CALCULUS WITH MAPLE® V

CALCULUS WITH MAPLE® V

John S. Devitt
University of Saskatchewan

Brooks/Cole Publishing Company
Pacific Grove, California

The trademark ITP is used under license.

Brooks/Cole Publishing Company
A Division of Wadsworth, Inc.

© 1993 Wadsworth, Inc., Belmont, California 94002. All rights reserved. No part of this book may be reproduced, stored in a retrieval system, or transcribed, in any form or by any means—electronic, mechanical, photocopying, recording or otherwise—without the prior written permission of the publisher, Brooks/Cole Publishing Company, Pacific Grove, California 93950, a division of Wadsworth, Inc.

Printed in the United States of America
10 9 8 7 6 5 4 3 2 1

Library of Congress Cataloging-in-Publication Data
Devitt, John S., [date]-
    Calculus With Maple V / John S. Devitt.
      p.   cm.
    Includes index.
    ISBN 0-534-16362-9
      1. Calculus–Data processing.   2. Maple (Computer file)
I. Title.
QA303.5.D37D48   1993
515'.0285'5369–dc20   93-10291
                                                                                              CIP

Sponsoring Editor: *Jeremy Hayhurst*
Editorial Assistant: *Elizabeth Barelli*
Production Coordinator: *Marlene Thom*
Manuscript Editor: *Carol Dondrea*
Cover Design: *Susan Haberkorn*
Printing and Binding: *Malloy Lithographing Inc.*

To Diane, Deanna, and Craig,
for their enthusiastic support

Preface

This text has arisen out of several semesters of using a computer algebra system (*Maple*) in the mathematics classroom. This discussion is based on the daily use of symbolic algebra during lectures. As much as 60 percent of the lecture time is spent involving students in developing ideas or solving mathematical problems directly at the computer terminal. The traditional blackboard has been relegated to a role of providing summaries, outlines, and overviews. We have been pleased by the effectiveness of this approach. The potential for focusing student energies on the concepts and ideas of mathematics instead of just mimicking routine computations is enormous. Although considerable work remains to be done before such tools become widely accessible, we hope that the problem solutions and observations presented here will help make others aware of the great promise that exists for these and similar methods.

Background

During the 1980s there was a significant rise in the level of faculty concern about the standard introductory calculus sequence. This concern came from a number of quarters, centering on

1. Lack of interest and preparedness of the students.
2. Apparent lack of relevance of the course content to the modern student.

At times it has seemed that while the rest of the world has moved on to take advantage of new tools and applications, mathematicians have been lost in a dream world. Each client discipline is upset by the students' lack of direct experience in their problem domain. Computer science instructors argue that the introductory mathematics sequence should involve more discrete mathematics. Their students don't need (want or have time for?) calculus as much as formal logic, the solution of difference equations, graph theory, and combinatorics. Physics instructors are concerned because the mathematics they want to use in the first term is not taught until the end of the first year. Mathematicians are concerned because of the perceived loss of good mathematics students to more glamorous disciplines.

Meanwhile, we praise the virtues of pencil and paper, lament the decline of the slide rule, and suggest that anything involving computation by machine must be unhealthy or at least not the work of a *true* mathematician.

Early attempts to modify the curriculum to accommodate numerical calculators did not have much success. Calculators seemed to be *tolerated* by assigning problem sets that allowed students to investigate numerically facts that could be easily deduced algebraically. Programming languages for mathematics courses failed because an enormous number of the programming and implementation details were not central to the *mathematics* we were trying to present. Interactive graphics software met with some success but still could not interact easily with the algebraic representation. Numerical and graphical tools could only be used to confirm or hint at facts that abstract reasoning told us must be true, but they did not assist with the direct manipulation and exploration of the underlying model itself.

Symbolic algebra systems address three crucial concerns:

1. No programming is involved.
2. The machine does not intrude on the underlying mathematical concepts. (An integer is an integer, rational numbers are rational numbers, and sets are sets.)
3. The algebraic formulas can be manipulated, not just tested.

When such a system includes graphics and extensive numerical capabilities the result is a single interactive environment in which you can build and manipulate mathematical models and use them. The need for algebraic computation is now being recognized in many traditionally numeric domains. For example, products such as MathCAD®, MATLAB®, and IMSL/IDL® all make use of *Maple* to address algebraic computation.

Algebraic computation systems have been in existence for more than thirty years; so why has it taken them so long to reach the classroom? Part of the answer lies in the enormous complexity of these systems and the minimum computing resources traditionally required to operate them. It is only in the last five years that this software has migrated from mainframe computers, installed in large air-conditioned rooms, to battery-powered portable computers affordable by students. A second part of the answer lies in the evolution of the user interface. During the past ten years, this has evolved from a sequence of typed commands and typewriter-like output with little or no graphics to "live" worksheets with three-dimensional graphics and animation, and the use of standard mathematical notation. There are still computations for which pencil and paper are more convenient, but even these problems are being addressed.

Advances in size and portability are only hinted at by the existence of calculators with some symbolic and graphic capabilities, such as Hewlett Packard's HP 48 - SX™, and there are similar developments in interface design on the horizon. But no matter how these evolve, a common thread is the notion of tackling mathematical problems with the aid of an expert assistant. Now is the time to start becoming familiar with such an approach.

An Electronic Blackboard

The derivation of formulas and the development of techniques for limits, derivatives, and integration often involves major computational components, as do many proofs. This side of mathematics is very much an experimental science and can be as computationally oriented, as the more traditional numerical work and applications. True, the domain of computation is different, but by using a symbolic algebra system as a primary delivery tool we can deal easily with computation from both worlds. We need not bog down on routine calculations of either sort.

From a technological point of view, we are already in a position to use a symbolic algebra system in an interactive session much as we would the blackboard and chalk. A live symbolic algebra session, projected to the class by means of an overhead projector, can be an effective part of every lecture. The blackboard may still be used for sketching diagrams, motivating concepts, and stating theorems. However, almost all computation can be done by the machine. Through this approach we can accomplish several goals.

- The majority of lecture time is spent on ideas rather than on messy but mundane computations done at the board to obtain evidence of your claims. If you spend 50 percent of your classroom time doing routine computations at the board, does this imply that 50 percent of the content of your course is routine computation?
- Students regularly observe how to carry out certain basic tasks within the system. This greatly reduces the task of familiarizing them with the computer system in labs.
- The instructor provides an important role model for students in terms of attitude. Students are given a mixed message if the instructor says "Use the computer!" but tacitly says "Computers aren't much use for what I do!" by avoiding the machine at every possible turn.

Dealing with the Black Box Phenomenon

Symbolic algebra systems are impressive in terms of the range and type of computations they can do. In fact, it is easy to feel that the systems are almost too powerful. Brute force use of symbolic algebra systems is about as pedagogically satisfying as an answer book that contains only the final answers.

We have addressed this issue by augmenting the basic system with routines that allow us to proceed in *student size* steps. These special routines are contained in the *student package* within *Maple* and are loaded by the command **with(student);**. They have been fully incorporated into the standard distribution of *Maple*, beginning with version 4.1.

Central to this approach is the provision of functions that are analogous to standard system functions but that remain unevaluated. These *unevaluated functions* can, in turn, be manipulated as algebraic entities to illustrate basic properties of the underlying operations. *Maple*'s standard versions of these functions always remain available to complete a computation.[1] For example, three typical commands and their results are

```
> Sum((i+5)^3,i= 1..n);
```

[1] In what follows, the actual *Maple* commands will occur on lines beginning with ">".

$$\sum_{i=1}^{n}(i+5)^3$$

```
> expand( " );
```

$$\sum_{i=1}^{n} i^3 + 15 \sum_{i=1}^{n} i^2 + 75 \sum_{i=1}^{n} i + 125 \sum_{i=1}^{n} 1$$

```
> value( " );
```

$$\frac{1}{4}(n+1)^4 + \frac{9}{2}(n+1)^3 + \frac{121}{4}(n+1)^2 + 90n - 35$$

Notes to Instructors and Students

No Major Attempt to Revise the Curriculum — Yet...

On close examination, instructors will note that although the method of delivery of material makes extensive use of the computer algebra system, little attempt as yet has been made to *revise* the curriculum. This has been deliberate. The intent here is to guide instructors and students through material that is familiar to the instructor, and to allow time for the instructors to examine and compare the capabilities of such a computer system with the more traditional presentation of calculus. These notes can be used either together with or even instead of a traditional calculus text without forcing a radical revision in terms of subject matter. The body of these notes provides worked solutions to familiar problems so that both the commands and the potential style of interaction with the computer system become apparent.

The two issues of symbolic algebra usage and effective calculus reform are in some sense separate. Understanding the tools and their capabilities is a necessary precursor to the latter. Furthermore, it is foolish to attempt such a major and critical task as revising calculus instruction to include use of a specific tool without a good understanding of that tool and its potential.

Not Just Another Language

If you find yourself thinking of a symbolic algebra system as simply another procedural programming language, then force yourself to look again. At the first sign of frustration or inability to get the system to do something, you will be tempted to return to the familiar pencil and paper. However, remember that specific mathematical knowledge is considerably more than the ability to carry out a particular computation by hand. Don't be afraid to help the system out with your own knowledge. Sequences of interactive steps, each carried out using the best tool[2] available, are often the best way to solve a particular problem. The process of identifying appropriate steps to solve a particular problem is at least as important as a detailed understanding of how to complete each step.

---

[2] As you develop more experience with the system, you will be better able to identify whether such instances are actual weaknesses of the system, or simply a lack of awareness of some of the functionality of the system.

Don't Forget Interaction

A symbolic algebra system is an extremely powerful tool, especially when used in an interactive *pencil and paper-like* style. It allows for extremely rapid progress to be made in formulating, testing, and verifying or refuting mathematical conjectures, and it is already having a major impact on how we carry out mathematical research in many branches of mathematics. These tools are still in their infancy but – if the last decade is any indication – they are evolving much more rapidly than is our ability to rethink the curriculum. We are still very much in the process of identifying the proper "primitives"[3] and functionality for an effective learning and working environment.

Many of the commands and the style of use presented here have been developed during the evolution of this course explicitly in response to a need for certain key primitives required in this interactive learning process. User interfaces, and particularly the manner in which these primitives are invoked, will continue to evolve at a very rapid pace. Thus (with time) it will become easier to carry out the necessary steps to solve a problem, though the actual steps will likely remain much the same. It is still important for you to use existing tools as early as possible in order to identify the direction in which such tools should evolve and to become aware of the possibilities opening up to you.

The Algebra System Works with the Student

Perhaps the single most important aspect of working and learning in such a mathematical laboratory is that, with some care on the part of the instructor, the tool can be used at a level of refinement that exactly matches the level of the student. As each concept or technique is mastered and understood, it becomes available as a single step in solving the next problem. This bundling of multiple steps into single higher-level steps is important because it allows both the student and instructor to recognize and deal with a concept as a unit. It becomes easier to identify the concept, to identify when such a step is not fully understood, and to explore its applications. Through use of the algebra system, we allow for a conscious decision to use steps in a solution that are not intended to be mastered in detail. Thus, steps in a solution can be understood in terms of functionality and purpose without the need to go into the gory detail.

We may be seeing and working with the first generation of students who have an opportunity to see *the forest* without getting lost in *the trees*.[4] We don't yet know the full impact of making accessible a top-down approach to problem solving (something that is implicit in these methods), but it may drastically affect how our society sees mathematics.

Specific Observations in the Classroom or Laboratory

The examples given in this section regarding the impact of the software on the mathematics course are only a sample of the kinds of things that can be and have been done

---

[3] These very primitives (that is, user-level commands for carrying out basic interactive manipulations) are relied on extensively in several interface projects that depend on *Maple*.
[4] This may be as important to society as doing away with the original implicit requirement that every car driver be a automobile mechanic.

in class. We emphasize that the tools are used extensively both in class and for assignments. As things have progressed, it has become more and more natural to respond to questions by using the software to work through suggestions made by the students. Lab assignments are becoming more exploratory. Some specific observations are:

- The treatment of inequalities takes on a whole new light if regarded as statements about the graphs of real-valued functions. By transforming each question into one about a plot, it is easy to get students to understand the goals. Because the system solves many forms of inequalities, we can also verify answers directly.
- Limits play a much more active role when they can be easily computed. In our presentation, discussion of the rules for computing derivatives has been delayed for an extra three weeks. In the interim, limits and the system **diff( )** function are used extensively. This helps avoid the rather common confusion of *knowledge of the rules for computing derivatives* with *the concept of a derivative*.
- When taught, the various ϵ, δ arguments seem to be better received. Each discussion of finding bounds on $x - a$ so that y falls into a given range amounted to defining an inequality for y, factoring both sides to *isolate* $(x - a)$, and using graphics to obtain an estimate for δ. These three steps can be repeated frequently and an intuition developed about how to establish a bound on one quantity to achieve certain restrictions on another. Or are we in too big a hurry to abandon theory?
- We can develop an intuition for the notion of asymptotes and asymptotic behavior of a function through graphical investigations. There is no need to restrict the study to vertical and horizontal asymptotes.
- For proofs and problem solving, the emphasis can be placed on the structure of the problem rather than on the tedious computations. Top-down problem solving becomes more natural, and the value and role of various transformations becomes much clearer. By the end of the course, good students can outline a solution by jotting down short sequences of commands as their rough solutions to problems. They then test and verify these solutions the next time they are in the lab.
- More students can and do try all the problems at the end of each section of the traditional text that have been associated with the course. In each case, the students seem to focus more on recognizing which type of problem they are dealing with and become more confident that their solutions are correct partly because of the increased opportunity for independent verification of the accuracy of each of the intermediate steps as well as of the final solution.
- The notion of easy versus hard computations arises naturally. Two or more ways can be tried, and students can discover the drastic difference that their approach can make to the computational complexity of a problem.
- Correctness and techniques for checking correctness are of much greater interest to the student. Their motivation goes beyond that of passing an exam to wanting to be able to determine that the results computed by the algebra system are reasonable and correct.
- A glimpse of the depth and breadth of mathematics becomes possible as we are no longer restricted to using only those techniques available to students. This can provide additional motivation and allow glimpses of techniques yet to be learned. Important

questions such as "Which equations can be solved?" or "Which integrals are easy to compute?" are raised naturally in this context.

The opportunity to tailor one's lectures and assignments to selected concepts has shown tremendous potential. Not only is it possible to work at a more abstract level, but students seem much more receptive to that abstraction. It would appear that with such an approach we can "get our hands dirtier quicker." The long-term potential is enormous and it is our belief that this approach can and will contribute seriously to a deeper understanding of mathematics at all levels.

Testing and Evaluation

Given that we are actively encouraging students to use computational tools to solve mathematical problems throughout the course, the issue of testing becomes very sensitive and important. Two things of particular concern to students are:

1. Direct comparisons (for credit) with students from traditionally taught sections of the same course.
2. Testing that does not take into account the shift in emphasis from *Carry out this computation!* to *What should be done and how should we do it?*.

When students compare the demands placed on them with those placed on students in a traditional section of the same course, they may feel that they are being asked to do everything their classmates in the more traditional sections are asked to do and more. In reality, after an initial familiarization period, the software becomes a true expert assistant and makes it easy to cover the required material in the remaining time. However, many students lack the self-confidence to take on this challenge and become paralyzed by the fear of a lower grade. Responsible testing must respect and address this concern.

The tests generally associated with the traditional courses do not acknowledge any shift in emphasis from the rote learning of certain computational techniques to the actual use of those same techniques for developing and using mathematical models. Responsible testing must reflect the style in which the material has been taught and the expectations of the instructor.

It is often difficult to hold traditionally scheduled tests in which the students have access to the software throughout the exam period. Instead, we have used questions that present information of the sort that might have been obtained by using the machine and ask how this information might be used to reach certain conclusions. The data for the questions is presented in a variety of ways including graphical representations, and in a manner that forces students to deal with general properties and strategies rather than with completion of a detailed computation. Many of the tests have been open book, taking the view that the student's role is to make use of available resources (perhaps including the assistance of software) to explain mathematical phenomena.

The details of computer language should never be tested in class. We have evaluated their familiarity with the tools by assigning student projects. The projects, completed by groups of two students or individually, generally involve comparing a number of methods for solving a particular problem. Students must prepare a recommendation of one of the methods for a ficticious employer and must justify their choices. The projects have proved very popular and effective.

To the Student

By the time you have reached college or university, you have spent a considerable amount of time learning a variety of computational techniques. Everything from testing to homework exercises has probably reinforced the notion that *knowledge* of the steps of a complicated computational procedure along with the ability to carry out those steps in detail is *success* in mathematics.

In a sense, we are now relegating most of that computation to the machine, and this can be very disconcerting at first. The very fact that the machine can do all the computations (to the level of an "A" student on most traditional final exams for introductory calculus) is troublesome when combined with the fact that you are confronted with the details of a new software system. This can cause insecurity at first. Initially your reaction may be "this is not mathematics." However, it is very important that you recognize that this course is deliberately deemphasizing computational skills and replacing them with the longer term goal of achieving a clear vision of the basic concepts and patterns of mathematics. Success will be measured here by your ability to recognize which mathematical procedures and tools are relevant to a problem at hand, what they do, what their limitations are, and how they may be combined to reach certain results. Persevere even though the course seems very different from what you are used to classifying as mathematics.

The details and power of *Maple* may seem overwhelming during the first couple of lectures, but it is important to realize that after the first week or so, and certainly by the end of the first hands-on lab session, the amount of new-software related details to be mastered in each class tails off drastically. At that point, the emphasis shifts substantially to the mathematical content of the course and any delays in addressing the mathematical content are soon made up for by progress elsewhere. By the end of the course, you should be working with concepts at a level substantially beyond where you would be in a traditional course. Such concepts are really no more difficult to deal with than the detailed computations of your earlier mathematical experiences; they are just different and essential if we are to have any hope of using the emerging computational tools effectively. Curiously, studies also suggest that the actual hand computations that we would traditionally regard as the content of the course are not lost.

Above all, keep in mind that our primary goal in this course is to understand concepts and relationships. The computing environment is very much a laboratory that allows us to experiment with ideas and clarify discovered relationships. When something is unclear to you, don't hesitate to explore on your own by asking the computing system to carry out computations while you look for underlying patterns. Once the concepts are clear, the methods and strategies for carrying out a particular computation tend to reveal themselves.

Acknowledgments

I thank Professor Keith Geddes and Professor Gaston Gonnet for having the vision to start the *Maple* project and for the thrill of having been allowed to participate fully in the project through the years. I also thank all my colleagues in the (extended) Symbolic Computation Group of the University of Waterloo, for their dedication to that vision. *Maple* is very much the result of a team effort in which every member of that team has, and continues to, put that vision first.

Thank you to Jeremy Hayhurst and Robert Evans from Brooks/Cole Publishing Company for their faith in *Maple* and this project, and their dedication to promoting the role of technology in mathematics education. There have been too many times that they have waited helplessly while the author was busy improving the software. I also thank Marlene Thom and Carol Dondrea from Brooks/Cole and Nora Sleumer and Maurizio Bianchi for the many improvements they have suggested for the manuscript, and the many students who have participated in classes based on this approach. In addition, I thank the following reviewers for their helpful comments: Maurino Bautista, Rochester Institute of Technology; Robert Eslinger, Hendrix College; Thomas Judson, University of Portland; Robert Lopez, Rose-Hulman Institute of Technology; and John Mathews, California State University, Fullerton.

I am grateful to the Alfred P. Sloan Foundation for the financial support that made possible the initial classroom experimentation with these ideas, and to Dr. Samuel Goldberg, Dr. Zaven Karian, Dr. Ram Manohar, and Dr. Robert Moody, and my parents and family, for their moral support and encouragement.

Finally, this book would not have been possible without the generous support of the University of Waterloo. Thank you for creating an environment that encourages exploration and development of ideas, and for providing the facilities that were used to develop the majority of this manuscript.

J. S. Devitt
Waterloo

Contents

1 The Basics and Interactive Computing — 1

1.1 Introduction   1
    1.1.1 Some Pointers to Better User–Machine Communication   2
    1.1.2 A Sample Session   3
    1.1.3 Basic Objectives   6
    1.1.4 Expressions and Equations   6
    1.1.5 Graphical Representations   9
1.2 Plotting as an Aid to Understanding Inequalities   12
    1.2.1 More on Inequalities   15
    1.2.2 Type 1 Inequalities   15
    1.2.3 Type 2 Inequalities   20
1.3 Equations Involving Two Variables   22
    1.3.1 Equations of Straight Lines   23
    1.3.2 Distances Between Points   26
    1.3.3 Solving an Algebraic Equation Step by Step   27
    1.3.4 Quadratic Equations and Their Graphs   29
    1.3.5 Solving a Quadratic   30
    1.3.6 Equations Involving x^2 and y^2 31
1.4 Functions   36
    1.4.1 Defining a Function   36
    1.4.2 Using Functions   37
    1.4.3 Restricted Domains   39
    1.4.4 Finding Domains of Functions   40
    1.4.5 Absolute Values   42
    1.4.6 Step Functions   43
    1.4.7 Even Functions   44
    1.4.8 Odd functions   44
    1.4.9 Constant Functions   45
1.5 New Functions from Old   45

1.5.1 Arithmetic Operations   46
1.5.2 Composition of Functions   46
Exercise Set 1   48

2 Limits   57

2.1 Introduction   57
2.2 Limit Computations in *Maple*   61
2.3 Some Limit Computations   62
    2.3.1 Ratios of Polynomials   62
    2.3.2 Obscure Common Factors   66
    2.3.3 Limits That Are Bounded but Do Not Exist   69
    2.3.4 Unbounded Functions   69
2.4 A Formal Definition of a Limit   70
    2.4.1 Computing δ 71
2.5 Properties of Limits   77
2.6 The Squeeze Theorem   81
2.7 Continuity   83
    2.7.1 One-Sided Limits   85
    2.7.2 Combining Continuous Functions   88
2.8 The Intermediate Value Theorem   89
2.9 Exact Computations Versus Approximations   92
Exercise Set 2   93

3 Derivatives   98

3.1 Introduction   98
    3.1.1 Interpretations of the Derivative   101
    3.1.2 Leibnitz Notation   110
3.2 Differentials   112
3.3 Differentiation Formulas   114
    3.3.1 Some Fundamental Examples   115
    3.3.2 Constants   116
    3.3.3 Pure Powers   117
    3.3.4 Laws for Addition and Multiplication   122
    3.3.5 Derivatives of $\frac{1}{g(x)}$ 126
    3.3.6 Summary   128
3.4 Expressions versus Functions   130
3.5 Trigonometric Functions   131
    3.5.1 Limits of Trigonometric Functions   131

	3.5.2 Derivatives of Trigonometric Functions 135	
3.6	The Chain Rule 137	
	3.6.1 Compositions Involving More Than Two Functions 146	
	3.6.2 Summary 147	
3.7	Derivatives of Exponentials and Logarithms 148	
	3.7.1 Logarithms as Inverse Exponential Functions 150	
3.8	Implicit Derivatives 152	
	3.8.1 Treating x and y as Functions of t 156	
	3.8.2 Related Rates 158	
3.9	A Derivation of Newton's Formula 163	
	Exercise Set 3 167	

4 Optimal Solutions and Extreme Values 173

- 4.1 Maximums and Minimums 173
  - 4.1.1 Optimizations in *Maple* 179
  - 4.1.2 Local Maximums and Minimums 179
  - 4.1.3 The Extreme Value Theorem 187
  - 4.1.4 Summary 188
- 4.2 The Mean Value Theorem 188
  - 4.2.1 The Average Slope 190
- 4.3 Monotonic Functions 197
  - 4.3.1 The First Derivative Test 198
- 4.4 Concavity and Points of Inflection 202
- 4.5 Asymptotes 209
- 4.6 Applied Maximum and Minimum Problems 213
  Exercise Set 4 223

5 Integration 227

- 5.1 Introduction 227
- 5.2 Summations 228
  - 5.2.1 Summations and Area Under a Curve 231
  - 5.2.2 Rules for Combining Sums 232
  - 5.2.3 Formulas for Specific Sums 234
  - 5.2.4 Discovering Formulas 235
- 5.3 Area 239
  - 5.3.1 An Underestimate of the Area 239
  - 5.3.2 An Overestimate of the Area 242
  - 5.3.3 Better Estimates 243
- 5.4 The Definite Integral 245

        5.4.1 Curves Above the Axis   246
        5.4.2 Curves Below the Axis   247
        5.4.3 Curves That Cross the Axis   249
5.5   Shortcuts in Computation   250
        5.5.1 The Basic Manipulations   250
        5.5.2 Order Relationships   254
5.6   The Fundamental Theorem of Calculus   260
        5.6.1 Indefinite Integrals   265
5.7   Applications of the Fundamental Theorem   268
Exercise Set 5   273

6 Applications of Integration   278

6.1   Areas Between Curves   278
6.2   Volume   285
6.3   Solids of Revolution   288
6.4   Generalized Cross Sections   294
6.5   Cylindrical Shells   298
        6.5.1 Visualizing Cylinders   301
        6.5.2 Cylindrical Decompositions Off the Axis   303
6.6   Work   305
Exercise Set 6   309

7 Integration Techniques   313

7.1   Introduction   313
7.2   Changing Variables   314
        7.2.1 Using a Change of Variables to Integrate   316
        7.2.2 Summary   316
        7.2.3 Additional Examples   317
        7.2.4 The Effective Use of Change of Variables   318
        7.2.5 Definite Integrals   321
7.3   Integration by Parts   326
        7.3.1 Applications of Integration by Parts   328
        7.3.2 Definite Integrals   331
        7.3.3 Reduction Formulas   333
7.4   Trigonometric Substitutions   334
        7.4.1 Mixtures of Sines and Cosines   338
        7.4.2 Identities for Secant and Tangent   339
        7.4.3 Integrating Secant   342
        7.4.4 Sums and Differences of Angles   344

7.5 Square Roots of Quadratics   344
7.6 Partial Fraction Decompositions   352
    7.6.1 Computing a Partial Fraction Decomposition   352
    7.6.2 Patterns for Partial Fractions   354
    7.6.3 Integrals with Quadratics in the Denominator   356
    7.6.4 Partial Fractions in Action   357
7.7 Numerical Approximations   361
    7.7.1 The Trapezoidal Rule   361
    7.7.2 Simpson's Rule   366
    Exercise Set 7   *373*

8 More Applications of Integration   380

8.1 Volumes Through Integration   380
    8.1.1 Visualizing Stacks of Disks   385
    8.1.2 Variations on Volume   388
8.2 Cylindrical Shells   396
8.3 Arc Length   398
8.4 Surface Area   405
    Exercise Set 8   *409*

9 Parametric Equations   413

9.1 Introduction   413
9.2 Parametric Curves   413
    9.2.1 Finding Cartesian Representations   415
9.3 Tangents and Areas Revisited   420
    9.3.1 Second Derivatives   424
    9.3.2 Areas   426
9.4 Arc Length and Surface Area Revisited   427
    9.4.1 Arc Length   428
    9.4.2 Surface Area   429
9.5 Polar Coordinates   430
    9.5.1 Curve Sketching in Polar Coordinates   431
    9.5.2 Tangent Lines and Polar Coordinates   432
9.6 Areas in Polar Coordinates   435
9.7 Arc Lengths in Polar Coordinates   436
    Exercise Set 9   *440*

10 Sequences and Series — 446

- 10.1 Introduction 446
- 10.2 Sequences 446
  - 10.2.1 Recurrence Relations 447
  - 10.2.2 Asymptotic Behavior of Sequences 449
- 10.3 Series 451
  - 10.3.1 Arithmetic on Series 456
- 10.4 Testing for Convergence and Divergence 457
  - 10.4.1 The Integral Test 457
  - 10.4.2 Comparison Tests 460
  - 10.4.3 Ratio Tests 463
  - 10.4.4 The Root Test 465
- 10.5 Alternating Series 465
- 10.6 Power Series 468
- 10.7 Constructing Power Series 471
  - 10.7.1 Algebraic Manipulations of Power Series 472
  - 10.7.2 Constructing Coefficients 477
  - 10.7.3 Taylor Series 480
- 10.8 Approximations 481
  - 10.8.1 Error Analysis 483
- *Exercise Set 10* 488

A The Computing Environment — 495

- A.1 The Student Package 495
- A.2 Production Notes 498

1 The Basics and Interactive Computing

1.1 Introduction

At the beginning of most calculus courses, it is customary to review some of the basic mathematical skills and ideas that you are expected to be familiar with. We have an even more important reason to do this here as we want you also to become familiar with *Maple* – our computing environment for automatic algebraic computation – in a situation where we will all know what to expect. A computer system can be intimidating to the casual user, and it is essential that this does not get in the way of our mathematical objectives. By beginning with a review of familiar material, we will be better able to predict the results generated by the computing system.

Numerous manuals have been written on how to use a particular computer system or software package. All too often the bulk of the information these manuals provide is wasted on the first-time user. When presented with a collection of computer systems manuals, especially in your mathematics course, a likely (and perhaps reasonable) reaction is to say, "I haven't got time for all that!" In practice you will be using only a very small part of the total system, and it is foolish to wade through *all* the system details and manuals before even starting to use the system.

A main objective in this section is to provide just enough information to allow you to start using *Maple* for simple computations. We introduce more advanced topics (both mathematical and computational) later as they are needed.

Our emphasis throughout this course is on mathematics. Your first encounter with specific commands and syntax[1] will be, as much as possible, by actually using them in their mathematical context. We rely heavily on this proper context to provide useful information on how to understand the computer notation (syntax) used. Many of the commands in use have optional parameters. However, wherever possible we avoid or postpone the use of these. Doing so helps us avoid cluttering the mathematical discussion with long explanations of details you may never have to know. The next section (1.1.1) contains specific system pointers that may save you some grief during those first few sessions. The actual details of each command we use, including examples, are also available through help facilities built into the program.

The *Maple* program we are using is best thought of as an algebraic calculator. The details of how to get the program started vary depending on how the machines and software are set up at your particular installation. For our present purposes we assume that such details have been taken care of and that you have just started up the symbolic algebra program.

---

[1]The *syntax* of a computing language consists primarily of details such as key words, how they can be combined into commands, and the proper punctuation.

1.1.1 Some Pointers to Better User–Machine Communication

A symbolic algebra system is designed to help us evaluate mathematical expressions. We make some concessions to the fact that the only reasonable means of entering data at present on most systems is a typewriter keyboard. The main restrictions are:

1. Ordinary names are used instead of Greek letters (or other unusual symbols) to represent variables, unknown quantities, or special actions. For example, summations are denoted by **sum(i, i = 1 .. n)** or **Sum(i, i = 1 .. n)** rather than ($\sum_{i=1}^{n} i$), and α is represented by *alpha*. The distinction between capitals and lowercase letters is important, as explained later.
2. Multiplication is indicated by the symbol " * ". We know $2a$ means 2 *times a*, but what does *alphab* mean? Is it a strange variable name, or is it *alpha times b*? What about $f(x + y)$ or $D^2 y$? Sometimes it is easier to give the software (and the student) a few hints. The explicit "*" (meaning "multiply") accomplishes this.
3. $a \div b$ is entered as **a/b**. Similarly, $\frac{a}{b+c}$ is entered as **a/(b+c)**.
4. Subscripts, such as in x_n, are written **x[n]**.
5. Exponents or powers such as x^n are written **x^n** or **x**n**.
6. Command lines appear in the text as lines beginning with the symbol >, as in

   ```
   > a + b;
   ```

   The results generated by the computing system usually appear immediately following the command.
7. Complicated expressions may result in the need to enter a single command across several lines of input. To accommodate this, *Maple* does *not* complete and evaluate an expression until it sees a ";" somewhere on a line. This need for a ; indicating "I'm finished" can lead to some confusion for the first-time user[2] since the commands[3]

   ```
   > a + b;
   ```

   and

   ```
   > a
   > +
   > b
   > ;
   ```

   both cause the system to evaluate the expression $a + b$.

   The use of the semicolon also allows us to enter two or more expressions on the same line, as in

   ```
   > a + b;  1/c;
   ```

---

[2] Many people first interpret the lack of a result after accidentally entering a carriage return in the middle of an expression as an indication that they made a mistake. They respond incorrectly by starting the expression over again. The result is the same as typing part of the expression twice: a syntax error.
[3] The prompts (>) shown here are assumed to have been provided by *Maple*.

1.1.2 A Sample Session

In this section we present various *Maple* commands followed immediately by the results generated by the computer system.

To some extent, *Maple* is used like a traditional calculator. You can use *Maple* to perform basic arithmetic simply by entering any required expression. As mentioned earlier, a ; at the end of the command line indicates that the command is complete. After you cause the command to execute (usually by pressing the Enter or Return key, depending on the computer system you are on), the actual *value* of the expression appears immediately following the command line.

```
> (3 + 4)* 12;
```

$$84$$

We often need to refer to the *value of the previous expression*. We can do this with a minimal amount of typing by using the double quote symbol " " " " (chosen because of its similarity to "ditto marks") as a name for the last computed value. Thus, for example, the reciprocal of the the preceding expression is:

```
> 1/";
```

$$\frac{1}{84}$$

Observe that *Maple* supports true rational arithmetic. This alone is a major departure from traditional calculators. Things get even more interesting when we discover that there is also no practical limit on the size of the integers used. The product of the first one hundred integers, *one hundred factorial*, is computed as

```
> 100!;
```

$$93326215443944152681699238856266700490715968264381621468592\\963895217599993229915608941463976156518286253697920827223 75\\82511852109168640000000000000000000000000$$

and, for example, *Maple* simplifies $\frac{100!}{99!}$ to 100.

```
> "/(99!);
```

$$100$$

This departure from *machine-dependent* restrictions is important for it allows us to work more closely with the mathematical concepts.

Algebraic Expressions

Expressions may not always simplify this easily. If we know that we have a rational number but do not know which one specifically, we might want to write $\frac{p}{q}$ as an unevaluated expression.

```
> p/q;
```

$$\frac{p}{q}$$

This is our third major departure from traditional calculators. Here we intend that p and q be integers with no common divisor. We have no more information, so about the best the system (or anyone for that matter) can do is to leave the expression unevaluated in the manner shown. This contrasts sharply with the familiar error message that many computing systems would give at this point.

Although *Maple* prefers to work with exact rational numbers and symbols, there are often times when we need to approximate a given numerical quantity. Such approximations are represented using a decimal point.[4]

Numerical Approximations

By default, *Maple* includes a full ten digits in its floating-point approximations but makes provision for us to request more. The more digits we request, the more accuracy we are expecting for our answer. All of *Maple*'s exact numerical quantities can be approximated by using the command **evalf( )**. This command literally means "evaluate using floating-point numbers." Some simple examples of the use of this command follow:

```
> evalf( 5/17 );
```
$$0.2941176471$$

```
> evalf( 5/17, 30 );
```
$$0.294117647058823529411764705882$$

```
> evalf( 5/17, 100 );
```
$$.2941176470588235294117647058823529411764705882352941176470588235294117647058823529411764705882352941$$

Of course, not all real numbers are rational numbers. Familiar examples of such exceptions include $\sqrt{2}$ and π. *Maple* represents both of these exactly[5] as

```
> sqrt( 2 ), Pi ;
```
$$\sqrt{2}, \pi$$

Floating-point approximations for these can be obtained to an arbitrarily high degree of precision. For example, we have:

```
> evalf( sqrt(2) , 40 );
```
$$1.414213562373095048801688724209698078570$$

```
> evalf( Pi, 100 );
```
$$3.1415926535897932384626433832795028841971693993751058209749445923078164062862089986280348253421117068$$

---
[4]In a computing system, such numbers are traditionally called *floating-point* numbers.
[5]Note that we can evaluate two or more expressions in one command providing that we separate adjacent expressions by ",".

Among the real numbers, true *rational numbers* are recognizable by the fact that they have a *periodic*[6] representation. Some – such as integers – can even be represented exactly.

Example 1.1 An example of an *irrational* number (i.e., one that is not rational) is $\sqrt{2}$. Show that $\sqrt{2}$ is not rational.

Solution To see that $\sqrt{2}$ is not rational, pretend for a moment that it is rational. Then the following equation should hold, with p and q such that they have no common factors such as 2 (i.e., at most one of them is divisible by 2.).

```
> sqrt( 2 ) = p/q;
```

$$\sqrt{2} = \frac{p}{q}$$

We can simplify this equation by squaring both sides[7] and then multiplying both sides by q^2.

```
> sqrt( 2 )^2 = (p/q)^2;
```

$$2 = \frac{p^2}{q^2}$$

```
> "*(q^2);
```

$$2q^2 = p^2$$

In this form it is clear that p^2 must be an even number (i.e., divisible by 2) and so p is even. If we replace[8] p by, say, $2m$ to represent this fact, we can cancel one power of 2 from both sides of the equation.

```
> subs( p = 2*m, " );
```

$$2q^2 = 4m^2$$

```
> "/2;
```

$$q^2 = 2m^2$$

From the form of this equation we see that q and q^2 must be even as well. But this contradicts our assumption that at most one of p and q has 2 as a divisor or, in effect, our assumption that $\sqrt{2}$ can be represented by a rational number. Since our assumption about $\sqrt{2}$ being rational led to a logical inconsistency, it must not hold. ∎

The decimal expansions of irrational numbers never evolve into repeating patterns and so, in a genuine sense, they are never finished. However, algorithms exist and are built into *Maple* for computing as many digits as desired for many common irrational quantities such as π and $\sqrt{2}$.

---

[6] By a *periodic* decimal representation we mean *evolving into a repeating pattern*. The repeating pattern may consist only of 0's as for the integers, or fractions with denominator a power of 10.
[7] An easy way to do this will be shown later, but for now we will just reenter the modified equation.
[8] The **subs( )** command will be described in more detail in section 1.1.4.

1.1.3 Basic Objectives

The computer algebra system did not play an essential role in solving the preceding problem concerning $\sqrt{2}$. We probably all could carry out the steps faster on paper. Rather, we have used symbolic algebra here to demonstrate how the system supports even the primitive operations that arise in solving such a problem. The setting was chosen so you could *see* how this system might be used in much the same way that we currently use pencil and paper.

When we solve more involved problems, we will encounter steps in the solution process that, while conceptually straightforward, involve cumbersome (but hopefully well-understood) details. The power of this approach will begin to show when we are able to *describe* and *carry out* a solution step by step at this more conceptual level, leaving these cumbersome details to the algebra system. By letting the algebra system look after details, the overall structure of our approach to a given problem begins to stand out.

> "If I could carry out steps (a), (b) and (c), I would have a solution to this problem..."

Ideally, each step will correspond directly to a computation in the algebra system. Of course, understanding and completing each of these subtasks may require that we understand some concept, and that might be the topic of some other discussion or course in its own right. The point is that in this environment, once a certain concept or procedure has been mastered, it is immediately available to us as a general tool and we can turn to the task of understanding how to use it.

The problems in this chapter should be regarded both as a review of some essential mathematical concepts and as an introduction to a "style" of using a computer algebra system. Ideally, the use of the computer algebra system will become as natural as the use of pencil and paper. Although a full realization of this awaits further developments in technology, our current goal is to exploit the help that today's technology can give in this direction.

1.1.4 Expressions and Equations

The real power of a symbolic algebra system starts to show when we deal with expressions involving unknown quantities. We have already seen that variables can be used in expressions and can be manipulated much like numbers. Consider, for example, the expression

```
>  x + z;
```

$$x + z$$

We can raise this to a power, expand[9] the result fully, and factor it again – all using facilities provided by the algebra system.

```
>  "^6;
```

$$(x + z)^6$$

---
[9] Recall Pascal's triangle or the binomial theorem. Try expanding $(x + z)^{100}$.

```
> expand( " );
```
$$x^6 + 6x^5z + 15x^4z^2 + 20x^3z^3 + 15x^2z^4 + 6xz^5 + z^6$$

```
> factor( " );
```
$$(x+z)^6$$

It is often convenient to name an intermediate expression.[10] This is done as follows:

```
>   y :=   4*x  -  3;
```
$$4x - 3$$

The symbol ":=" is used to label the the expression $4x - 3$ with the name y. The value of y becomes $4x - 3$, and whatever we do with y, we actually do to its value. Examples of this include everything from simply referring to y, to manipulating y, or to using it as an argument to a *Maple* command.

```
> y;
```
$$4x - 3$$

```
> y^2 + 5*x + 3;
```
$$(4x - 3)^2 + 5x + 3$$

```
> expand( y^3 );
```
$$64x^3 - 144x^2 + 108x - 27$$

We frequently need to replace a particular unknown (or even an entire subexpression) in an existing expression. We do this via use of the **subs( )** command, as in

```
> subs( x=t, y );
```
$$4t - 3$$

The first argument to **subs( )** is always an equation in which the left-hand side is the object to be replaced and the right-hand side is its new value. Similarly, we can replace z by $s + t$ in the expression $(x + y + z)^6$:

```
> (x+z)^6;
```
$$(x + z)^6$$

```
> subs( z=(s+t), " );
```
$$(x + s + t)^6$$

We create equations simply by indicating that one expression be equal to another.

```
> 4*x + 3 = x + 7;
```
$$4x + 3 = x + 7$$

---

[10] An *intermediate expression* is any expression that arises during the course of a longer computation but that is usually needed only to go on to the next step.

Equations, like expressions, can be named by using :=, as in:

```
> eq1 := ";
```
$$4x + 3 = x + 7$$

Given an equation, we can extract the left and right sides of the equation by using the commands **lhs( )** and **rhs( )**

```
> y := lhs( eq1 );
```
$$4x + 3$$

```
> z := rhs( eq1 );
```
$$x + 7$$

Note that in these statements we actually associated names with the computed values through the use of :=. The values of the names such as $eq1$, y, and z remain unchanged until they or any symbols appearing in the associated expressions (i.e., x) are assigned different values through use of assignment statements.

In general, equations may be subtracted from one another, or each may be multiplied by nonzero constants without changing their meaning. In particular, the equation $eq1$ is equivalent to all of:

```
> y-z = 0;
```
$$3x - 4 = 0$$

```
> " + (4 = 4);
```
$$3x = 4$$

```
> "/3;
```
$$x = \frac{4}{3}$$

This sequence of manipulations corresponds to a "*solution of eq1 for x,*" as we can easily verify by trying the proposed value for x in the original equation. Verification is easy to achieve in *Maple* through use of the **subs( )** command, as in

```
> subs( ", eq1 );
```
$$\frac{25}{3} = \frac{25}{3}$$

In this case the result of the substitution is an equation that is easily seen to be true.

REMARK: The statement "$x = a$ is a solution to the equation" means that the substitution $x = a$ into the original equation must result in a valid equation.

Once we understand how to solve a given type of equation, we can reduce the process of isolating x to a single command.[11]

---
[11]See **?isolate**. This command is a a special variant of the **solve( )** command. It is available after executing the command **readlib(isolate):** or the command **with(student):**.

```
> isolate( y = z , x );
```

$$x = \frac{4}{3}$$

(This command can be read as "*isolate x in the equation* $4x + 3 = x + 7$.")

The *Maple* **solve( )** command can be used for finding exact solutions to systems of one or more equations in one or several unknowns. It can also be used in the case of a single equation, as in the preceding, though it often returns an answer in a slightly different form from the **isolate( )** command. Two such variations are:

```
> solve( eq1, x );
```

$$\frac{4}{3}$$

which just gives the value (or values) of x, and

```
> solve( eq1, {x} );
```

$$\left\{ x = \frac{4}{3} \right\}$$

This last variation (where the variable to be solved for is specified as the set $\{x\}$) always results in an answer that is a set containing one or more equations. Even so, the result can still be used directly with the **subs( )** command to verify the solution or to make use of it in a new expression.

```
> subs( ", x^2 + 3 );
```

$$\frac{43}{9}$$

1.1.5 Graphical Representations

A convenient way of understanding any equation is to reorganize it into an equation of the form "expression in $x = 0$." In this form, the question "Which values of x satisfy this equation?" becomes "For which values of x is this *expression in x* equal to 0?" This is often most easily understood as a question about the behavior of a graph.

Points on a line can be identified with real numbers by first choosing an arbitrary reference point on the line (called the *origin*) and labeling it 0. The point corresponding to any given real number other than zero is found a certain distance in the positive direction from 0 (usually to the right) measured using a convenient scale. The negative of x is found the same distance in the opposite direction.

In a similar manner, pairs of numbers $[x, y]$ can be associated with points in the plane. This time two number lines called coordinate axes are used, one for each of the numbers. They are called the x-axis and y-axis, respectively. These are placed at right angles and cross at their respective origins. Together they are referred to as a *Cartesian coordinate system*, and the plane they map out is called the *Cartesian plane*.

The point $P = [a, b]$ of the plane can be located using an $a \times b$ rectangle. The rectangle is placed parallel to both axes with one point at the origin. It is oriented so that

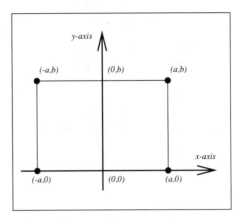

Figure 1.1 Locating a point on the Cartesian plane

the sides parallel to the *x*-axis and the *y*-axis are of lengths a and b respectively. The point P is the corner of the rectangle furthest from the origin ($[0, 0]$) (see Figure 1.1).

Given an expression y defined in terms of an unknown quantity x, it is natural to ask how the value of y changes as we try different values of x. One of the simplest ways to develop a feeling for how x and y are related, especially if we have the computational power to do so rapidly and accurately, is to compute the value of y for many different values of x.

By marking pairs of points $[x, y]$ on the Cartesian plane, with y computed in terms of x, we obtain a picture that represents the expression y. We call this picture *the graph of y*. Ideally, the graph shows a point for every pair of numbers $[x, y]$. In practice, we compute only as many pairs of points as we can in a reasonable amount of time and then guess at the position of other points on the basis of the points already computed. This latter process (called *interpolation*) is very useful but error prone and must be used with extreme caution. Mathematical knowledge that can be used to anticipate behavior of the plot can be useful in verifying results. We can use algebraic properties of the expression y to deduce essential features of the graph.

The process of drawing these graphs we refer to as *plotting*. We conclude this section with some sample plots obtained using the plotting function built into *Maple*.

```
> y := x^2 + 3;
```
$$x^2 + 3$$

```
> plot( y , x );
```

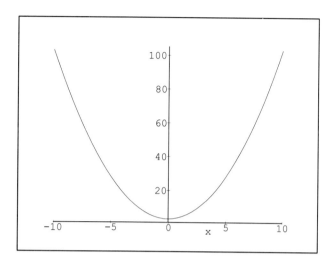

In this plot we did not specify a range for the values of either *x* or the expression *y*. By default, *x* is given values in the interval $-10\ldots 10$, and the range of *y* values is chosen to accommodate all the computed points. This is done automatically for us and sometimes results in a surprising choice of scaling for the vertical axis.

We can control the scaling of both the *x*- and the *y*- axis by giving the desired range of values explicitly.[12]

```
> plot( y , 'x'=-3 ... 3 , 'y'=0 ... 20 );
```

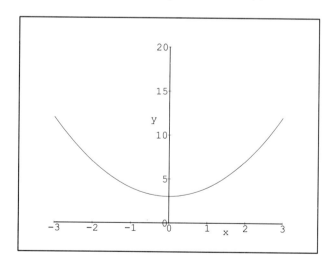

In this example, $'x' = -3\ldots 3$ indicates that the horizontal range will be from -3 to 3. Similarly, the expression $'y' = 0\ldots 20$ denotes the vertical range, and that the vertical axis will be labeled "y." We have added single forward-slanting quotes to the

---

[12]See **?plot,options** for a full list of available settings.

names *x* and *y* in the *Maple* command to guarantee that their names, rather than their *values*, are given to the **plot( )** command. To see why this is important, compare the difference in results for the following two statements:

```
> y;
```
$$x^2 + 3$$

```
> 'y';
```
$$y$$

1.2 Plotting as an Aid to Understanding Inequalities

It is traditional to spend a significant amount of time at the beginning of a calculus course verifying that students are proficient at the algebraic solution of inequalities. Instead of following this tradition, we will start by using graphics to develop an intuition for inequalities. Later this will evolve into a general technique that can be used to derive more traditional algebraic solutions.

Example 1.2 If *y* is given as

```
> y := 3*x - 5;
```
$$3x - 5$$

then determine the values of *x* for which the inequality $y < 4$ is valid.

Solution It is easy to interpret the meaning of this inequality by examining the graph of the expression $y - 4$.

```
> plot( y - 4, x );
```

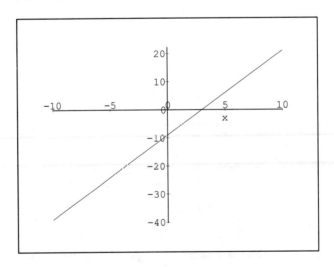

In this case the default plotting domain of $-10...10$ captured the interesting part of the graph (i.e., the part that allows determination of where the graph crosses the *x*-axis)

so there was no need to specify the domain. From the graph we can read off the algebraic solution: $\{x : x < 3\}$.

This solution can also be obtained algebraically by *Maple*.

```
> solve( y-4 < 0 , {x} );
```
$$\{x < 3\}$$

We have used **solve( )** here only to indicate that the mathematical environment does support such solution techniques, but we will not dwell on the actual algebraic technique used to derive this solution. ∎

Our emphasis here is on constructing a graphical interpretation of inequalities. To see how important it is to use a graphical interpretation, consider the problem of comparing two or more expressions.

Example 1.3 Find the values of x for which the expression y is greater than the expression z when y and z are defined as:

```
> y := 3*x - 11;
```
$$3x - 11$$

```
> z := -3*x + 11;
```
$$-3x + 11$$

Solution The plot command allows you to graph several expressions at once simply by specifying a set of expressions as the first argument.

```
> Set1 := { y , z };
```
$$\{3x - 11, -3x + 11\}$$

```
> plot( Set1, x );
```

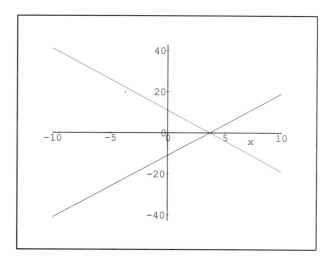

You still must deduce which of the lines on the graph correspond to which expression and you must look for the points where the two lines cross, but the basic process remains quite easy. In this case, the actual solution is *y is greater than z for all x such that* $x > \frac{11}{3}$.

The x value ($x = \frac{11}{3}$) corresponds to the intersection of the two lines and is the solution of the equation $y = z$.

```
> y=z;
```
$$3x - 11 = -3x + 11$$

```
> solve( ", {x} );
```
$$\left\{ x = \frac{11}{3} \right\}$$

Again, the **solve( )** command confirms our algebraic solution.[13]

```
> solve( y-z > 0 , {x} );
```
$$\left\{ \frac{11}{3} < x \right\}$$

An alternative graphical approach is to plot a graph of the difference of the two expressions (i.e., $y - z$), as in

```
> plot( y-z, x );
```

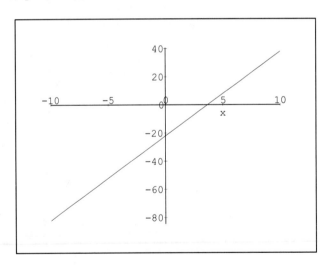

Here the solution to the inequality is the values of x for which this line is above the x-axis. ∎

The solutions generated by the **solve( )** command for inequalities are always described in terms of a sequence of one or more sets of inequalities depending on the nature of the problem at hand. An x value is a solution if at least one of these sets of inequalities is completely satisfied.

---

[13] *Maple* will always write $\{x > 11/3\}$ as $\{11/3 < x\}$ so that it need only deal with one type of strict inequality.

Maple supports set notation and set operations. This can help in setting up the initial graphs and manipulating the resulting solutions. The expression *Set1* is a *set* in the usual mathematical sense. There are no repeated elements and we can test for membership through use of the command **member( )**.

```
> member( y, Set1 );
```

true

REMARK: The variable *x* is not a member of the set $Set1 = \{3x - 11, -3x + 11\}$ even though it appears in both the expressions *y* and *z*.

```
> member( x, Set1 );
```

false

Commands also exist to compute the *union*, *intersection*, and *difference* (*minus*) of two sets. For example, *Set1* could have been constructed as

```
> { y } union { z };
```

$$\{3x - 11, -3x + 11\}$$

(Try the *Maple* help commands: **? sets** and **? solve**.)

1.2.1 More on Inequalities

Inequalities may be much more complicated than those shown. Still, our technique for solving them will remain much the same. In each case we can reorganize the inequality into the form of an algebraic expression compared with a constant and then examine the graph of the algebraic expression in detail. For convenience, we have elected to break these into two categories depending on whether or not the constant is zero.

1.2.2 Type 1 Inequalities

If the inequality is of the form of

"product involving *x*" compared with 0

we can proceed as in the following example.

Example 1.4 Find the values of *x* that satisfy

```
> (x-1)/(x+1) > 0;
```

$$0 < \frac{x-1}{x+1}$$

Solution As always, we first graph the expression $\frac{x-1}{x+1}$ in order to estimate where the expression is nonnegative. Some extra care is needed here since the denominator becomes 0. For example, the graph produced by the simple plot command

```
> plot( (x-1)/(x+1), x );
```

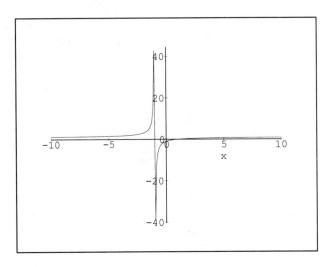

is not very informative near $x = -1$ because of the extremely small values found in the denominator. The **plot( )** command works by sampling the curve at various points and then using straight lines or other smooth curves in between the true sample points. Sometimes, as in this example, this leads to a very poor approximation for parts of the curve.[14] When this happens we just describe the graph more precisely, as in

```
> plot( (x-1)/(x+1), 'x'=-3..3, 'y'=-5..5 );
```

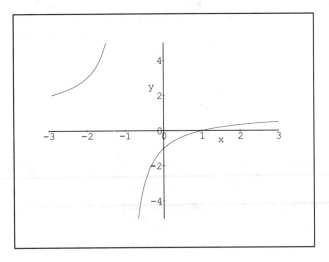

By specifying both the domain and the range of the plot, we force the **plot( )** command to ignore any large numbers resulting from division by the very small denominators near $x = -1$.

[14] The style option **style=LINE** can be used to force the **plot( )** command to use straight lines. The style option **style=POINT** shows the actual sample points.

1.2 Plotting as an Aid to Understanding Inequalities

The algebraic solution to this inequality is

```
> solve ( (x-1)/(x+1) > 0 , x );
```

$$\{x < -1\}, \{1 < x\}$$

which corresponds to the regions of the graph where the curve $y = \frac{x-1}{x+1}$ is nonnegative. A closer examination of the graph shows that the points where the subexpressions $x - 1$ and $x + 1$ are zero play a crucial role. The whole problem can be reformulated as one of determining the signs (positive or negative) of the algebraic subexpressions. If both $x - 1$ and $\frac{1}{x+1}$ have the same sign, their product will be positive; otherwise, their product will be negative. The sign of $x + 1$ and $\frac{1}{x+1}$ are the same and so the signs of $x - 1$ and $x + 1$ are all that need be checked. This information can be read off directly from the next plot:

```
> plot( {x+1,x-1}, x );
```

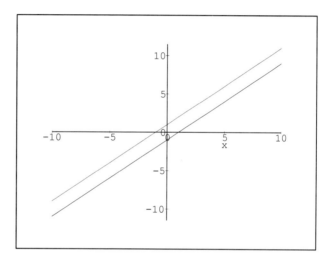

For example, both these straight lines are negative at $x = -5$, so the value of the original expression must be positive at $x = -5$. ∎

The general solution for this type of inequality involves carrying out this sort of analysis (i.e., the counting of negative factors) systematically. Consider the inequality

```
> (x^2- 5*x + 6)/(x^2 + 3*x + 2) < 0;
```

$$\frac{x^2 - 5x + 6}{x^2 + 3x + 2} < 0$$

```
> f := lhs( " );
```

$$\frac{x^2 - 5x + 6}{x^2 + 3x + 2}$$

First we form a set of factors[15] of both the numerator and the denominator of f. The numerator is given by the command **numer( )** and can be factored.

```
> numer( f );
```
$$x^2 - 5x + 6$$

```
> p := factor( " );
```
$$(x - 2)(x - 3)$$

Similarly, the denominator can be obtained and factored.

```
> denom( f );
```
$$x^2 + 3x + 2$$

```
> q := factor( " );
```
$$(x + 2)(x + 1)$$

The combined set of factors of both the numerator and the denominator are given by

```
> e2 := convert( p , set ) union convert( q , set );
```
$$\{x + 1, x + 2, x - 3, x - 2\}$$

and this is precisely the set of expressions whose signs for a given x value determine the sign of the entire expression. The following is a graph of this set of expressions.

```
> plot( e2, x=-5..5 );
```

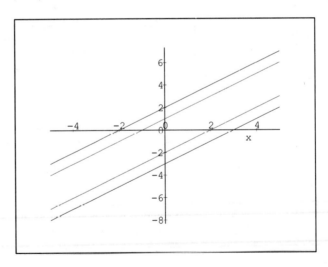

From the graph we see that, for example, at $x = 0$ exactly two of the expressions are negative and two are positive, so the overall value of f is positive at $x = 0$ (i.e., the solution to this inequality should not include $x = 0$).

To find the locations where the sign of a particular expression can change, we must find the zeros of the corresponding graph. For example, the first expression in $e2$ is

---

[15]It is important here that a given factor appear only in the numerator or the denominator if the counting of negative factors is to come out right. Later we will use the **normal( )** command to ensure that this is the case.

```
> e2[1];
```
$$x+1$$

and it is zero at

```
> solve( "=0, {x} );
```
$$\{x = -1\}$$

REMARK: The locations where the actual *number of positive expressions* can change in f are precisely at the zeros of the various terms of the numerator and denominator of f.

Also, the changes in sign and even the sign of the expression

$$\frac{x^2 - 5x + 6}{x^2 + 3x + 2}$$

exactly match those for the graph of the product of all the individual factors

```
> e2;
```
$$\{x+1, x+2, x-3, x-2\}$$

as in

```
> f2 := convert( e2, '*' );
```
$$(x+1)(x+2)(x-3)(x-2)$$

```
> plot( f2, 'x'=-5..5, 'y'=-5..15 );
```

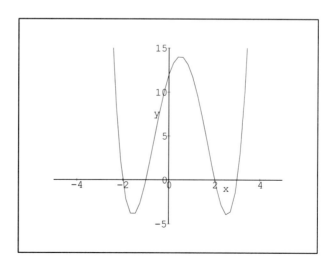

We see that the inequality $f < 0$ is satisfied for $-2 < x < -1$ and for $2 < x < 3$.

1.2.3 Type 2 Inequalities

Type 2 inequalities are of the form *product compared to c* where $c \neq 0$. We can still graph the expression to obtain a direct comparison with the graph of the constant function. However, to determine the exact location of where the graph of the expression crosses the line given by the constant, we can no longer just plot the factors appearing in the expression. Instead, we proceed by first converting our Type 2 inequality to a Type 1 inequality as in the following example.

Example 1.5 Solve the inequality $\dfrac{x+3}{x^2+2x+1} < 2$ for x.

Solution The given inequality is

```
> expr := (x+3)/(x^2 + 2*x + 1) < 2;
```

$$\frac{x+3}{x^2+2x+1} < 2$$

As always, we could approximate the solution by examining the graph of the expression

```
> lhs( expr ) - rhs( expr );
```

$$\frac{x+3}{x^2+2x+1} - 2$$

to determine where it is positive. This can be seen as follows:

```
> plot( ", 'x'=-5..5, 'y'=-5..5 );
```

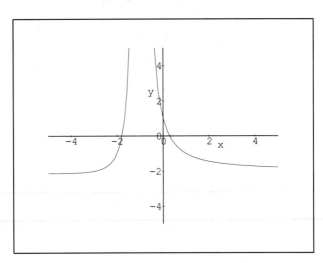

A more detailed solution comes from rearranging the problem as a *Type 1* problem. First, we move all the terms to the left-hand side, as in

```
> lhs( expr ) - rhs( expr ) < 0;
```

$$\frac{x+3}{x^2+2x+1} - 2 < 0$$

We are interested primarily in the left-hand side.

```
> lhs( " );
```

$$\frac{x+3}{x^2+2x+1} - 2$$

We can rearrange the left-hand side further by placing everything on the left over a common denominator and removing common factors. This is precisely the job of the **normal( )** command.

```
> expr2 := normal( " );
```

$$-\frac{3x - 1 + 2x^2}{x^2 + 2x + 1}$$

The structure of the new problem, **expr2 < 0**, is that of a *Type 1 problem,* and so it can be solved directly by working with the zeros of the numerator and denominator.

```
> f1 := numer( expr2 ):
> f2 := denom( expr2 ):
> solve( f1*f2 = 0, {x} );
```

$$\left\{x = -\frac{3}{4} - \frac{1}{4}\sqrt{17}\right\}, \left\{x = -\frac{3}{4} + \frac{1}{4}\sqrt{17}\right\}, \{x = -1\}, \{x = -1\}$$

These zeros are where the graph of both $f_1 f_2$ and f_1/f_2 can change sign. The graph

```
> plot( f1*f2, x );
```

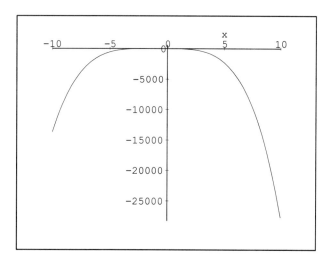

does not help us much in finding exactly where the graph of the product crosses the *x*-axis since *Maple* has automatically adjusted the scale to include all the sample points generated. However, we obtain a more informative result by overriding the vertical scaling, as in

```
> plot( f1*f2, 'x'=-5..5, 'y'=-4..2 );
```

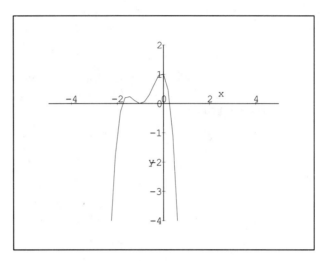

We can now see that *expr2* is negative for $x < -3/4 - \frac{\sqrt{17}}{4}$ and for $x > -3/4 + \frac{\sqrt{17}}{4}$. ∎

1.3 Equations Involving Two Variables

In previous examples we used *y* as the name of an expression whose value was dependent on *x*. In fact, we used the system command := to specify such name assignments, and once a value had been assigned to the name *y*, then every time we used *y* in an expression we were in effect using the associated value (generally an expression *x*). For example, the assignment

```
> y := x^2 + 3;
```
$$x^2 + 3$$

allowed us to conveniently write other expressions such as

```
> y^2 + y;
```
$$\left(x^2 + 3\right)^2 + x^2 + 3$$

If, for some reason, as in the following discussion, we require that *y* cease to have any associated value, we can accomplish this by the assignment

```
> y := 'y':
```

We can verify that this *unassignment* has had the desired effect simply by reevaluating *y* in familiar expressions such as

```
> y, y^2;
```
$$y, y^2$$

1.3 Equations Involving Two Variables

We wanted to redefine y as just a name again because often we have a genuine need to be able to work with equations that specify more general relationships between two variables x and y. Examples of such equations include

```
> e1 := (y + 5 = 5*x + 7);
```
$$y + 5 = 5x + 7$$

```
> e2 := (x^2 + y^2 = 1);
```
$$x^2 + y^2 = 1$$

```
> e3 := (y = x^2 -3*x + 2);
```
$$y = x^2 - 3x + 2$$

The resulting equations can be referred to by name and have no effect on the values of x or y.

DEFINITION 1.1 An equation has been *solved for* y if it is in a form where y appears isolated on the left-hand side of the equation.[16]

The last equation (*e3*) is somewhat special in that it has already been *solved for* y. Though all these equations can be associated with a *graph* of the pairs of points $[x, y]$ that satisfy the specified equation, when an equation has been solved for y it is much easier to determine a value of y that corresponds to a given value of x and to generate *the graph of the equation*. When presented with a relation in this *solved* form, we often say that y *depends* on x. Even if the equation has not been solved for y, we may choose to think of y as *implicitly defined in terms of x* by the given equation.

1.3.1 Equations of Straight Lines

Slope-Intercept Form

A common and useful form for equations of straight lines is called the *slope-intercept* form. It is characterized by the fact that it has been solved for the dependent variable y and also expanded (multiplied out fully) on the right-hand side.

An example of such an equation is

```
> y = m*x + b;
```
$$y = mx + b$$

where both m and b are to be interpreted as numerical constants. This form of the equation is particularly useful because the values of m and b correspond to specific geometric properties of the associated graph. The *slope* of the line is given by m while the y-intercept (i.e., the point on the line representing the function for which $x = 0$) is b.

This knowledge of the correspondence between the algebraic and the geometric representations is useful as it can be used to construct the algebraic equation for a line from its geometric properties.

---

[16]We generally use the **isolate( )** command to solve a single equation though sometimes we need the more comprehensive **solve( )** command.

The slope *m* is computed by taking any two points on the line (i.e., two distinct pairs of *x-y* values satisfying the equation) and computing the ratio of the change in *y* values to the change in *x* values.

```
> slope ( [x1,y1],[x2,y2] );
```

$$\frac{y1 - y2}{x1 - x2}$$

Knowledge of the slope can be used to determine which line corresponds to which *y* values in the following graph.

```
> fns := { 3*x + 5 , 4*x + 5, 5*x + 5 ,3*x + 5 ,6*x + 5};
```

$$\{3x + 5, 4x + 5, 5x + 5, 6x + 5\}$$

```
> plot( fns , 'x'=-5..5, 'y'=-10..10 );
```

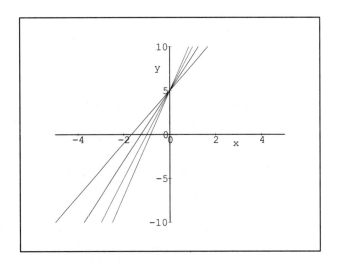

Similarly, knowledge of the *y*-intercept can be used to determine which line corresponds to which expression for *y* in the next graph.

```
> fns2 := { 3*x + 5 , 3*x + 7 , 3*x };
```

$$\{3x + 5, 3x, 3x + 7\}$$

```
> plot( fns2, 'x'=-5..5, 'y'=-10..10 );
```

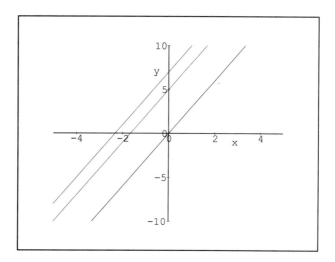

A particular equation may still represent a straight line even if it is not in the *slope-intercept* form. In fact, depending on the geometric properties we have been given, we may even prefer to write the equation in a different form.

Example 1.6 If we know that the line goes through the point [4, 5] and has slope 12, the corresponding equation can be found by comparing the slope (as determined by using the point [4, 5] and an arbitrary point [x, y]) to 12.

```
> slope( [x,y], [4,5] ) = 12;
```

$$\frac{y-5}{x-4} = 12$$

We can convert such an equation into the slope-intercept form by isolating y.

```
> isolate( ", y );
```

$$y = 12x - 43$$

Sometimes it may be necessary to expand this further using **expand( )**.

Example 1.7 If we know the slope of the line is 8 but the specific point is unspecified (e.g., $[x1, y1]$), we simply compare the point (as determined by the point $[x1, y1]$ and an arbitrary point $[x, y]$) to 8.

```
> slope( [x,y], [x1,y1] ) = 8;
```

$$\frac{y - y1}{x - x1} = 8$$

By specifying different points $[x1, y1]$ we may choose a different line, but all such lines will have the same slope and are said to be *parallel*.

Standard Form

Sometimes we may be given the equation of the line in the form

> a*y + b*x = c;

$$ay + bx = c$$

where a, b, and c are constants. This form has the advantage that it is still valid even when one of a or b is 0. Of course, we may not be able to isolate y in all cases. For example, the equation $4x = 12$ corresponds to the vertical line at $x = 3$.

1.3.2 Distances Between Points

The distance of a point $[a, b]$ from the origin can be expressed in terms of a and b. It is

> c := distance([a,b], [0,0]);

$$\sqrt{a^2 + b^2}$$

Since the coordinate axes are at right angles, this is merely an interpretation of the familiar rule for right-angle triangles that $a^2 + b^2 = c^2$. By measuring along lines parallel to the axes, we can extend this rule to compute the distance between any two points.

> distance([x2,y2], [x1,y1]);

$$\sqrt{(x1 - x2)^2 + (y1 - y2)^2}$$

This can be reexpressed in the forms

> expand(");

$$\sqrt{x2^2 - 2x2\,x1 + x1^2 + y2^2 - 2y2\,y1 + y1^2}$$

and

> completesquare(", [x2]);

$$\sqrt{(x2 - x1)^2 - 2y2\,y1 + y1^2 + y2^2}$$

The midpoint of the line segment drawn between these two points is computed as

> midpoint([x1,y1], [x2,y2]);

$$[\frac{1}{2}x1 + \frac{1}{2}x2, \frac{1}{2}y1 + \frac{1}{2}y2]$$

We conclude this section with some examples of problems involving distance.

Example 1.8 Find an equation for the set of points equidistant from $[2, 5]$ and $[8, 12]$.

Solution Let $[x, y]$ be an arbitrary point in the prescribed set. What do we know about this point? Just that it is the same distance from each of $[2,5]$ and $[8,12]$.

> P := [2,5]: Q := [8,12]:
> d := distance([x,y], P) = distance([x,y], Q);

$$\sqrt{(x-2)^2 + (y-5)^2} = \sqrt{(x-8)^2 + (y-12)^2}$$

This equation is already an answer, though it is certainly not in the form that we want it. We can simplify it by isolating y.

```
> isolate( ", y );
```

$$y = -\frac{6}{7}x + \frac{179}{14}$$

From this new form of the equation we can deduce that the set of points we are trying to describe forms a straight line. ∎

1.3.3 Solving an Algebraic Equation Step by Step

Until now we have relied very heavily on *Maple* to solve most of the equations that we have encountered. We can also look at the equation-solving process in more detail. The basic manipulations required to complete the immediately preceding solution for y include the following operations:

1. Square both sides of the equation.
2. Move all terms to one side (say the right) of the equation by subtracting one side from the other. This has the effect of canceling common terms and collecting multiples of x and y.
3. Move all terms involving y to the left side.
4. Divide both sides by the coefficient of y.

Example 1.9 Simplify the equation

```
> d;
```

$$\sqrt{(x-2)^2 + (y-5)^2} = \sqrt{(x-8)^2 + (y-12)^2}$$

into the slope-intercept form for a line.

Solution   To complete this task we find it convenient to add to the functionality provided by *Maple*. For example, to simplify the task of squaring both sides of the equation, we first define our own *Maple* function that can square any algebraic quantity.

```
> sq := (x) -> x^2;
```

$$x \mapsto x^2$$

We can easily test that we have created the desired function by actually trying it on sample values, as in

```
> sq( y ), sq( 3 );
```

$$y^2, 9$$

We usually need to modify both sides of an equation in the same manner. The **Maple** command **map( )** can be used for this and (as we shall see) a wide variety of other tasks. The notations `""` and `"""` can be used to refer to the second and third last expression values, as required, while the **select( )** command and **lhs( )** and **rhs( )** allows us to single out parts of the given expressions.

Armed with these tools, we are ready to begin our solution. The first step is to *apply* the function **sq** to both sides of the original equation.

```
> map( sq, d );
```
$$(x - 2)^2 + (y - 5)^2 = (x - 8)^2 + (y - 12)^2$$

The terms common to both sides will automatically disappear if we rearrange the equation as

```
> 0 = expand( rhs( " ) - lhs( " ) );
```
$$0 = -12x + 179 - 14y$$

The terms involving x^2 and y^2 have canceled so we know the result is a straight line. The term on the right that involves y is

```
> select( has, rhs( " ), y );
```
$$-14y$$

It can be subtracted from both sides of the previous equation by the command

```
> ("") - ("=");
```
$$14y = -12x + 179$$

Finally, we can divide both sides of the equation by the coefficient of y.

```
> "/coeff(lhs("),y);
```
$$y = -\frac{6}{7}x + \frac{179}{14}$$

We could have also used **isolate( )** to complete this last step. ∎

A Solution That Begins by Assuming a Straight Line

There is more than one approach to most problems. For the previous problem, people often start by assuming that the answer is a straight line and then use this knowledge to construct its equation.

Solution Assume that the solution is the straight line perpendicular to the line segment PQ and passing through its midpoint. The specified points are

```
> P, Q;
```
$$[2, 5], [8, 12]$$

The negative reciprocal of the slope of the line segment PQ is

```
> m := -1/ slope( P, Q );
```

The solution line must pass through the midpoint of the line segment PQ, which is

```
> M := midpoint( P, Q );
```

$$[5, \frac{17}{2}]$$

An equation for this line of known slope and passing through a known point M is

```
> ( y - M[2] ) / ( x - M[1] ) = m;
```

$$\left(y - \frac{17}{2}\right)(x - 5)^{-1} = -\frac{6}{7}$$

The slope-intercept form of this line is

```
> isolate( ", y );
```

$$y = -\frac{6}{7}x + \frac{179}{14}$$

∎

1.3.4 Quadratic Equations and Their Graphs

Our next goal is to graph equations that involve either x^2 or y^2 (or possibly both). In each instance, the equation in x and y may correspond to one or more values of y for each x, or vice versa. These multiple solutions arise from the solutions to the quadratic equation, as in

```
> eq := a*x^2 + b*x + c = 0;
```

$$ax^2 + bx + c = 0$$

```
> sol := solve( eq , {x} );
```

$$\left\{x = \frac{1}{2}\left(-b + \sqrt{b^2 - 4ac}\right)a^{-1}\right\}, \left\{x = \frac{1}{2}\left(-b - \sqrt{b^2 - 4ac}\right)a^{-1}\right\}$$

These two formulas together are often referred to as *the quadratic formula*.
In a specific instance such as

```
> 0 = x^2+2*x-3;
```

$$0 = x^2 + 2x - 3$$

the quadratic formula gives rise to the solutions

```
> solve( ", {x} );
```

$$\{x = -3\}, \{x = 1\}$$

Integer solutions such as these can also be read directly from the factored form of the equation

> `factor("");`

$$0 = (x+3)(x-1)$$

When we are working with real numbers, we can take square roots only of non-negative numbers. Thus, depending on the value of the quantity $d = b^2 - 4ac$, the formula may give rise to zero, one, or two solutions. If $d = 0$, then both solutions are identical, whereas if it is negative, no solutions exist at all.

1.3.5 Solving a Quadratic

You may not recall exactly how the quadratic formula is derived but you should find the case with no linear x term manageable. This is just the previous equation *eq* with $b = 0$, as in

> `b := 0; eq;`

$$ax^2 + c = 0$$

To solve this for x we simply move the constant to the right side and take square roots![17]

> `" - (c=c);`

$$ax^2 = -c$$

> `"/a;`

$$x^2 = -\frac{c}{a}$$

> `map( sqrt, " );`

$$x = \sqrt{-a^{-1}}\sqrt{c}$$

Although only one solution appears here, there are really two since $(+x)^2 = (-x)^2$.

Exactly the same approach works in the situation where there is both a linear and quadratic term involving x, but to see this we have to complete the square[18] prior to beginning the whole operation.

> `b := 'b': eq;`

$$ax^2 + bx + c = 0$$

> `completesquare( eq, x );`

$$a\left(x + \frac{1}{2}\frac{b}{a}\right)^2 - \frac{1}{4}\frac{b^2}{a} + c = 0$$

---

[17] For real solutions to exist, a and c should have opposite signs.
[18] See **?student,completesquare**.

1.3 Equations Involving Two Variables

The rest is fairly direct algebraic simplification. The term $x + \frac{b}{2a}$ is isolated as in

```
> isolate( ", (x+b/2/a)^2 );
```

$$\left(x + \frac{1}{2}\frac{b}{a}\right)^2 = \left(\frac{1}{4}\frac{b^2}{a} - c\right)a^{-1}$$

This can be further simplified by placing terms over a common denominator using the **normal( )** command.

```
> lhs(") = normal(rhs("));
```

$$\left(x + \frac{1}{2}\frac{b}{a}\right)^2 = -\frac{1}{4}\frac{-b^2 + 4ac}{a^2}$$

If we take square roots on both sides

```
> map(sqrt,");
```

$$x + \frac{1}{2}\frac{b}{a} = \frac{1}{2}\sqrt{b^2 - 4aca^{-1}}$$

then we can move the constant part back over to the right-hand side by isolating x.

```
> isolate( ", x );
```

$$x = \frac{1}{2}\sqrt{b^2 - 4aca^{-1}} - \frac{1}{2}\frac{b}{a}$$

Again there are really two possible solutions depending on whether we choose the positive or the negative square root.

1.3.6 Equations Involving x^2 and y^2

Quadratic equations might involve both x and y. For example, we can have equations such as

```
> e1 := y^2 = 2*x + 5;
```

$$y^2 = 2x + 5$$

Such equations still express a relationship between values of x and values of y, but there is generally more than one pair $[x, y]$ that satisfies the equation and can be plotted on our Cartesian coordinate system.

Example 1.10 Obtain a graph[19] corresponding to the expression $e1$.

Solution We first consider the equation as a quadratic in y, with a *constant* x. We solved such a quadratic in the previous section albeit with different constants ($a, b,$ and c). As expected, there are two solutions, one corresponding to each root (positive and negative).

```
> solve( e1, {y} );
```

$$\left\{y = \sqrt{2x+5}\right\}, \left\{y = -\sqrt{2x+5}\right\}$$

[19]Later we will make use of the command **implicitplot( )**.

A more useful form of the **solve( )** command, especially when it comes to plotting, is as follows:

> solve(e1, y);

$$\sqrt{2x+5}, \ -\sqrt{2x+5}$$

In this form, the **solve( )** command returns a sequence of values rather than a sequence of equations. We can plot these two expressions simultaneously on the same coordinate system as follows. First we place the expressions for y in a set.

> {"};

$$\left\{-\sqrt{2x+5}, \sqrt{2x+5}\right\}$$

Then we invoke **plot( )** on this set.

> plot(", 'x'=-10..10, 'y'=-10..10);

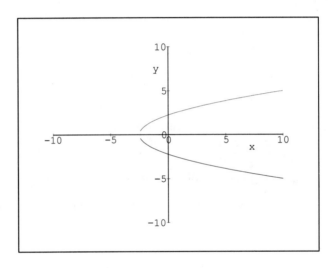

Example 1.11 Obtain a graph corresponding to the equation

> e2 := x^2 + y^2 = 16;

$$x^2 + y^2 = 16$$

representing a circle centered at the origin.

Solution We proceed exactly as before.

> solve(e2, y);

$$\sqrt{-x^2+16}, \ -\sqrt{-x^2+16}$$

```
> plot( { " }, 'x'=-10..10, 'y'=-10..10 );
```

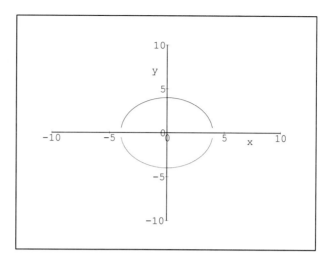

The result is a circle of radius 4, but the graph is distorted by the fact that we used a different scaling[20] for the x and y axes. By *scaling*, we mean the actual length of one unit in the direction of a given axis. Small gaps appear in the graph because the curve has been approximated by two separate curves, each of which is a bit short because of the way the **plot( )** command has chosen sample points for the curves. ∎

REMARK: The approach just outlined works for all quadratics involving both x and y, and generally, for any equation that **solve( )** is able to handle. For all these types of examples, we can obtain the graph by reducing the problem to that of graphing a set of expressions simply by doing

```
> solve( ", y );
> plot( { " } );
```

Thus, the problem of graphing such relations can be reduced to that of solving equations.

Remember to Think!

At this point we might be tempted to think that future thoughts on these topics are wasted. Each problem has been reduced to graphing or solving equations and the same approach works for a large class of problems. If we know how to solve the equations, then what more is there for us to do?

Observe the following:

1. A slight change in our point of view can often greatly simplify things. Equations such as $y^2 = 2x + 5$ are much more easily handled if we are willing to obtain the graph

---

[20]The axis can be constrained by making use of the plot option **scaling=constrained**.

by computing *x* values in terms of *y* values. In short, we can avoid the complicated **solve( )** step altogether. This small observation is typical of mathematical problem solving. Always be ready to view your problem in a different light.

```
> solve( e1, x );
```

$$\frac{1}{2}y^2 - \frac{5}{2}$$

```
> plot( ", y );
```

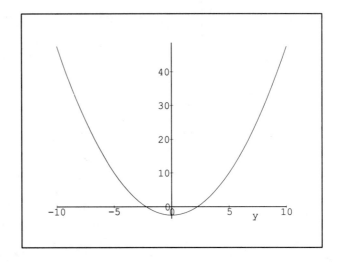

2. We will always be able to find examples where the limitations of the computational power available to us means that we need to come up with a better way to solve the problem.
3. We may need to solve a similar problem many thousands of times, in which case it would be well worth our while to use the best method available to us.

The key here is to use all the available tools to investigate alternative approaches. The thinking goes on, but at the level of combining the various techniques and developing an intuition for the computational costs associated with each approach.

Example 1.12 The graph of the hyperbola

```
> eq := 4*x^2 - 9*y^2 = 36;
```

$$4x^2 - 9y^2 = 36$$

can be obtained by the same technique as above.

Solution We proceed exactly as before.

```
> solution := solve( ", y );
```

$$-\frac{2}{3}\sqrt{x^2 - 9}, \frac{2}{3}\sqrt{x^2 - 9}$$

```
> plot( { " }, 'x'=-10..10 );
```

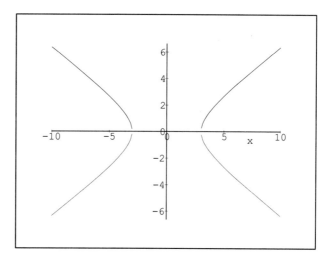

Such *hyperbolic equations* are more often presented in the form

```
> eq/(4*9);
```

$$\frac{1}{9}x^2 - \frac{1}{4}y^2 = 1$$

The form

```
> x^2/a^2 - y^2/b^2 = 1;
```

$$\frac{x^2}{a^2} - \frac{y^2}{b^2} = 1$$

helps us identify two lines corresponding to the expressions $\{\frac{bx}{a}, \frac{-bx}{a}\}$. These two lines are examples of what are known as *asymptotes*, which will be studied in section 4.5. They are important since they describe in simple terms the approximate behavior of the original graph. You can see this effect by graphing the asymptotes and the original graph together.

```
> { solution, 2/3*x , -2/3*x };
```

$$\left\{\frac{2}{3}x, -\frac{2}{3}x, \frac{2}{3}\sqrt{x^2-9}, -\frac{2}{3}\sqrt{x^2-9}\right\}$$

```
> plot( ", x=-10..10 );
```

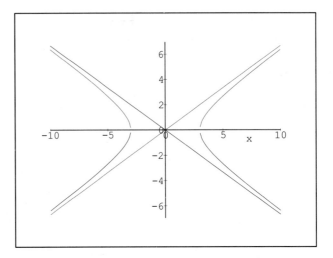

Any time we can approximate a complicated expression accurately with a simpler one, we have gained some insight to the behavior of a mathematical object.

1.4 Functions

Functions play an important role in mathematics and in *Maple*. We include here a few examples of function definitions and ways in which they can be manipulated and combined to make new functions.

1.4.1 Defining a Function

A function can be thought of as a machine. Given an *input*, the machine uses this input as required to compute exactly one *output*. There are three distinct points to note when defining a function.

1. A symbolic name is used to represent a typical input value (i.e., the value that is to be transformed by the rule into a new value).
2. The computations necessary to obtain the corresponding output value must be described. This is usually given as a sequence of one or more formulas that use the given symbolic name as necessary to refer to the current input value.
3. (*optional*) The rule or function is often given a name. We do not want to completely redescribe the rule every time we refer to the function or use it.

In mathematical discussions functions are frequently defined informally – by example. For example, we might say f is the function defined by $f(x) = x^2$. What we really mean by this statement is that x is a sample of the input to the function, x^2 is the rule for computing a new value from the input x, and f is the function name.

Sometimes an equation such as $f(x) = x^2$ may occur as the result of solving an equation for $f(x)$, as in

```
> eq := x^2 + 3 = 3*f( x );
```
$$x^2 + 3 = 3f(x)$$

```
> isolate( ", f( x ) );
```
$$f(x) = \frac{1}{3}x^2 + 1$$

To avoid any possible confusion between definitions and equations involving $f(x)$, we will always use the following special notation when defining a function. The statement

```
> f := (x) -> x^2;
```
$$x \mapsto x^2$$

defines the function named f as the function for which $f(x) = x^2$. The simulated *arrow* indicates that the specified *argument* on the left is to be transformed into the expression appearing to the right of the arrow.

1.4.2 Using Functions

The notation $f(x)$ represents the result of *applying* the function f to the input value x whatever x is. For example we can have each of

```
> f( x ), f( 2 ), f( x+1 ), f( a+b );
```
$$x^2,\ 4,\ (x+1)^2,\ (a+b)^2$$

The function named f is *applied* to a specific input value and if possible the result is actually computed.

REMARK: If no rule has been associated with a name g, then the value at x is the *unevaluated function call $g(x)$*.

Unevaluated function calls still stand for (i.e., are names for) the values that would have been computed had we had a way of doing so. Often we are able to compute the value of an unevaluated expression later in a computational session by using additional information uncovered by our investigations. For example, the equation

```
> 3*g( x ) + x^3 = x^2;
```
$$3g(x) + x^3 = x^2$$

allows us to discover the relation

```
> isolate( ", g( x ) );
```
$$g(x) = \frac{1}{3}x^2 - \frac{1}{3}x^3$$

REMARK: A function is still a function even if it has no name. A name is merely a convenience.

The notion of a function without a name is especially useful if we are going to use the function only once. Instead of defining the function g as $g := (x) \to x^3 + x$ and then applying it to u, we write

```
> ((x) -> x^3 + x)( u );
```
$$u^3 + u$$

Here the function definition appears as part of the expression itself in place of a name. We still know to apply the function to its argument by the positioning of the definition in front of the parenthesized argument.

To apply a function to both sides of an equation use the **map( )** command, as in

```
> map( f, a = b ), map( g, a = b );
```
$$a^2 = b^2, \ g(a) = g(b)$$

Similarly, given an ordered *list* of values

```
> L := [1,2,3,4];
```
$$[1,2,3,4]$$

the result of mapping a function f onto a list of arguments is the list of function values $[f(1), f(2), f(3), f(4)]$, as in

```
> map( f, [1,2,3,4] );
```
$$[1,4,9,16]$$

Neither the function definition nor the equation or list appearing in the arguments to **map( )** need have names.

```
> map( (x) -> x^2, a=b );
```
$$a^2 = b^2$$

```
> map( g, [a,b] );
```
$$[g(a), g(b)]$$

A distinct advantage of the symbolic environment is that we can discover the definition of simple functions or operations just by trying them with unevaluated names. For example, by mapping an unknown function g onto $a = b$, we discover that the result of such a map is $g(a) = g(b)$. We can then define g after the fact, or replace it in the previous command by the substitution

```
> subs( g=f,");
```
$$[f(a), f(b)]$$

Such substitutions are very much in the spirit of editing instructions. They construct a new expression much as if you had typed it in, but the result is not evaluated. To see how it evaluates, you must refer to it again

```
> ";
```
$$[a^2, b^2]$$

1.4.3 Restricted Domains

Domains of functions (i.e., those values for which the rule is truly defined) may be restricted to intervals such as $0 \leq x \leq 3$. This is accomplished in *Maple* by using a slightly longer form of function definition, as illustrated here.

```
> g := proc( x )
> if 0 <= x and x <= 3 then
>     x^2
> else
>     ERROR( x, `not in domain' )
>     fi
> end:
```

We can use this function inside the domain

```
> g( 0 ), g( 2 ), g( 1.5 );
```

$$0, 4, 2.25$$

but when we use it outside the domain we obtain an error.

```
> g( 4 );
Error, (in g) 4, not in domain
```

This approach also allows us to define functions that use different rules on different subintervals. These are often referred to as *piecewise defined functions*.

```
> h := proc( x )
> if -3 <= x and x <= 3 then
>     x^2 - 2
> elif 3 < x and x <= 5 then
>     x-3
> elif x > 5 then
>     -x^2 + 34
> else
>     undefined
>     fi
> end:
```

The result can be used to generate a plot,[21] as in

```
> plot( h ,-10...10 );
```

---

[21] We cannot use the command **plot(h(x),x);** because the expression $h(x)$ is not defined for arbitrary x.

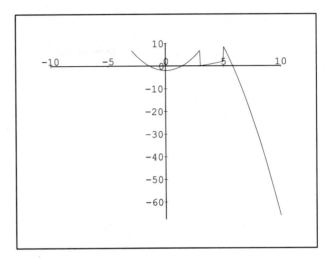

The actual domain of this function is from $x = -3$ to ∞. The **plot( )** command[22] is clever enough to ignore the points outside of its domain and still continue with the plot. However, another weakness that is inherent in graphing by sampling shows up here as unwanted lines connecting portions of the graph that should not be connected. (Can you find them?) Without knowing and working with the mathematical properties of a function there is no way of being sure of such interpretations.

1.4.4 Finding Domains of Functions

Formulas need not be defined for all real numbers. Problems usually arise because of zeros in the denominators of the defining formulas, or because of restrictions arising from operations such as taking the square root.

Example 1.13 Find the domain of the function f as defined by

```
> f := (x) -> 1/(x^2 -x);
```

$$x \mapsto (x^2 - x)^{-1}$$

Solution To solve this problem we must look closely at the expression used to define the function f.

```
> f( x );
```

$$(x^2 - x)^{-1}$$

This expression is undefined where the denominator is zero. Sometimes we can discover where this is simply by factoring the denominator.

```
> factor( " );
```

[22] You specify the domain for the graph as an optional extra argument, as in **plot(h,a..b)**.

$$\frac{1}{x(x-1)}$$

This shows that the denominator has a zero when $x = 0$ or $x = 1$, as verified by any attempt to evaluate f at these two points.

```
> f( 0 );
Error, (in f) division by zero
```

```
> f( 1 );
Error, (in f) division by zero
```

Apart from values of x that cause the denominator to become zero, this particular formula for computing values of $f(x)$ is valid everywhere. The domain is the entire real line except for $x = 0$ and $x = 1$. This is because the formula is constructed from expressions like polynomials that are defined everywhere. ■

Example 1.14 Find the domain of f when f is defined by

```
> f := (x) -> sqrt( 2 - x - x^2 );
```

$$x \mapsto \sqrt{2 - x - x^2}$$

Solution Again we look at the formula defining the function.

```
> f( x );
```

$$\sqrt{2 - x - x^2}$$

The quantity under the square root must be nonnegative if we are to get a real number.

```
> "^2;
```

$$2 - x - x^2$$

```
> factor( " );
```

$$-(x + 2)(x - 1)$$

```
> plot( ", x=-3..3 );
```

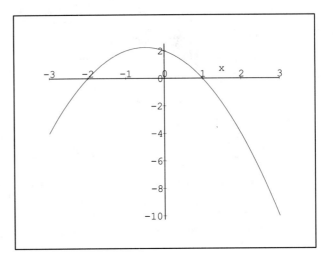

We see from this graph that the domain of x must be restricted to the values of x between the two roots of the polynomial if we are to get a real number as a function value. ∎

1.4.5 Absolute Values

When we use absolute values, we are in effect using two different formulas: one when the argument to the **abs( )** function is positive and another when the argument is negative. Consider the graph

```
> plot( abs(x^3 - 3*x ), x = -2..2 );
```

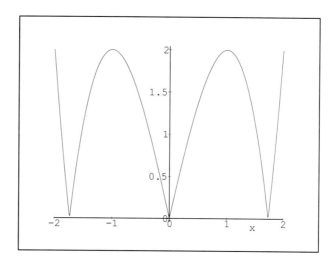

On the intervals $x < -\sqrt{3}$ and $0 < x < \sqrt{3}$, the formula $-x^3 + 3x$ is used; for the rest of the graph, the formula $x^3 - 3x$ is used.

1.4.6 Step Functions

Step functions are functions that are constant on selected subintervals. This kind of behavior occurs when we truncate to the greatest integer less than a given value, as in

```
> plot( trunc( x ), x=0...5, style=LINE );
```

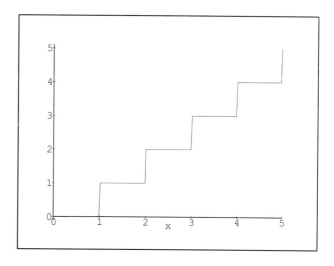

Although this plot incorrectly includes vertical straight lines between each of the steps because of the sampling technique being used, it leaves no doubt as to the origin of the name *step function*.

1.4.7 Even Functions

Even functions are characterized by the fact that $f(x) = f(-x)$. They are symmetric about the y-axis. Examples usually are based on the exclusive presence of even powers of x.

```
> f := (x) -> x^2 + 3*x^4 + 1;
```
$$x \mapsto x^2 + 3x^4 + 1$$

```
> f( x ) = f( -x );
```
$$x^2 + 3x^4 + 1 = x^2 + 3x^4 + 1$$

```
> f( 2 ) = f( -2 );
```
$$53 = 53$$

An exception is the trigonometric expression $\cos(x)$, but even this can be described in terms of even powers of x. Later we shall see how it can be approximated by a polynomial-like expression[23] of the form

```
> taylor( cos( x ), x, 10 );
```
$$(1 - \frac{1}{2}x^2 + \frac{1}{24}x^4 - \frac{1}{720}x^6 + \frac{1}{40320}x^8 + O(x^{10}))$$

which involves only even powers of x.

1.4.8 Odd functions

Odd functions are characterized by the fact that $f(x) = -f(-x)$. Examples of these usually are based on the exclusive use of odd powers of x.

```
> f := (x) -> x^3 + 3*x;
```
$$x \mapsto x^3 + 3x$$

```
> f( 2 ) = - f( -2 );
```
$$14 = 14$$

```
> f( 3 ) = - f( -3 );
```
$$36 = 36$$

```
> f( x ) = - f( -x );
```
$$x^3 + 3x = x^3 + 3x$$

An exception to this rule is the expression $\sin(x)$, which it turns out can still be approximated by odd powers through the polynomial-like expression

```
> taylor( sin( x ), x, 10 );
```
$$(x - \frac{1}{6}x^3 + \frac{1}{120}x^5 - \frac{1}{5040}x^7 + \frac{1}{362880}x^9 + O(x^{10}))$$

which involves only odd powers of x.

---
[23] This is an example of *taylor series* and will be dealt with in chapter 10.

1.4.9 Constant Functions

For these functions, the argument to the function can be completely ignored. For example, the function

```
> f := (x) -> 2;
```
$$2$$

results in the following evaluations

```
> f( a ), f( b ), f( 2 ), f( 100000000000 );
```
$$2, 2, 2, 2$$

REMARK: *Maple* understands an expression to be a function by the manner in which you use it. If you apply one expression to another, then the former must be a function. This can lead to some surprising results. For example,

```
>   3(x^2 + 3);
```
$$3$$

must be understood as the result of applying the constant function defined by $t \to 3$ to the argument $x^2 + 3$. Do not confuse it with `3*(x^2 + 3)`.

1.5 New Functions from Old

The following examples hint at the extent to which *Maple* supports the standard mathematical notation for some of the common ways of combining old functions to create new functions. Sometimes care must be taken in determining the new domains and ranges. The main problem that arises is that if divisions are performed, the denominator may be zero.

```
> f( x ), g( x );
```
$$f(x), g(x)$$

```
> (f+g);
```
$$f + g$$

```
> (f+g)( x );
```
$$f(x) + g(x)$$

```
> (f-g)( x );
```
$$f(x) - g(x)$$

```
> (f*g)( x );
```
$$f(x)g(x)$$

```
> (f/g)( x );
```
$$\frac{f(x)}{g(x)}$$

1.5.1 Arithmetic Operations

Consider the following two functions:

```
> f := (x) -> sqrt( x+1 );
```
$$x \mapsto \sqrt{x+1}$$

```
> g := (x) -> sqrt( 9-x^2 );
```
$$x \mapsto \sqrt{9-x^2}$$

Three examples of the use of functions derived from these are

```
> (f+g)( x ), (f-g)( x ), (f/g)( x );
```
$$\sqrt{x+1} + \sqrt{9-x^2}, \; \sqrt{x+1} - \sqrt{9-x^2}, \; \sqrt{x+1}\frac{1}{\sqrt{9-x^2}}$$

The valid domains of these new functions are determined from the valid domains of the original functions. Since for now we are not allowing you to take square roots of negative numbers, the effective domain of f is $x \geq -1$. Similarly, the effective domain of g is $-3 \leq x \leq 3$. In all the combined cases shown, the new domain is essentially $-1 \leq x \leq 3$. The only exception is for the f/g case, where zeros in the denominator may exclude certain points.

```
> plot( { f( x ), g( x ), (f+g)( x ) }, 'x'=-5..5 );
```

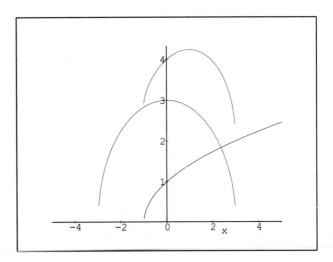

Which of the preceding functions correspond to which line on the graph?

1.5.2 Composition of Functions

The essential idea of functional composition is that of applying functions one after another, each to the result produced by the previous functional application. The usual mathematical notation for the *composition* of f and g is $f \circ g$, but largely because of the limitations of our typewriter keyboard we use `f@g`. The actual value of $(f \circ g)(x)$ is defined as $f(g(x))$, as in

```
> (f@g)(x);
```
$$f(g(x))$$

Example 1.15 Compute all the possible compositions of the two functions shown below.
```
> f := (x) -> x^3;
```
$$x \mapsto x^3$$
```
> g := (x) -> 3*x-2;
```
$$x \mapsto 3x - 2$$

Solution Using *Maple*'s notation we compute the various compositions as
```
> (f@g)( u );
```
$$(3u - 2)^3$$
```
> (g@f)( u );
```
$$3u^3 - 2$$
```
> (f@f)( u );
```
$$u^9$$
```
> (g@g)( u );
```
$$9u - 8$$

■

Note the effect of applying such a composition to a more complicated argument, as in
```
> (f@g)( a+b );
```
$$(3a + 3b - 2)^3$$

Example 1.16 Compute the result of applying the compositions $f \circ g \circ h$ and $f \circ h \circ g$ to x when f, g, and h are defined by
```
> f := (x) -> x/(x+1):
> g := (x) -> x^10:
> h := (x) -> x+3:
```

Solution These compositions each involve three functions:
```
> (F@G@H)(x);
```
$$F(G(H(x)))$$

For the specific functions and compositions given in this example we obtain
```
> (f@g@h)( x );
```

$$\frac{(x+3)^{10}}{(x+3)^{10}+1}$$

and

> (f@h@g)(x);

$$\frac{x^{10}+3}{x^{10}+4}$$

∎

You should experiment with each of these operations on simple examples until you can accurately predict what the outcome will be in each case.

EXERCISE SET 1

Consult your instructor for details about how to start a Maple session on a specific machine. Look over the following section on how to get started with Maple V, Release 2 and on preliminaries. Then use the indicated commands to solve the problems listed on the following pages.

Using *Maple*

Maple is a mathematical manipulation language developed by the Symbolic Computation Group at the University of Waterloo. It differs from traditional high-level computing languages in that it can manipulate symbols according to various algebraic rules as well as do the usual operations on numbers and variables.

At this point, start your *Maple* session. Your terminal or computer should respond with something like

```
     |\^/|        MAPLE V  Release 2
._|\|   |/|_.  Copyright (c) 1981-1992 by the University of Waterloo.
 \  MAPLE  /   All rights reserved.  MAPLE is a registered trademark of
 <____ ____>   Waterloo Maple Software.
      |        Type ? for help.
>
```

The "prompt" ($>$, or some similar symbol) indicates that *Maple* is ready to receive a command.

It is important to know how to end a session. When you are ready to end your current session, you need only type the *Maple* command **quit** followed by a carriage return. On some machines you will be able to select a "Quit" button using a mouse.

The commands you must type are all prefaced by the prompt symbol $>$. The results of those commands appear immediately after. For example, assigning an algebraic value to *p*, inspecting its value, and then cubing that value are accomplished by entering the commands

> p := x^2+ 3*x + 2;

$$x^2 + 3x + 2$$

```
> p;
```
$$x^2 + 3x + 2$$

```
> p^3;
```
$$(x^2 + 3x + 2)^3$$

Try it! You can always restore *Maple* to the initial state used in most of these chapters by entering the commands

```
> restart; with(student):
```

This clears all assignments that you have made to variables during your session and then reloads a package of extra routines that are used extensively throughout this book.

Preliminaries

Maple is intended to be used interactively. Instead of writing programs, you will be issuing commands to proceed one step at a time through the algebraic operations you require. There are actually thousands of commands in *Maple*, but the only ones you will require will be illustrated on a *need to know* basis. You can (and should) ignore the rest of *Maple*'s capabilities until much later. Tables 1.1 and 1.2 summarize the commands used in this chapter.

As you start to enter commands, recall the following technical points already covered in this chapter.

- All *Maple* commands must end with a semicolon (;) or a colon (:). A colon is used to suppress the printing of the output or result that is produced by the command, but the result is still there. (Type the command "; to see it.)
- If you forget the semicolon, just type it on the next line.
- Extra semicolons are perfectly acceptable and can be typed any time between commands.
- All arithmetic operations, including multiplication, must be specified. Thus, you must type x*y rather than x y.
- The symbol = is used to denote equations.
- The symbol := is used to assign names to expressions.
- Use exact rational numbers (e.g., 2/3) instead of decimal expansions (e.g., 0.66) as much as possible. Many algebraic operations such as factoring are only defined for *exact* quantities.

Finding Help

All the available commands are documented in an online help facility. For example, to get help on the command **factor** just type "? factor". If you do not know the name of the command, try using the first two or three characters of a likely sounding name, perhaps"? fac", or use the help index.

Depending on the interface to *Maple* that you are using, you may also have access to a *help browser*. This allows you to access help files directly by using a mouse to click on a button appearing on your screen.

Command	Description
";	The previous expression
"";	The expression two back
""";	The expression three back
(f+g)(x);	Compute $f(x) + g(x)$
(f*g)(x);	Compute $f(x) * g(x)$
(f@f)(u);	Compute $f(f(u))$
(x−1)/(x+1) > 0;	An inequality
100!;	Compute 100 factorial
4*x + 3 = x + 7;	An equation
P := [1,2];	Name a list of two elements as P
P;	The value of P
Pi;	A symbol representing π
x^2+x+3;	A polynomial of degee 2
abs(x);	The absolute value of x
assign(a=b);	Execute the command **a:=b;**
b := 'b';	Unassign (or remove) the value of b
completesquare(x^2 + 2*x,x);	Complete the square
convert(e,'*');	Convert to a product
convert(p,list);	Convert p to a list
convert(p,set);	Convert p to a set
denom(f);	The denominator of f
distance([a,b],[c,d]);	Distance between points
e[1];	The first element of a list
evalf(5/17);	Approximate to 10 digits
evalf(5/17, 100);	Approximate to 100 digits
expand(s);	Expand s by multiplying out in full

Table 1.1 Some of the Commands Used in Chapter 1

Command	Description
f := (x)-> x^3;	Define the function f
f(x+3);	Evaluate f at $x+3$
factor(p);	Factor the polynomial p
isolate(eq, x);	Isolate x in the equation eq
lhs(eq);	The left hand side of the equation eq
map(g, [a,b]);	Construct the list $[g(a), g(b)]$
member(x, Set1);	Test if x is in $Set1$
midpoint(P,Q);	The midpoint of a line segment PQ
numer(f);	Find the numerator of f
p:=";	Name the previous expression p
plot(f(x) , x);	Plot an expression in x
plot(f(x) , x=-3...3);	Plot the expression $f(x)$ over the specified domain
plot(h);	Plot the function h
plot({f(x),g(x)},x=-10..10,y=-10..10);	Plot a set of expressions with specified domain and range
restart;	Reinitialize your *Maple* session
rhs(eq);	The right hand side of an equation
select(has, f , y);	Choose the terms in f which have y
slope(P, Q);	Slope of the line segment PQ.
solve(eq, y);	Solve the equation for y returning the result as an expression.
solve(eq, x);	Solve the equation for x returning the result as an equation
sqrt(2);	Compute the square root of 2
subs(a=3, eq);	Replace a by 3 in eq
taylor(f(x), x, 10);	Compute a series in x to order 10
trunc(1.5);	Truncate to an integer
with(student);	Load the routines from the student package
{3*x + 5=0 , 6*x=0};	A set of equations
{2*y+3,z};	A set of expressions

Table 1.2 More of the Commands Used in Chapter 1

Course Specific Commands

The *Maple* solutions to problems in this text often rely on commands that are defined only after you have loaded the *student* package. This package of routines is part of every distribution of *Maple* and is loaded by the command

```
> with(student);
```

For the material covered in this course, this command should be placed in a special *Maple* initialization file[24] so that it is executed every time you start *Maple*.

Exercises

The following exercises are intended to provide you with some experience carrying out basic computations within the algebra system. They use some of the basic *Maple* commands to carry out what should be familiar mathematical calculations.

REMARK: Ask questions early, rather than late. In the beginning stages of becoming familiar with *Maple*, a timely hint from your instructor or tutor, or even a fellow student, can avoid a lot of unnecessary frustration.

REMARK: Above all else, *experiment*! Try variations on the following commands and try to anticipate how the system will respond to your own requests.

1. An algebra system allows you to manipulate undefined quantities. To accomplish this, the system must also support a lot of other useful concepts such as large integers and rational numbers. Use *Maple* to execute each of these commands and then read the comments that follow.

    ```
    > 1000!;
    > 2/3 + 7/8;
    > (a+b)^10;
    > expand( " );
    > factor( " );
    > expand( (x+y)^3 ) / (x+y);
    > normal( " );
    ```

    - Big integers are exact.
    - Rational numbers are exact.
    - The double-quote symbol " is used as a name for *the previous expression*.
    - Expressions can contain undefined symbols such as *x* and *y*.
    - Commands such as **factor( )**, **expand( )**, and **normal( )** can be used to manipulate algebraic expressions; **expand( )** multiplies out expressions in full, **factor( )** finds polynomial factors, and the command **normal( )** is used to regroup quotients of polynomials over a commmon denominator and then remove common factors.
    - To find out more about the use of these commmands, use the help commands **?factor, ? expand**, and **?normal**.

---

[24] Ask your local system administrator to find out how to do this.

- The most important part of these help pages is the worked examples at the end of each page.
2. Experiment with each of the following commands:

   ```
   > restart;
   > x^12;
   > 3/11;
   > "^12;
   > p;
   > p := t^2 + 3*t - 2;
   > p;
   > answer := solve( p = 0 , {t} );
   > answer[1];
   > evalf( " );
   > p := 'p';
   > p;
   ```

   - The command **restart** restores *Maple* to its initial state.
   - The *assignment* := makes p the *name* of the expression $t^2 + 3t - 2$.
   - The current value of p is given by the command **p;** .
   - There is a **solve( )** command that can be used to solve exactly a large variety of equations, systems of equations, and inequalities.
   - An individual name such as p can be "unassigned" by assigning it itself in quotes as a value, as in '**p**'. The single quotes prevent evaluation.

3. The **evalf( )** command is used to produce decimal expansions of numbers. Compare the decimal expansions of the following two expressions. Can you see a repeating pattern? What does the presence or absence of a repeating pattern tell you about a number?

   ```
   > Pi;
   > evalf( Pi );
   > evalf( Pi, 100 );
   > 317/23;
   > evalf( ", 100 );
   ```

   - The special number π is represented in *Maple* as **Pi**.
   - The results of the **evalf( )** command usually have decimal points. Such numbers are always regarded by *Maple* as approximate.
   - *Maple* ordinarily gives you ten digits of precision. To extend this, specify the required precision as part of the command.

4. Use the plotting techniques developed in this chapter to investigate the following inequality.

   ```
   > r1 := (x + 5)/(x + 1) > (x+2)/(x -2 );
   > lhs( r1 ) - rhs( r1 );
   > plot( ", x = -10 ... 10, y = -20 ... 20 );
   ```

   - If you do not specify a domain, the **plot( )** command uses the range $-10...10$.
   - The commands **lhs( )** and **rhs( )** are used to grab parts of an equation or inequality.

5. Experiment with the **solve( )** command on the following problems.

```
> eq1  :=   a*x^2 + b*x + c = 0;
> eq1;
> solution := solve( eq1 ,{x} );
> solution[1];
> solution[2];
> solve( eq1, {c} );
> solve( eq1 );
> eq2 := y = 3*x/( 2 - x^2 );
> solve( eq2, {x} );
> e1 := 3*x + 4*y = 12;
> e2 := 4*x - 3*y = 10;
> solve( {e1,e2} );
```

- The first of the two named solutions has the name **solution[1]**.
- The **solve( )** command nearly always tries to do something, but if you do not tell it what to solve for, it may choose to solve a different problem from what you intended.
- The answers given by **solve( )** are exact.
- The **solve( )** command can solve systems of equations.

6. *Maple* works with unknown functions. Try each of the following:

```
> f := 'f'; g := 'g'; h := 'h';   #unassign the names
> h := f + g;
> h( x );
> h := f*g;
> h( x );
> h := f@g;
> h( x );
```

- Functions can be combined to form new functions.
- The notation **f@g** denotes the composition of the two functions f and g.

7. The following commands define some functions. Apply them to various values as shown; try them on some numeric values as well.

```
> f := 'f'; g := 'g'; h := 'h';   #unassign the names
> h := f@g;
> h( t );
> f := (x) -> x^2 + 3;
> g := (x) -> x^3;
> f( t );
> f( t+a );
> g( t );
> g( t+1 );
> h;
> h( t );
```

- Notice the difference in the value of $h(t)$ once the functions f and g have been defined.

- What would happen if we now undefine g? (i.e., `g := 'g';`) Try it!
8. We can use *Maple* to solve for unkown parameters. The following sequence of commands illustrates how to solve the problem: *Given a function f defined by* $(x) \to \frac{x^2-2x+2}{x^2+s}$ *find a value of s such that $f(f(t)) = t$.*

    ```
    > f := (x) -> ( x^2 -2*x + 2 ) / ( x^2 + s );
    > f( t );
    > (f @ f)( t );
    > " = t;
    > sol := solve( ", {s} );
    > subs( sol[1], f( t ) );
    ```

    - Once a function is defined it is easy to use in compositions.
    - We can substitute the result from **solve( )** directly into an expression.
    - Try this substitution with the other solutions.

    How would you have solved this equation?

9. Try *Maple*'s solution to the following problem. *Find a polynomial of degree 2 whose graph passes through the three points* $[0, 1]$, $[2, 4]$, *and* $[3, 1]$. *Sketch a graph of the polynomial.*

    ```
    > restart;
    > f := (x) -> a*x^2 + b*x + c;
    > eqns := { f(0)=1, f(2)= 4 , f(3)=1 };
    > solve( eqns );
    > assign(");
    > plot(f,-5..5);
    ```

    - The **assign( )** command is useful for transforming equations of the form **name = something** into assignments of the form **name := something**.

10. Even if you do not know how to solve a system of three equations in three unknowns, you can still use *Maple* to solve them one at a time and substitute the results back into the other equations.

    ```
    > solve( f(0)=1 , {c});
    > subs(", eqns );
    ```

    Use this technique to find the previous solution provided *Maple*.

11. Use the **plot( )** command to generate a graph of the function defined by

    ```
    > f := (x) -> 40*x^8-700*x^7+5250*x^6-21985*x^5+56065*x^4
    > -88890*x^3+85320*x^2-45225*x+10125;
    ```

    $$x \mapsto$$
    $$40x^8 - 700x^7 + 5250x^6 - 21985x^5 + 56065x^4 - 88890x^3 + 85320x^2 - 45225x + 10125$$

    Try to include all the places where $f(x) = 0$. It is okay to simply try different domains and ranges, but you may want to use **factor( )**, **solve( )**, and **fsolve( )** to try and locate these zeros.

12. Given a circle $c1 := (x - 2)^2 + (y - 3)^2 = 16$ and a line $l1 := 3x - 2y = 3$, find the two points of intersection of the circle and the line.

a. Start by drawing a graph of the circle and the line. This can be done in *Maple* by solving each equation for y and plotting the three expressions representing the top and bottom of the circle and the line.

```
> solve( c1 , y ); solve( l1 , y );
> plot( {","""} );
```

b. You can use the **solve( )** command to find intersections of pairs of curves. Solve the equation $l1$ for y and substitute this value for y in the equation $c1$ to produce a quadratic equation in x. Then solve this quadratic equation.

c. If *eqns* $= c1, l1$, then the commands

```
> fsolve( eqns , {x,y} , x=-1...1);
> fsolve( eqns , {x,y} , x=4...6);
```

will give approximate solutions. Compare these approximate solutions with your exact solutions by using **evalf( )**.

d. Solve the previous system of equations in one step by using the command

```
> solve( eqns );
```

The result should involve something that looks similar to

$$\text{RootOf}(13Z^2 - 66Z - 54)$$

This expression is *Maple*'s way of describing an exact quantity that could be either root of the equation $13Z^2 - 66Z - 54 = 0$. Each choice yields a solution. They are recovered by the command

```
> allvalues(" , 'd' );
```

The option **'d'** indicates that the same root should be used to define both x and y.

2 Limits

2.1 Introduction

The notion of a limit is a fundamental mathematical concept. It arises in an essential way in two main contexts in this course. First, it helps us estimate values for functions that are defined in terms of quotients at points where the denominator becomes zero. Second, it helps us approximate the behavior of functions of x for very large values of x.

Evaluating Rational Functions at Zeros of the Denominator

A rational expression such as $(x^3 + 2*x - 3*x^2 - 6)/(x - 3)$ cannot be evaluated directly at $x = 3$ because of division by zero.

```
> e1 := (x^3+2*x-3*x^2-6)/(x-3);
```

$$\frac{x^3 + 2x - 3x^2 - 6}{x - 3}$$

```
> value( subs( x=3, e1 ) );
Error, division by zero
```

Limits are used to discover what the expression value should have been, had it been defined. In this example, the missing value can be represented by

```
> Limit( e1 , x=3 );
```

$$\lim_{x \to 3} \frac{x^3 + 2x - 3x^2 - 6}{x - 3}$$

which evaluates to

```
> value(");
```

$$11$$

In most cases the missing value is easily predicted by examining a graph of the function in a region near the point in question. Consider

```
> plot( e1, 'x'=1..4 );
```

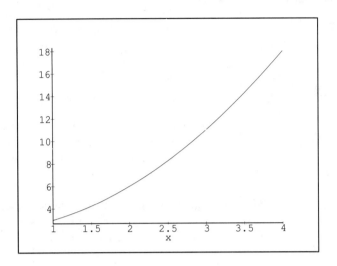

If you look closely you will see that there is a point missing from the graph at $x = 3$. Except for that single troublesome point, the resulting graph is identical to the graph of the expression

```
> normal( e1 );
```

$$x^2 + 2$$

a new expression that is the result of simplifying *e1* by removing a common factor of $x - 3$ from both the numerator and the denominator.

The result of such a limit computation is not always finite. For example, the expression

```
> e1/(x-3);
```

$$\frac{x^3 + 2x - 3x^2 - 6}{(x - 3)^2}$$

gets infinitely large as x approaches 3.

```
> plot(",x=1..4,y=-100..100);
```

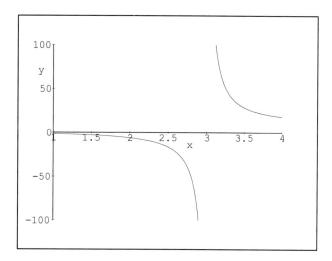

Note that it is still easy to predict this fact by examining the graph in a region near the point $x = 3$.

Determining the Behavior of $f(x)$ as x Gets Large

Limits are also used to estimate the value of expressions for very large values of x. Consider the expression $e2$, defined by

```
> e2 := (4*x^2 + 1/x)/(3*x^2 - x);
```

$$\frac{4x^2 + x^{-1}}{3x^2 - x}$$

The graph of the function defined by this expression on the domain $[1, 10]$ descends rapidly as x increases and then seems to level off at just less than 1.4.

```
> plot(e2,x=1..10);
```

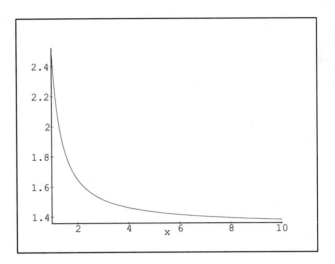

This leveling off shows even more dramatically if we extend the domain out to $x = 100$.

```
> plot(e2,x=1..100);
```

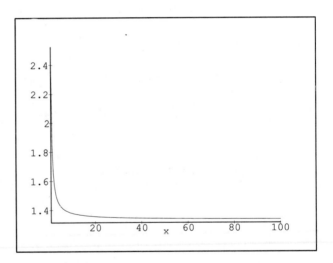

The curve actually approaches the horizontal line corresponding to the y value represented by

```
> Limit(e2,x=infinity);
```

$$\lim_{x \to \infty} \frac{4x^2 + x^{-1}}{3x^2 - x}$$

which evaluates to

```
> value(");
```

$$\frac{4}{3}$$

As x gets closer to ∞, the value of the expression gets closer to $\frac{4}{3}$.

The algebraic structure of the expression gives us clues as to why this happens. In both the numerator and the denominator, the x^2 term grows much more rapidly than either x or $\frac{1}{x}$ as x tends to ∞. If we make the rather bold assumption that all the smaller terms can be ignored and replace them by zeros, the remaining expression can actually be simplified to $\frac{4}{3}$.

2.2 Limit Computations in *Maple*

The notion of estimating the value of an expression at a point $x = a$ by looking at the graph of the function for smaller and smaller regions near that point is referred to as *taking the limit of $f(x)$ as x approaches $x = a$.*

DEFINITION 2.1 The limit of $f(x)$ as x approaches a is denoted by

$$\lim_{x \to a} f(x)$$

If the limit is to be a number L, then $f(x)$ must get close to L and remain close to L as x approaches a. If we interpret *increasing without bound* as *getting close to ∞*, we can also use limits to denote unboundedness near $x = a$.

$$\lim_{x \to a} f(x) = \infty$$

Limits are specified in *Maple* using the **Limit( )** command. We must indicate both the expression in x and the x value of interest. For example, we write

```
> Limit( e1, x=3 );
```

$$\lim_{x \to 3} \frac{x^3 + 2x - 3x^2 - 6}{x - 3}$$

to represent the idea of computing the estimate for $e1$ at $x = 3$.

The **Limit( )** command is inert; that is, *Maple* makes no attempt to carry out the computation. The actual value of this limit can be computed by applying the *Maple* command **value( )** to this expression.

```
> value( " );
```

$$11$$

Rather than entering two commands (as above), we can use the single command **limit( )** (with a lowercase "l") to ask *Maple* to compute the value of the limit immediately.

```
> limit( e1 , x=3 );
```
$$11$$

One of the goals of this chapter is to develop techniques that will enable you to carry out such computations. Throughout, we rely heavily on plotting to guide our intuition.

REMARK: The value of the limit
$$\lim_{x \to a} f(x)$$
can always be estimated by examining an accurate graph of f near $x = a$ for smaller and smaller regions surrounding a.

2.3 Some Limit Computations

For well-behaved functions f, such as those defined by polynomials, the result of taking a limit at $x = a$ is always the same as evaluating the function at a (i.e., $f(a)$). For example, given

```
> f := (x) -> x^3 + 5*x^2 + 2;
```
$$x \mapsto x^3 + 5x^2 + 2$$

the limit

```
> Limit( f(x), x=a );
```
$$\lim_{x \to a} \left(x^3 + 5x^2 + 2 \right)$$

evaluates to

```
> value(");
```
$$5a^2 + a^3 + 2$$

The concept of a limit starts to play an essential role for functions defined by a quotient in which the denominator is zero at $x = a$. Consider some of the typical ways in which this arises.

2.3.1 Ratios of Polynomials

In the introduction to this chapter we encountered a function that was defined by a quotient of polynomials. Many of our important examples have this form.

Example 2.1 For the function f defined by

```
> f := (x) -> (x^2-1)/(x+1):
```

compute the limit

```
> e1 := Limit(f(x),x=-1);
```

$$\lim_{x \to -1} \frac{x^2 - 1}{x + 1}$$

Solution The limit concept plays an essential role here since if we attempt to evaluate $f(x)$ directly at $x = -1$, we commit the error of attempting to divide by zero.

```
> f( -1 );
Error, (in f) division by zero
```

The actual value of the limit $e1$ is

```
> value(e1);
```

$$-2$$

To see this, examine the graph of f near $x = -1$.

```
> plot(f(x),x=-3..1);
```

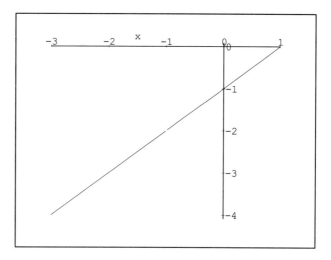

There is a tiny "hole" in the curve at $x = a$ reflecting the fact that $f(-1)$ is undefined. Apart from this, the graph clearly suggests that the correct value for $f(-1)$ is -2.

Often such answers can be obtained algebraically by simplifying the quotient of polynomials to remove common factors from the numerator and denominator. For

```
> f(x);
```

$$\frac{x^2 - 1}{x + 1}$$

a common factor of $x + 1$ can be canceled, yielding

```
> normal(f(x));
```

$$x - 1$$

This linear polynomial is defined at $x = -1$ but agrees with $f(x)$ everywhere else. It has the value of -2 at $x = -1$, as predicted by the graph of f.

```
> subs(x=-1,");
```
$$-2$$

∎

REMARK: Whenever f is defined in terms of a quotient of polynomials, a good first step in evaluating the limit is to simplify $f(x)$ in order to eliminate as many factors as possible from the denominator.

Approximating the Rate of Change at $t = a$

Quotients of polynomials arise naturally in rate-of-change problems. A function f is used to indicate a position (perhaps as a distance from some reference point) as a function of time. At time a, we are at position $f(a)$ and at time b we are at position $f(b)$. The average rate of change from time a to time b is just the slope of the straight line from the starting point on the curve to the finishing point on the curve.

```
> with(student):
> m := slope( [b,f( b )], [a,f( a )] );
```

$$\frac{f(b) - f(a)}{b - a}$$

As the point b is taken closer and closer to a, we obtain a better and better estimate of the instantaneous rate of change at time $t = a$. The formal definition of this instantaneous slope at $t = a$ is

```
> Limit( m , b=a );
```

$$\lim_{b \to a} \frac{f(b) - f(a)}{b - a}$$

This evaluates to the slope of a line tangent to the graph of f at $t = a$.

Example 2.2 Suppose that a ball is dropped from the Calgary Tower, a free-standing restaurant that is about 185 meters high. What is its velocity after one second?

Solution Let $s(t)$ be the distance traveled downward after time t. A study of the physics involved tells us that $s(t)$ is essentially $-\frac{1}{2}at^2 + s(0)$, where a is the acceleration due to gravity (9.8 m/sec^2) and $s(0)$ is the position of the object at time $t = 0$.

```
> s := (t) -> -4.9*t^2 + 185:
> s( t );
```

$$-4.9 t^2 + 185$$

An approximation to the instantaneous rate of change at $t = 1$ is:

```
> slope( [1+1,s(1+1)],[1,s(1)]);
```

$$-14.7$$

2.3 Some Limit Computations

A graph showing the secant line defined by these two points and the curve s is shown below.

```
> Line := makeproc( [1+1,s(1+1)],[1,s(1)] );
```

$$x \mapsto -14.7x + 194.8$$

```
> plot( { s(t),Line(t) },t=0..3);
```

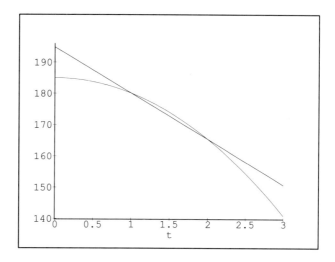

If we move the second point closer to $t = 1$, we get an even better approximation to this rate of change.

```
> slope( [1.2,s(1.2)],[1,s(1)]);
```

$$-10.78000000$$

and the line becomes closer to a tangent line. The true slope at $t = 1$ is

```
> Limit( ( s( t ) - s(1)) / (t - 1), t = 1 );
```

$$\lim_{t \to 1} \frac{-4.9 t^2 + 4.9}{t - 1}$$

To compute this limit, we assume that the numbers appearing in this expression are all exact. This is to avoid any complications that might arise from trying to factor approximate polynomials. We make them exact in *Maple* by converting all the numbers in the expression to rational numbers.

```
> convert(",rational);
```

$$\lim_{t \to 1} \left(-\frac{49}{10} t^2 + \frac{49}{10} \right) (t-1)^{-1}$$

The quotient of polynomials then simplifies to a polynomial. An equivalent limit computation is

```
> simplify(");
```

$$\lim_{t \to 1} \left(-\frac{49}{10} t - \frac{49}{10} \right)$$

which is easily computed by direct function evaluation.

```
> value( " );
```

$$-\frac{49}{5}$$

A graph showing both the function and the tangent line at $t = 1$ is generated by the following command:

```
> showtangent(s(t),t=1,t=0..3);
```

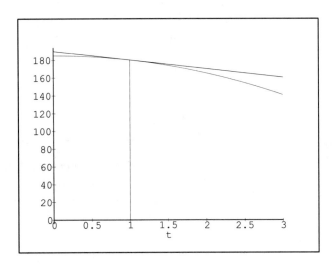

2.3.2 Obscure Common Factors

Algebraic simplifications can be difficult to detect. Consider the following function:

```
> f := (t) -> (sqrt( t ) - 3)/ ( t - 9 );
```

$$t \mapsto (\sqrt{t} - 3)(t - 9)^{-1}$$

The graph of this function

```
> plot( f( t ), t=6..12);
```

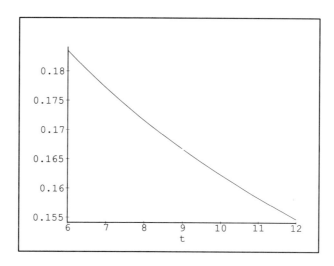

suggests that the limit at $t = 9$ should be

```
> limit( f( t ), t=9 );
```

$$\frac{1}{6}$$

```
> evalf(");
```

$$0.1666666667$$

To obtain this result algebraically, we need to simplify $f(t)$. Ordinarily we would use commands such as **factor( )** and **normal( )** in an attempt to cancel out the zero factor in the denominator. However, these commands do not help here since they are designed to work primarily on polynomials that have rational number coefficients. Neither

```
> factor( numer( f( t ) ) );
```

$$\sqrt{t} - 3$$

nor

```
> factor( denom( f( t ) ) );
```

$$t - 9$$

has any effect.

It turns out that we can factor the expression if we regard t as x^2 and $x = \sqrt{t}$. Then $t - 9$ becomes $x^2 - 9$. A simple way to accomplish this transformation is by the substitution

```
> subs( t=x^2, f( t ) );
```

$$\frac{x - 3}{x^2 - 9}$$

The factored form of this new expression involves polynomials in x that are equivalent to polynomials in $t^{1/2}$.

```
> factor( " );
```

$$(x+3)^{-1}$$

An appropriate limit of this new expression can also be computed. Both x^2 and x go to zero when t goes to zero, so we use the limit

```
> Limit( f(t),t=9) = Limit( "", x=3 );
```

$$\lim_{t \to 9} \left(\sqrt{t} - 3\right)(t-9)^{-1} = \lim_{x \to 3} \frac{x-3}{x^2 - 9}$$

which evaluates to

```
> value(rhs("));
```

$$\frac{1}{6}$$

In general, $x^2 \to a$ as $t \to a$, so that we would need to compute the new limit

```
> subs(t=x^2,f(t));
```

$$\frac{x-3}{x^2 - 9}$$

```
> Limit(",x=sqrt(a));
```

$$\lim_{x \to \sqrt{a}} \frac{x-3}{x^2 - 9}$$

instead of $\lim_{t \to a} f(t)$.

A Change of Variables for Limits

In *Maple* the computations required to replace t by x^2 in a limit computation can be done all at once. The limit

```
> restart: with(student):
> Limit(f(t),t=a);
```

$$\lim_{t \to a} f(t)$$

is transformed, using the identity $t^2 = x$, to a limit in terms of x by the command

```
> changevar( x^2=t,",x);
```

$$\lim_{x \to \sqrt{|a|}} f(x^2)$$

The last parameter in this command is there to make explicit the fact that the result is to be a limit that is specified in terms of x. Also, the use of $\sqrt{|a|}$ instead of $\sqrt{a}$ makes it clear that for now we are really thinking about positive values of a.

2.3.3 Limits That Are Bounded but Do Not Exist

Sometimes no unique number exists as a limit even though the function value always lies in a narrow range such as $[-1...1]$.

```
> f := (x) -> sin( 3*Pi/x );
```

$$x \mapsto \sin(\frac{3\pi}{x})$$

```
> plot( f( x ), x=0... 1. );
```

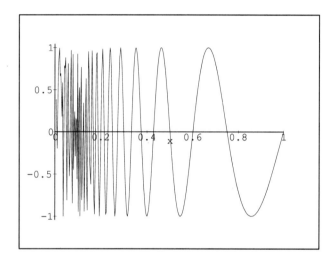

```
> limit( f( x ), x=0 );
```

$$-1\ldots 1$$

The magnitude of $f(x)$ will never exceed 1 near 0, but neither will it remain close to 1. Providing that we include 0, no matter how small we make the domain of the function, its values will still range over all of $[-1, 1]$.

2.3.4 Unbounded Functions

Consider

```
> f := (x) -> 1/x^2;
```

$$x \mapsto x^{-2}$$

It is apparent from the following graph that something unusual happens at or near $x = 0$.

```
> plot( f( x ), x=-1..1, y=0..20 );
```

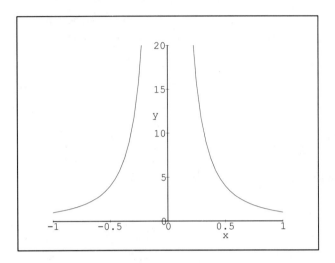

A *finite limit* does not exist here as there is no specific number L that $f(x)$ stays close to for all the values of x near 0. Since $f(x)$ does get closer to ∞ as x gets closer to 0, this behavior can still be described as

$$\lim_{x \to 0} f(x) = \infty$$

2.4 A Formal Definition of a Limit

As expressions become more complicated we ultimately need a means of formally defining the concept of a limit. This notion of *staying near L* whenever x is near a is made more precise as follows.

DEFINITION 2.2 We say that $\lim_{x \to a} f(x) = L$ if for any $\epsilon > 0$ there exists a δ such that $|x - a| < \delta$ implies $|f(x) - L| < \epsilon$.

We will use this definition primarily to prove that a chosen limit value is correct. The way we apply this definition is like a game between two people.

The first person chooses a value for ϵ. Their goal is to choose an ϵ that is as nasty as possible (usually very small). The second person – you – has to choose a suitable value of δ (as requested by the definition) for use in conjunction with the specified ϵ.

Because the ϵ value chosen by the first person is completely arbitrary, the only possible way for the second person to win the game is to find a rule that can be used to produce a suitable value of δ for any given value of ϵ. The rule must work for every x in the interval $(a - \epsilon, a + \epsilon)$.

REMARK: In terms of graphs, the first person is looking for a rule that indicates how narrow to make the plot so that the total height of a plot that is automatically scaled to include every point generated is no bigger than the limit set by the second person.

Of course, you may be unable to find a suitable rule. This may be simply because you did not look hard enough, or it may be because there really is no possible way to specify δ in terms of ϵ. You will always fail to find such a rule if the true limit value is something other than the proposed L or if the limit does not exist. This occurs at points where the graph of f is discontinuous.

In the next example, we use this very formal procedure to *prove* that we have a correct answer for a specific limit computation. Our main goal is to study the process of proving that the given limit value is correct rather than to compute the actual limit value. Thus we use a very simple example.

Example 2.3 Prove that

$$\lim_{x \to 7} x - 4 = 3$$

Solution As the limit is defined in terms of a polynomial, there is no real doubt about the answer being 3. The proof that this is the correct limit value is as follows. For any given ϵ we must show how to compute a bound δ on $|x - 7|$ so that

$$|x - 7| < \delta \Rightarrow |(x - 4) - 3| < \epsilon$$

This will be true if we choose $\delta = \epsilon$. That is, for any x within a distance ϵ of 7, the value of $x - 7$ at that x will be within ϵ of the limit value $L = 3$. We have found a rule for defining δ in terms of ϵ so this completes the proof. ∎

REMARK: *Any* rule for choosing δ that ensures that $|f(x) - L| < \epsilon$ for all x satisfying $|x - 7| < \delta$ serves to complete this proof. In this example, we could just as easily choose δ to be $\delta = \frac{\epsilon}{2}$ or $\delta = \frac{\epsilon}{3}$; however, $\delta = 2\epsilon$ will not work.

2.4.1 Computing δ

To use the limit definition effectively we must become proficient at finding a rule to define δ as a function of ϵ. Here we show how to do this systematically for all functions defined in terms of polynomials.

Example 2.4 Prove that $\lim_{x \to 1} x^2 + 1 = 2$.

Solution The function is defined by

```
> f := (x) ->  x^2+1;
```

$$x \mapsto x^2 + 1$$

and the proposed limit value is

```
> L := 2;
```

$$2$$

We must find a suitable restriction on $|x - 1|$ so that the following inequality is satisfied when those values of x are used:

```
> abs( f( x ) - L ) < epsilon;
```

$$|x^2 - 1| < \epsilon$$

In the absence of any obvious way of proceeding, it is always a good idea to try rearranging the expression into a different form. We need an inequality involving $x - 1$. In this case factoring provides the necessary insight since the expression $x - 1$ turns out to be a factor of the left-hand side.

```
> factor( " );
```

$$|x - 1||x + 1| < \epsilon$$

Our first step at constructing a usable bound on $|x-1|$ in terms of ϵ is by rearranging this factored inequality.

```
> isolate( ", abs( x-1 ) );
```

$$|x - 1| < \frac{\epsilon}{|x + 1|}$$

The right-hand side of this inequality

```
> bound1 := rhs(");
```

$$\frac{\epsilon}{|x + 1|}$$

is a bound of sorts. The problem with it is that its value depends on the value of x. The type of bound we are looking for must work independent of our choice of x. That is, it must depend only on ϵ. The following crucial observations allow us to eliminate any reference to x in **bound1**.

1. We can always choose the smaller of two bounds. The rule we are seeking need only specify what happens near $x = a$. Thus, we can choose to ignore any values of x outside of an arbitrary region, say, $|x - 1| < \frac{1}{2}$ even if such x values place $f(x)$ near $L = 2$.
2. In the region $|x - 1| < \frac{1}{2}$, we can systematically underestimate the value of **bound1** by using instead a single value of **bound1** that is the worst we can find (i.e., the smallest) over this entire interval.

Underestimating the Bound

The numerator of **bound1**, ϵ, is a constant, having already been chosen by our opponent earlier in the game. Thus we can estimate **bound1** by examining the behavior of

```
> e1 := subs( epsilon=1, bound1 );
```
$$|x+1|^{-1}$$

on the interval $|x - 1| \leq \frac{1}{2}$ and scaling the result by ϵ. It is clear from the graph of $e1$ on this interval

```
> plot( e1, x=-1/2...1/2 );
```

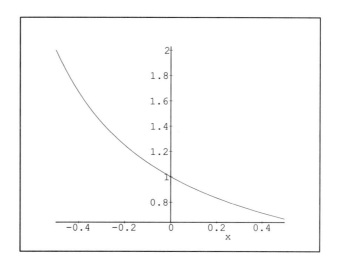

that its minimum value occurs at $x = 1/2$. This worst case becomes our second candidate for a bound.

The actual bound we use is the smaller of the two candidates

```
> delta := min( 1/2 , subs( x=1/2, bound1 ) );
```
$$\min(\frac{1}{2}, \frac{2}{3}\epsilon)$$

For this bound we have $|f(x) - L| < \epsilon$ whenever $|x - 1| < \delta = \min(\frac{1}{2}, \frac{2\epsilon}{3})$, since

$$|x - 1| < \min(\frac{1}{2}, \delta) \Rightarrow |x - 1| < \epsilon/|x + 1|$$
$$\Rightarrow |x - 1||x + 1| = |x^2 + 1 - L| < \epsilon$$

We have found a rule for computing δ so this concludes our proof. ∎

This approach generalizes to all polynomial examples.

Example 2.5 Prove $\lim_{x \to 2} 23x^5 + 105x^4 - 10x^2 + 17x = 2410$.

Solution The expressions involved are a bit more complicated, but we can proceed in exactly the same manner.

The function is

```
> f := (x) -> 23*x^5 + 105*x^4 - 10*x^2 + 17*x ;
```

$$x \mapsto 23x^5 + 105x^4 - 10x^2 + 17x$$

The limit value at $x = 2$ should be

```
> L := f(2);
```

$$2410$$

To prove that this answer is correct, we must use the inequality

```
> abs( f( x ) - L ) < epsilon;
```

$$\left|23x^5 + 105x^4 - 10x^2 + 17x - 2410\right| < \epsilon$$

to construct a rule for how close x should be to 2 to achieve this.

To construct the rule for computing δ from a given value of ϵ, we factor the preceding expression

```
> factor( " );
```

$$|x - 2|\left|23x^4 + 151x^3 + 302x^2 + 594x + 1205\right| < \epsilon$$

and rearrange it so that the crucial term $x - 2$ is isolated.

```
> isolate( ", abs( x-2 ) );
```

$$|x - 2| < \frac{\epsilon}{\left|23x^4 + 151x^3 + 302x^2 + 594x + 1205\right|}$$

The first approximation to a bound on $x - 2$ is just

```
> bound1 := rhs(");
```

$$\frac{\epsilon}{\left|23x^4 + 151x^3 + 302x^2 + 594x + 1205\right|}$$

Next we choose any small interval around $x = 2$. For example, $|x - 2| < 3$ will do. The important thing is that it be finite. In the interval $|x - 2| < 3$, we observe that the expression **bound1** is minimized at $x = 5$.

```
> plot( subs( epsilon=1, bound1 ), x=2-3..2+3);
```

2.4 A Formal Definition of a Limit 75

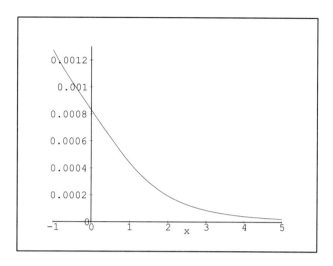

Thus, a suitable value of δ (anything smaller will do) is

```
> delta := min( 2, subs( x=5, bound1 ) );
```

$$\min(2, \frac{1}{44975}\epsilon)$$

```
> abs( x-2 ) < delta;
```

$$|x - 2| < \min(2, \frac{1}{44975}\epsilon)$$

Since we have found a rule for computing δ in terms of ϵ, we have proved that this limit is 2410. ∎

A Limit That Does Not Exist

Example 2.6 Prove that $\lim_{x \to 3} \frac{1}{x^2 - 5x + 6}$ does not exist.

Solution To *prove* that a limit does not exist [1] requires more than just saying, "I have been unable to find a rule for computing δ from ϵ."

To see how we might prove that a specific value of L is not the correct value for $\lim_{x \to 3} \frac{1}{(x-3)^2}$ let us look carefully at the steps used in the previous example to compute a bound δ in terms of ϵ. For

```
> L := 'L':
```

---

[1] Imagine simply telling your boss that you have been unable to do something. They might conclude that "if you can't do it nobody can," but more likely your boss will just hire somebody else to try again. A proof that it cannot be done by anyone might just save your job.

and

> f := (x) -> 1/(x^2-5*x+6);

$$x \mapsto (x^2 - 5x + 6)^{-1}$$

we attempt to find a region for x so that the inequality

> abs(f(x) - L) < epsilon;

$$\left|(x^2 - 5x + 6)^{-1} - L\right| < \epsilon$$

is satisfied. As before, factoring helps produce the necessary factor $x - 3$

> factor(");

$$\frac{\left|-1 + Lx^2 - 5Lx + 6L\right|}{|x - 2| |x - 3|} < \epsilon$$

only this time the inequality is structured quite differently.

No matter what the value of L is, the left-hand side of this inequality is bounded away from 0 in any small interval near 3 such as $|x - 3| < \frac{1}{4}$. Consider the case $L = 10$.

> plot(abs(f(x)-10) , x=3-1/4..3+1/4,y= 0...5);

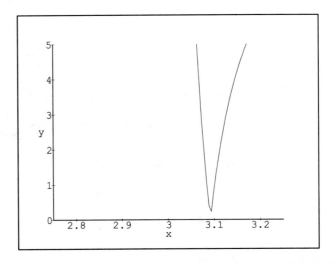

As x approaches 3, the left-hand side gets bigger instead of smaller. To win the game, all the person choosing ϵ need do is choose an ϵ smaller than the minimum value of $|f(x) - L|$ in the graph over the interval $|x - 3| < \frac{1}{4}$.

This same approach can be used to prove that the inequality cannot be satisfied throughout any entire region including 3, for any other choice of L. ∎

REMARK: In examples such as this one we say that *the limit does not exist*. This example is another instance of

$$\lim_{x \to 0} f(x) = \infty$$

2.5 Properties of Limits

When computing limits we often make use of their general algebraic properties to simplify computations. These properties are used to break large problems down into smaller, more manageable ones. The validity of each of these general rules is established by using the formal definitions discussed in section 2.4.

REMARK: You can choose to apply these rules or not as the case may be. In each case the decision to reformulate the problem using one of the rules will be based on whether or not it helps break down the computation into manageable pieces.

We begin by using *Maple* to recall the general properties of limits. This is best done by starting a new *Maple* session.

```
> restart: with( student ):
```

In what follows, c is a constant, $f(x)$ and $g(x)$ represent arbitrary expressions in x, and the command **expand( )** is used to invoke various transformations. For example, the limit of a constant is that constant.

```
> Limit( 3, x=a );
```

$$\lim_{x \to a} 3$$

```
> expand( " );
```

$$3$$

and constants can be factored right out of a limit computation.

```
> Limit( c*f( x ), x=a );
```

$$\lim_{x \to a} cf(x)$$

```
> expand( " );
```

$$c \lim_{x \to a} f(x)$$

Limits can also "map" onto each of sums, products, and exponents, as in

```
> Limit( f( x ) + g( x ), x=a );
```

$$\lim_{x \to a} (f(x) + g(x))$$

```
> expand( " );
```

$$\lim_{x \to a} f(x) + \lim_{x \to a} g(x)$$

> Limit(f(x) * g(x), x=a);

$$\lim_{x \to a} f(x)g(x)$$

> expand(");

$$\lim_{x \to a} f(x) \lim_{x \to a} g(x)$$

> Limit(f(x) ^ g(x), x=a);

$$\lim_{x \to a} f(x)^{g(x)}$$

> expand(");

$$\lim_{x \to a} f(x)^{\lim_{x \to a} g(x)}$$

The next rule is really the rule for powers in disguise.

> Limit(sqrt(f(x)), x=a);

$$\lim_{x \to a} \sqrt{f(x)}$$

> expand(");

$$\sqrt{\lim_{x \to a} f(x)}$$

Example 2.7 Use the preceding rules to rewrite the following limit in terms of limits of simpler expressions:

> eg01 := (x^3 + 2*x^2 - 1) / (5 - 3*x) :

> Limit(eg01, x=-2);

$$\lim_{x \to -2} \frac{x^3 + 2x^2 - 1}{5 - 3x}$$

Solution This limit can be rewritten as

> expand(");

$$\frac{\lim_{x \to -2} x^3 + 2 \lim_{x \to -2} x^2 - 1}{5 - 3 \lim_{x \to -2} x}$$

The value of the resulting expression is

2.5 Properties of Limits

```
> value( " );
```
$$-\frac{1}{11}$$

∎

REMARK: Generally, an expansion is not appropriate if it leads to a division of zero by zero.

Example 2.8 Rewrite the following limit in terms of limits of simpler expressions:

```
> eg02 := (x^2 -x)^(1/5) + ( x^3 + x)^9 :
> Limit( eg02, x=1 );
```

$$\lim_{x \to 1} \left(\sqrt[5]{x^2 - x} + \left(x^3 + x\right)^9 \right)$$

Solution This limit can be rewritten using the rules for expansion of limits as

```
> expand( " );
```

$$\sqrt[5]{\lim_{x \to 1} x^2 - \lim_{x \to 1} x} + \left(\lim_{x \to 1} x^3 + \lim_{x \to 1} x \right)^9$$

The value of the resulting expression is

```
> value( " );
```
$$512$$

∎

We usually cancel common factors from the numerator and denominator before applying these rules. However, such factors may not always be obvious.

Example 2.9 Use the rules for expansion to compute the following limit:

```
> eg03 := ((1+x)^(1/3) - 1) / x ;
```

$$\left(\sqrt[3]{1+x} - 1 \right) x^{-1}$$

```
> Limit( eg03, x=0 );
```

$$\lim_{x \to 0} \left(\sqrt[3]{1+x} - 1 \right) x^{-1}$$

Solution The numerator and denominator of *eg03* are

```
> p := numer( eg03 );
```

$$\sqrt[3]{1+x} - 1$$

```
> q := denom( eg03 );
```
$$x$$

If we let $1+x = a^3$, then p and q can be rewritten as:
```
> p = a - 1;
```
$$\sqrt[3]{1+x} - 1 = a - 1$$
```
> q = a^3 - 1;
```
$$x = a^3 - 1$$

In this form, q can be factored as
```
> factor( " );
```
$$x = (a-1)(a^2 + a + 1)$$

and this can be used to simplify the limit computation. The expression becomes
```
> subs( p=a-1, q=a^3-1, p/q );
```
$$\frac{a-1}{a^3 - 1}$$
```
> factor( " );
```
$$(a^2 + a + 1)^{-1}$$

and since $a \to 1$ as $x \to 0$, the value of the limit can be rewritten as
```
> Limit( ", a=1 );
```
$$\lim_{a \to 1} (a^2 + a + 1)^{-1}$$

which evaluates to
```
> value(");
```
$$\frac{1}{3}$$

■

The example
```
> e3 := Limit(eg03,x=0);
```
$$\lim_{x \to 0} \left(\sqrt[3]{1+x} - 1\right) x^{-1}$$

can be evaluated directly by *Maple*, as in
```
> value(e3);
```
$$\frac{1}{3}$$

or reformulated all at once by the command
```
> changevar( 1+x=a^3,e3, a );
```
$$\lim_{a \to 1} \frac{a-1}{-1 + a^3}$$

2.6 The Squeeze Theorem

The *squeeze theorem* allows us to estimate the value of limits by comparing them with nearby functions.

THEOREM 2.1 If $f(x) \leq g(x) \leq h(x)$ for all x near a, and

$$\lim_{x \to a} f(x) = L = \lim_{x \to a} h(x)$$

then

$$\lim_{x \to a} g(x) = L$$

If a particular function is too messy to allow us to compute its limit directly, we may still be able to estimate it by using nicely behaved functions that bound it.

Example 2.10 Prove that

$$\lim_{x \to 0} x \sin(1/x) = 0$$

Solution We first examine a plot of this function near 0.

```
> f := (x) -> x*sin( 1/x );
```

$$x \mapsto x \sin(x^{-1})$$

```
> plot( f(x) , x=-0.5 ... 0.5);
```

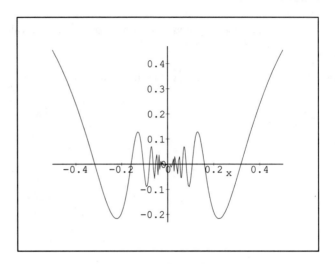

The resulting graph is only approximate (an inherent danger in any process based on sampling function values), but it would appear that the limit is 0 at $x = a$.

We can compute this limit exactly[2] at $x = 0$ because $-|x| \leq x \sin(x^{-1}) \leq |x|$ for all x near 0. The plot

```
> plot( {abs(x),f(x)},x=-1.0..1.0 );
```

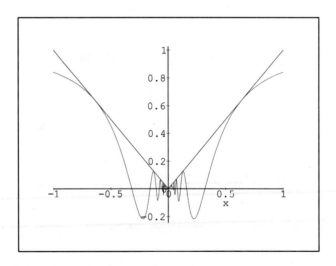

shows how $|x|$ is an upper bound. Similarly $-|x|$ is a lower bound.

Since $x \sin(x^{-1}) \leq |x|$, we have

$$\lim_{x \to 0} x \sin(x^{-1}) \leq \lim_{x \to 0} |x|$$

---

[2]Note that $\sin(x^{-1})$ does not have a limit at $x = 0$.

Similarly
$$\lim_{x \to 0} -|x| \leq \lim_{x \to 0} x \sin(x^{-1})$$

This leads to the conclusion
$$\lim_{x \to 0} x \sin(x^{-1}) = 0$$

The overall effect is dramatically summarized by plotting all three functions together.

```
> plot( {-abs(x),f(x),abs(x)},x=-1.0..1.0);
```

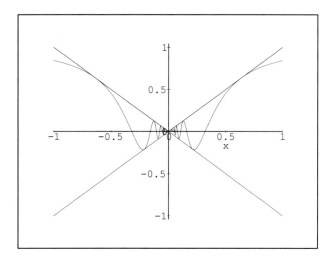

Note that the upper and lower bounds meet at $x = 0$. ∎

2.7 Continuity

There are many ways in which $\lim_{x \to a} f(x)$ may not exist. Some typical ways are as follows:

- *Step discontinuities:* There may be more than one reasonable limit value. This will occur at the boundary between two pieces of a piecewise-defined function if the adjacent pieces don't agree on the boundary.
  The piecewise-defined function obtained by using $f1$ for $x > 3$ and $f2$ otherwise with
  ```
  > f1 := (x) -> if x>3 then sqrt(x)+2 else undefined fi:
  > f2 := (x) -> if x<=3 then -sqrt(x)+2 else undefined fi:
  ```
  is shown in
  ```
  > plot({f1,f2},0..5);
  ```

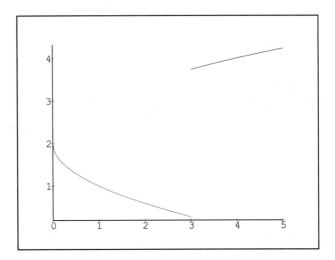

Which limit value you would expect at $x = 3$ depends entirely on which side of $x = 3$ you are approaching from.

- *One-sided limits:* The function may be undefined on one side of the point a. This is the situation for a function such as

```
> g := (x) -> sqrt(x+2)+1;
```

$$x \mapsto \sqrt{x+2} + 1$$

```
> plot(g,-3..10,0..5);
```

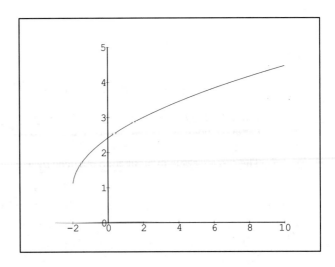

- *Jump discontinuities:* The function may be specifically redefined at $x = a$ to something inconsistent with the computed limit value.

```
> h := (x)-> x^2 + 3;
```
$$x \mapsto x^2 + 3$$

Ordinarily $f(2) = 7$, but we can force it to be any other value by directly assigning a specific value to $f(2)$. The following command forces $h(2)$ to evaluate to 1.

```
> h(2) := 1;
```
$$1$$

These three situations all result in *discontinuities* in the graph of the function. The **plot( )** command will often miss discontinuities because it generates a curve primarily by sampling function values and *assuming continuity* between sample points. However, the graphs of discontinuous curves can always be drawn properly by drawing the disconnected pieces separately and combining them, perhaps by using the **display( )** command.

The concept of a limit enables us to make this notion of continuity more precise.

DEFINITION 2.3   A function f is said to be *continuous* at a if

$$\lim_{x \to a} f(x) = f(a)$$

A function f will be discontinuous at a if it is undefined, if its limit at $x = a$ is undefined, or if $f(a)$ is different from the limit value at a.

2.7.1 One-Sided Limits

The limit concept is especially of interest at those places in the domain of a function where unusual things happen. To deal with these limits at boundaries we introduce the notion of a *one-sided limit*. Such a limit allows us to ignore points of the curve that would fall outside the domain of a particular function or that belong to the domain of the wrong piece of a piecewise-defined function.

DEFINITION 2.4   The one-sided limit of $f(x)$ as x approaches a from above is denoted by

$$\lim_{x \to a^+} f(x)$$

The one-sided limit of $f(x)$ as x approaches a from below is denoted by

$$\lim_{x \to a^-} f(x)$$

In *Maple* this notation becomes

```
> Limit( f( x ), x=a, right );
```
$$\lim_{x \to a^+} f(x)$$

Similarly, the one-sided limit approaching from below is denoted in *Maple* by

> Limit(f(x), x=a, left);

$$\lim_{x \to a^-} f(x)$$

The following example shows how the directional limits at $x = a$ may sometimes both exist but be different.

Example 2.11 Investigate the continuity of f defined by $(x) \to \frac{|x-1|}{x-1}$.

Solution The following plot shows an actual a jump discontinuity[3] at $x = 1$.

> f := (x) -> abs(x-1) / (x-1);

$$x \mapsto \frac{|x-1|}{x-1}$$

> plot(f, -2..2);

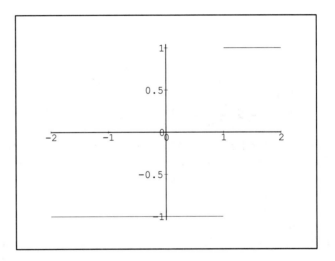

This is reflected in the fact that the one-sided limits at $x = 1$ both exist but are different. They are

> limit(f(x), x=1, left);

$$-1$$

and

> limit(f(x), x=1, right);

$$1$$

---

[3] In many instances the sampling technique used by **plot( )** will result in a picture that assumes continuity across this gap. The **plot( )** command gets this one right.

They are different so there is a discontinuity at $x = 1$. When we attempt to compute the ordinary (two-sided) limit

```
> Limit( f( x ), x=1 );
```

$$\lim_{x \to 1} \frac{|x - 1|}{x - 1}$$

we obtain

```
> value(");
```

$$\text{undefined}$$

■

Example 2.12 Consider the functions

```
> f := (x)-> if is(x >= 1) then sqrt(x) fi:
> g := (x)-> if is(x < 1) then x^2 - 2 fi:
```

Find the one-sided limits of the piecewise-defined function h defined by using $f(x)$ if $x > 1$ and $g(x)$ otherwise.

Solution By using the **is( )** command to define f and g we allow for f and g to return different formulas depending on the assumptions placed on symbolic arguments. We can still plot them, as in

```
> plot( {f,g}, -2..2, -4..4 );
```

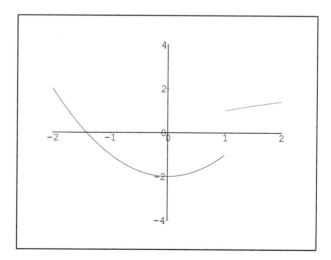

This is the graph of the piecewise-defined function h on the domain $[-2, 2]$. We can also apply these functions to symbolic arguments. If no assumptions are placed on x, then $g(x)$ returns unevaluated, as in

```
> g(x);
```

However, for the case[4]

```
> assume( x < 1 ):
```

we find that $g(x)$ evaluates as

```
> g(x);
```

$$x^2 - 2$$

and that the directional limit

```
> Limit( g(x), x = 1 , left );
```

$$\lim_{x \to 1^-} (x^2 - 2)$$

evaluates to

```
> value(");
```

$$-1$$

To force the domain to be $x \geq 1$ we make the assumption

```
> assume(x >= 1 ):
```

Then the limit

```
> Limit( f(x),x = 0,right);
```

$$\lim_{x \to 0^+} \sqrt{x}$$

evaluates to

```
> value(");
```

$$0$$

■

2.7.2 Combining Continuous Functions

Combinations of continuous functions usually result in continuous functions. For example, each of $f \circ g$ (written **f@g** in *Maple*), $f + g, fg$ and f/g are continuous everywhere they are defined. The only problems that arise are division by 0 for f/g, and mismatched domains for $f \circ g$.

Some of these combinations of functions are illustrated here. Addition, multiplication, and so on of functions are defined by

```
> (f+g)( x );
```

$$f(x) + g(x)$$

---

[4]To remove all such assumptions on x use the command **x := 'x';**.

```
> (f*g)( x );
```
$$f(x)g(x)$$

If the function c is defined so as to evaluate to a constant, as in

```
> c := (x) -> k;
```
$$x \mapsto k$$

then we have

```
> (c*f)( x );
```
$$kf(x)$$

These combinations are always continuous when f and g are continuous. The next example is continuous except where $g(x) = 0$.

```
> (f/g)( x );
```
$$\frac{f(x)}{g(x)}$$

We also have functional composition $h = f \circ g$ defined by $(f \circ g)(x) = f(g(x))$. In *Maple* this is just

```
> (f@g)(x);
```
$$f(g(x))$$

Note that if g is continuous and f is continuous at $g(a)$, then $f \circ g$ is continuous at a. An important consequence of this is as follows:

```
> Limit( (f@g)( x ), x=a ) = Limit( f( y ), y=g( a ) );
```
$$\lim_{x \to a} f(g(x)) = \lim_{y \to g(a)} f(y)$$

2.8 The Intermediate Value Theorem

THEOREM 2.2 If f is continuous on the closed interval $[a, b]$ and $f(a) \leq N \leq f(b)$, then there exists a number c in $[a, b]$ that satisfies $f(c) = N$.

This theorem helps us answer the following kinds of questions.

Example 2.13 Find out whether $f(x) = 0$ has a solution between $x = 1$ and $x = 2$ if f is defined by

```
> f := (x) -> 4*x^3 - 6*x^2 + 3*x - 2 ;
```
$$x \mapsto 4x^3 - 6x^2 + 3x - 2$$

Solution The function f is continuous since it is defined in terms of a polynomial, so we can use the theorem. We can also compute

```
> f( 1 );
```
$$-1$$

and

```
> f( 2 );
```
$$12$$

Since $f(1) < 0 < f(2)$, the theorem guarantees that there must be a c in $[1, 2]$ that satisfies the equation $f(c) = 0$ exactly. ∎

The theorem did not produce the exact solution, but at least we know roughly where to look to find one.[5] In this particular example we are fortunate to be able to compute an exact solution algebraically. A list of three distinct solutions is given by the command

```
> sol := [solve( f( x )=0, x )];
```

$$[\frac{1}{8}\sqrt[3]{3}\,8^{\frac{2}{3}} + \frac{1}{2}, -\frac{1}{16}\sqrt[3]{3}\,8^{\frac{2}{3}} + \frac{1}{2} + \frac{1}{16}\sqrt{-1}\,3^{\frac{5}{6}}8^{\frac{2}{3}}, -\frac{1}{16}\sqrt[3]{3}\,8^{\frac{2}{3}} + \frac{1}{2} - \frac{1}{16}\sqrt{-1}\,3^{\frac{5}{6}}8^{\frac{2}{3}}]$$

the values of which are approximately

```
> evalf( " , 5 );
```

$$[1.2211, 0.13945 + 0.62450\sqrt{-1}, 0.13945 - 0.62450\sqrt{-1}]$$

Not all polynomial equations can be so easily solved. Then the information provided by the mean value theorem becomes even more important.

Example 2.14 Solve the polynomial equation $f(x) = 0$ where f is defined by

```
> f := (x) -> 79*x^14+56*x^12+49*x^9+63*x^8+57*x^7-59*x^3;
```
$$x \mapsto 79x^{14} + 56x^{12} + 49x^9 + 63x^8 + 57x^7 - 59x^3$$

Solution First we attempt to solve this using *Maple*'s built-in solve( ) command.

```
> solve( f( x )=0, x );
```
$$0, 0, 0, RootOf(79\_Z^{11} + 56\_Z^9 + 49\_Z^6 + 63\_Z^5 + 57\_Z^4 - 59)$$

We find that *Maple* was able to find only three roots explicitly, all of them at $x = 0$. The **RootOf( )** expression indicates that the remaining roots are all roots of the indicated polynomial.

We can still find approximations to more of these roots if we know the locations of both positive and negative function values. In this case we have

```
> f( 1/2 );
```
$$-\frac{107633}{16384}$$

---

[5] The bisection method of root finding involves repeatedly subdividing the interval into subintervals and selecting the one containing the root as in the next example.

and

> f(1);

$$245$$

so that, by the intermediate value theorem, we know that there is a root between $x = 1/2$ and $x = 1$.

By continually breaking the interval size in half, we can get as close as we like to a solution to $f(x) = 0$. For example, at the midpoint of the interval $[\frac{1}{2}, 1]$ we have

> f(3/4);

$$-\frac{1104420393}{268435456}$$

which is still negative so we know that there is a root somewhere in $[\frac{3}{4}, 1]$. Similarly, since

> f(7/8);

$$\frac{187795842463999}{4398046511104}$$

is positive, we know that there is a solution in the interval $[\frac{3}{4}, \frac{7}{8}]$.

After ten more repetitions of this subdivision process, our interval will have narrowed from a width of $\frac{1}{2}$ to approximately $\frac{1}{2^{11}} = 0.0004$ units wide, which provides a fairly accurate estimate of the value of the root.

In *Maple* an accurate estimate of this root can be obtained in one step by the command[6]

> fsolve(f(x)=0,x,3/4..7/8);

$$0.7744769991$$

Ordinarily ten digits of accuracy is used but this can be changed. For example, the command **Digits := 100** can be used to request 100 digits of accuracy in subsequent computations. ■

REMARK: How many subdivisions would be required before we knew the location of the solution to this equation to within 0.00001 of the true value?

---

[6]If no range is given **fsolve( )** will attempt to find approximations to all real roots of the polynomial. If the option **complex** is given, **fsolve( )** will attempt to find approximations to all roots of the polynomial including those which involve complex numbers.

2.9 Exact Computations Versus Approximations

When a limit is to be computed, we may need to know exactly the x value of the point at which you are taking the limit. For example, if we are near a discontinuity, being even a fraction off could lead to drastically different values.

Example 2.15 Consider the function defined by $f(x) = x^2$ if $x > 3$, and $f(x) = -x^2$ if $x \leq 3$. The limit of $f(x)$ at $x = 2.9999$ is not even close to the limit of $f(x)$ at 3.0001. ∎

A crucial design decision for the designers of a symbolic algebra system is that of how to interpret floating-point numbers such as 5.0. The *Maple* system has opted to treat all floating-point numbers as approximations.

We can always force the computation involving floating-point approximations to proceed *as if the numbers were exact* by converting all floating-point numbers to rational numbers and then solving the transformed *exact* problem.

```
> v := x^2 - 2.0*x + 1.0;
```
$$x^2 - 2.0x + 1.0$$

```
> convert( v, rational );
```
$$x^2 - 2x + 1$$

In part because of these problems with approximations, it is frequently useful to solve the more general problem – for example, finding the slope exactly at an *arbitrary point* $t = a$ – and then substituting a numeric value for a in the resulting general formula.

Example 2.16 Estimate the slope of the tangent to the curve f at $x = 3.75$ when f is defined by

```
> f := (x) -> x^2 -2*x + 3:
```

by using the slope of a straight line passing through two points close to $x = 3.75$ and then taking the limit of this slope as the one point approaches the other.

Solution If we take one of these points to be at $x = a$ and the other to be at $x = b$, then the slope of the tangent line will be

```
> Limit( (f( b ) - f( a ))/ (b - a), b=a );
```
$$\lim_{b \to a} \frac{b^2 - 2b - a^2 + 2a}{b - a}$$

This evaluates to

```
> value( " );
```
$$2a - 2$$

At $a = 3.75$ this is

```
> subs( a=3.75, " );
```
$$5.50$$

∎

Command	Description
Limit(f(x), x=1);	Represent $\lim_{x \to 1} f(x)$
Limit(f(x), x=a, left);	A directional limit from below
Limit(f(x), x=a, right);	A directional limit from above
assume(x< 1);	Indicate domain constraints on x.
assume(x >= 1);	
is(x > 3);	Verify domain constraints on x
convert(1.5 , rational);	Change 1.5 to 3/2
expand(f(x));	Rewrite f(x) by expanding
f1 := (x) -> if is(x > 3) then sqrt(x) + 2 fi;	A function with a restricted domain
fsolve(f(x)=0,x,3/4..7/8);	Numerical root finding
limit(f(x), x=0);	Compute $\lim_{x \to 1} f(x)$
limit(f(x), x=1, left);	A directional limit from below
limit(f(x), x=1, right);	A directional limit from above
restart;	Reset internal state of *Maple*
showtangent(s(t),t=5,t=4..6);	Graph a curve and its tangent line
simplify(");	Perform general algebraic simplifications
value(");	Force inert expressions to evaluate

Table 2.1 New Commands Used in Chapter 2

EXERCISE SET 2

When attempting each of these exercises do not hesitate to use the capabilities of Maple to help you explore them. Look for opportunities to use **plot( )** *to plot a graph of the function near the points of interest. Commands such as* **factor( )** *and* **normal( )** *or* **simplify( )** *can be used to help simplify expressions.*

The new commands that have been used in this chapter are shown in Table 2.1. All these commands are documented in an online help facility. To discover how a particular command works, complete with examples, just type a **?** *followed by the name of the command you are interested in. Even if you don't know the name, simply type the first two or three characters of a likely sounding name and Maple will report any near matches. You can also use the help browser through mouse actions and buttons that form part of the particular interface you are using to Maple.*

Your primary focus should be on identifying what it is that you need to do to solve a particular problem and to understand why something works the way it does. Recognizing that you do not understand a particular step is half the battle in mastering the concepts and techniques.

The Maple command **limit( )** *can always be used to evaluate a particular limit. Your real task is to understand why the value of the limit computation is what it is.*

1. Investigate each of the given limits by carrying out the following steps.
   a. Plot a graph of the indicated function near the point where you are computing the limit. Use this graph to help predict when a limit will or will not exist at the

indicated point. You may need to specify both a domain and a range in your plot, as in the command

```
> plot( f(x), x=2..5, y = -10..10);
```

b. Evaluate the indicated expression at the indicated point. If evaluation fails try again after simplifying the expression prior to evaluation. You may use the commands **simplify( )** or more specifically **normal( )**, **factor( )**, and **expand( )** to assist with these simplifications. How does the outcome relate to the graph of the corresponding expression?

c. Reexpress the limit computation of the simplified expression in terms of limits of subexpressions. The **expand( )** command can be used to help construct such results. Note that this step will not always help as it may lead to expressions of the form $\frac{0}{0}$, or $\frac{\infty}{\infty}$.

d. Compare the results of evaluating the limits (as computed by hand or using **value( )**) with the alternative formulations of the problem just outlined (i.e., graphical, algebraically simplified, or limits of subexpressions).

i. $\lim\limits_{x \to 2} (x^2 + 1)(x^2 + 4x)$ ii. $\lim\limits_{x \to -1} \dfrac{x - 2}{x^2 + 4x - 3}$

iii. $\lim\limits_{t \to 16} t^{-1/2}(t^2 - 14t)^{3/5}$ iv. $\lim\limits_{t \to 1} \dfrac{t^3 - t}{t^2 - 1}$

v. $\lim\limits_{h \to 0} \dfrac{(h - 5)^2 - 25}{h}$ vi. $\lim\limits_{x \to 2} \dfrac{\frac{1}{x} - \frac{1}{2}}{x - 2}$

2. If $-x^2 + x + 3 < f(x) < x^2 + 2x + 2$ for all x where $f(x)$ is defined, then find the one-sided limit of $f(x)$ as x approaches -1 from above.

3. The slope of a tangent line can be computed by taking the limit of the slope of a *secant line* defined by two points on the curve, as the two points come closer and closer together. To see a visualization of this process for the function sin at $x = 1$, execute the following steps in *Maple*.

a. Let $a = 1$ and choose b to be a point at distance $\frac{1}{i}$ away from a. We can generate a sequence of x-values and points on the curve, one for each choice of $b.i$ as

```
> with(student):
> n := 10; a := 1; f := sin;
> for i to n do b.i := a + 1/i; pt.i := [b,f(b.i)] od:
```

b. Generate the functions corresponding to the ten secant lines formed by the pairs $[a,f(a)]$ and $[b.i,f(b.i)]$. This is done by

```
> for i to n do
>    l.i := makeproc( f(1)
>            + slope([a,f(a)], pt.i ])*(x-1) , x ) ;
> od;
```

c. Generate a vertical line segment from the x-axis to the curve as a list of two points.

```
>    vertical := [[a,0],[a,f(a)]]:
```

d. Generate the graph objects corresponding to each of these secant lines by using the commands

```
> for i to n do
>    g.i := plot( {vertical, sin(x),l.i(x)},
>         x=0.5..1.5, y=-1..1 ,color=green ):
> od:
```

Note that the last line ends with colon (:). This is so that none of the graph objects g.i show on the screen. You can still check that these really are plots by entering, for example, the command

```
> g7;
```

e. Display all the preceding graphs simultaneously.

```
> with(plots,display):
> display( [g.(1..n)]);
```

f. Animate the graphs by using the command

```
> display( [g.(1..n)] , insequence=true );
```

4. In Example 2.2 we investigated the instantaneous rate of change. The slope produced by taking the line defined by the two points $[1,f(1)]$ and $[b,f(b)]$ on the curve changes as b changes. Plot this slope as a function of b for the function f given by

```
> f := (t)-> -49/50*t^2 + 185;
```

$$t \mapsto -\frac{49}{50}t^2 + 185$$

You may use the command **slope( )**. By simplifying the algebraic expression for the slope, explain why this is a straight line.

5. a. Plot a graph of the expression defined by

```
> e1 := (t^(1/7) - 3)/(t-2187);
```

$$(\sqrt[7]{t} - 3)(t - 2187)^{-1}$$

Use it to estimate the value of the limit $\lim_{t=2187} \frac{\sqrt[7]{t}-3}{t-2187}$.

b. Try to compute this limit using *Maple*.[7]

c. Use **changevar( )** to transform this into an algebraic limit that can be computed by *Maple* as was done in Section 2.3.1. Obtain this solution algebraically.

6. The accuracy of a given plot can be increased by indicating that more sample points should be used. This is done by specifying an option such as **numpoints=200** to the **plot( )** command. Use this to investigate the graph shown in section 2.2.3 for

```
> f := (x) -> sin(3*Pi/x);
```

$$x \mapsto \sin(\frac{3\pi}{x})$$

in smaller intervals near $x = 0$.

a. Find an x value in the region $[0, \frac{1}{1,000,000}]$ for which $f(x) = 1$.

---

[7]*Maple* gets this one wrong as it fails to recognize that $\sqrt[7]{2187} - 3 = 0$.

b. Find an x value in the region $[0, \frac{1}{1,000,000}]$ for which $f(x) = -1$.
c. Extend the techniques you used to solve the two previous parts of this exercise to find a similar x value in the interval $[0, \frac{1}{1,000,000,000}]$.

7. Investigate the limit

```
> f := (x) -> 1/(x-1)^2/(x+1);
```

$$x \mapsto \frac{1}{(x-1)^2 (x+1)}$$

```
> Limit(f(x),x=-1);
```

$$\lim_{x \to -1} \frac{1}{(x-1)^2 (x+1)}$$

using directional limits, and draw an appropriate graph illustrating this limit. Does the limit exist? Do the directional limits exist? If so, what are their values?

8. Prove that $\lim_{x \to 3} x^2 + 11 = 20$ by setting up the inequality $|f(x) - L| < \epsilon$ and using factoring and plotting to find bounds for $|x - 3|$.

9. Prove that

$$\lim_{x \to 23} -85x^5 - 55x^4 - 37x^3 - 35x^2 + 97x + 50 = -562946823$$

using the same technique as in the previous question.

10. Prove that $\lim_{x \to 2} x^2 - 7x + 10 \neq 20$ by setting up the inequality $|f(x) - L| < \epsilon$ and then using factoring and plotting to bound $|f(x) - L|$ away from 0.

11. a. Investigate the behavior of $\frac{\sin(x^2)}{x^2}$ near $x = \infty$ by using graphics, limits, and upper and lower bounds.
    b. Investigate the behavior of $x^2 \sin(\frac{1}{x^2})$ near $x = 0$ by using graphics, limits, and upper and lower bounds.
    c. Use **changevar( )** to transform one problem into the other. Note what happens to the location at which we are computing the limit.

12. Investigate the behavior of $\frac{\sin(x^2)}{3x^3}$ by using graphics and limits as $x \to \infty$. Use upper and lower bounds and the squeeze theorem to prove the result.

13. a. Use graphics and limits to investigate the behavior of **exp(x)** and **exp(1/x)** as x gets large. What are these limits at $x = \infty$?
    b. By computing the limit of $\frac{f(x)}{g(x)}$ as $x \to \infty$, we can decide if $f(x)$ or $g(x)$ grows faster. If the limit is 1, they grow at the same rate. If the limit is less than 1, then the function on the bottom is growing faster, whereas if the limit is greater than 1 then the function on the top is growing faster. Use this technique to compare **exp(x)** with each of x, x^{10}, and with x^{100}.
    c. The above results should be consistent with the fact that **exp(x)** grows faster than any power of x.

14. a. Use graphics and limits to investigate the behavior of $\ln(x)$ and $\ln(1/x)$ as x gets large and as $x \to 0$.
    b. By using graphics and computing the limit of $\frac{f(x)}{g(x)}$ as $x \to \infty$, compare the growth rate of $\ln(x)$ to each of x, $\sqrt{x}$, and $\sqrt[10]{x}$.

15. Use the intermediate value theorem to prove that there is a c such that $f(c) = 10$ when f is defined by

    ```
    > f := (x) -> x^3 - 5*x^2 + x;
    ```
    $$x \mapsto x^3 - 5x^2 + x$$

16. Explain why the intermediate value theorem cannot be used to find a solution to $f(c) = 3$ for

    ```
    > f := (x) -> (x-2)^4 + 5;
    ```
    $$x \mapsto (x-2)^4 + 5$$

17. The displacement of a particle moving in a straight line is given by s where $s : (t) \to t^2 - 7t + 19$. Find the average velocity over the time intervals $[2, 5]$ and the instantaneous velocity at $t = 2$.

18. Investigate the continuity of the function g defined piecewise by

    $$g(x) = \begin{cases} 2x - x^2 & \text{if } 0 \le x \le 2 \\ 2 - x & \text{if } 2 < x \le 3 \\ x - 4 & \text{if } 3 < x < 4 \\ \pi & \text{if } 4 \ge x \end{cases}$$

19. a. Use limits to find a formula for the instantaneous slope of the tangent lines for each of f and g defined by

    ```
    > f := (x) -> sqrt(2-(x-1)^2) + 3::
    ```
    and
    ```
    > g := (x) -> (x-4)+3;
    ```
    $$x \mapsto x - 1$$
    at the point a.

    b. Find a point where the slope of f and g are equal. (Hint: You may want to use **fsolve( )**.)

20. According to Boyle's law, when the temperature of a confined gas is held constant, then the product of pressure P and volume V is a constant.

    a. Find the average rate of change of P in terms of V if $PV = 500$ and V changes from 100in^3 to 150in^3.

    b. Show that when V is regarded as a function of P, the instantaneous rate of change V at $V = a$ is proportional to $\frac{1}{a^2}$.

3 Derivatives

3.1 Introduction

Consider the equation $y = f(x)$. The tangent line to the curve defined by this equation at $x = a$ can be approximated by taking the line defined by $[a, f(a)]$ and a second point on the curve $[b, f(b)]$ and letting the second point approach the first along the curve.

```
> with(student):
> P := [a,f(a)];
```

$$[a, f(a)]$$

```
> Q := [b,f(b)];
```

$$[b, f(b)]$$

The slope of this secant line is

```
> slpe := slope(Q,P);
```

$$\frac{f(b) - f(a)}{b - a}$$

and this slope approaches the slope of the tangent line.

Example 3.1 Let f be defined by

```
> f := (x) -> (x-1)*(x-2)*(x);
```

$$x \mapsto (x - 1)(x - 2)x$$

Approximate the slope of f at $x = 1/2$.

Solution Let $a = 1/2$ and consider a sequence of points $b = a + h$ for small values of h.

```
> a := 1/5; b := a+h;
```

$$\frac{1}{5} + h$$

```
> h := .1; slope( [b,f(b)], [a,f(a)]);
```

$$0.6900000000$$

```
> h := .01; slope( [b,f(b)], [a,f(a)]);
```

$$0.8961000000$$

```
> h := .001; slope( [b,f(b)], [a,f(a)]);
```
$$0.9176010000$$

∎

To compute the true slope of the tangent line at a, we compute the limit of the preceding slope as $h \to 0$. It is

```
> h := 'h':
> Limit( slope( [b,f(b)],[a,f(a)]) ,h=0);
```

$$\lim_{h \to 0} \left(\left(-\frac{4}{5}+h\right)\left(-\frac{9}{5}+h\right)\left(\frac{1}{5}+h\right) - \frac{36}{125}\right) h^{-1}$$

which has the value

```
> m := value(");
```

$$\frac{23}{25}$$

The tangent line at $[a, f(a)]$ is then given by the equation

```
> y - f(a) = m * (x - a );
```

$$y - \frac{36}{125} = \frac{23}{25}x - \frac{23}{125}$$

This tangent line and the curve f are shown together in the next graph.

```
> showtangent(f(x),x=1/5,x=-1..1,y=-1..1);
```

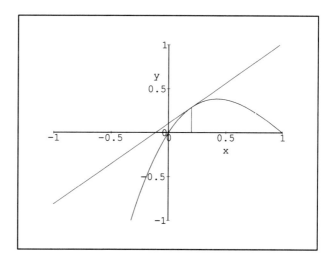

The result of this limit computation (the slope of the tangent line at $x = a$) is referred to as *the derivative of $f(x)$ at $x = a$*.

A General Formulation of the Derivative

For the ensuing discussion we want both f and a to be unspecified. Thus, we begin the discussion by unassigning them.

```
> f := 'f':   a := 'a':  h := 'h':
```

If we choose the second point b to be

```
> b := a + h:
```

then the slope of the secant line becomes

```
> slpe;
```

$$\frac{f(a+h) - f(a)}{h}$$

As b approaches a, h approaches 0, so the slope of the tangent line is given by

```
> m := Limit( slpe , h = 0);
```

$$\lim_{h \to 0} \frac{f(a+h) - f(a)}{h}$$

The result of actually computing this limit (once f has been defined) would be the slope of the curve $y = f(x)$ at $x = a$. If f is not yet defined, the value of this limit computation is written as

```
> value(");
```

$$D(f)(a)$$

REMARK: Because this rule for computing the slope of f at $x = a$ can be attempted for any value of a in the domain of the function f, the rule defines a new function. The name of this new function is $D(f)$. We will sometimes write $D(f)$ as f' and $D(f)(a)$ as $f'(a)$.

DEFINITION 3.1 The derivative of a function f, evaluated at a, is L, where

```
> L := Limit( (f(a+h)-f(a))/h , h = 0 );
```

$$\lim_{h \to 0} \frac{f(a+h) - f(a)}{h}$$

This limit computation can also be described in terms of the points a and b.

```
> a := 'a': b := 'b':
> m := Limit( slope([b,f(b)],[a,f(a)]) , b = a);
```

$$\lim_{b \to a} \frac{f(b) - f(a)}{b - a}$$

```
> value(");
```

$$D(f)(a)$$

To see that these two limits are indeed the same, make the change of variables defined by $h = b - a$.

```
> changevar(h=b-a,m,h);
```

$$\lim_{h \to 0} \frac{f(a+h) - f(a)}{h}$$

Note that h tends to 0 as b tends to a.

REMARK: We have not yet concerned ourselves with whether h is positive or negative — or equivalently, whether b is bigger than or smaller than a. Occasions will arise where this distinction will be important.

3.1.1 Interpretations of the Derivative

Although it may not be immediately obvious, the fact that we have an algebraic way of computing the slope of a tangent line for every point on a curve is extremely important.

Example 3.2 Investigate the relationship between the graph of the function f defined by

```
> f := (x) -> (x-1)*(x-2)*(x-3)*(x-4);
```

$$x \mapsto (x-1)(x-2)(x-3)(x-4)$$

and that of its derivative $f' = D(f)$.

Solution The function $f' = D(f)$ is[1]

```
> D(f);
```

$$x \mapsto (x-2)(x-3)(x-4) + (x-1)(x-3)(x-4) + (x-1)(x-2)(x-4) + (x-1)(x-2)(x-3)$$

and a graph of them together is

```
> plot( {f,D(f)},-1..5,-5..5);
```

---

[1] We deliberately postpone any discussion of how to compute this derivative to focus attention on general properties. The actual limit computation that defines $f'(x)$ can be completed by simplifying the expression to remove a common factor of h. This, and other techniques for computing derivatives, will be discussed later in this chapter.

Among other things, you should note that

- where f is increasing (left to right), the value of $f'(a)$ is positive
- where f is decreasing (left to right), the value of $f'(a)$ is positive
- where f peaks[2] or bottoms out, $f'(a)$ is 0.

∎

If we can construct an algebraic formula for a derivative, we can use it to investigate the behavior of the original function and to address questions such as where the maximums and minimums occur.

Example 3.3 Find where the function f defined by

```
> f := (x) -> x^2 - 8*x + 9 ;
```

$$x \mapsto x^2 - 8x + 9$$

is minimized.

Solution Because the tangent line will be horizontal at the (smooth) peaks and valleys occurring in the graph of f, we can find maximums and minimums by finding where the derivative is 0. The derivative of the function f is the function

```
> D(f);
```

$$x \mapsto 2x - 8$$

and we require a value a such that

```
> D(f)(a) = 0;
```

$$2a - 8 = 0$$

[2] For now we shall assume that the curve is smooth and well behaved.

This is given by

> isolate(",a);

$$a = 4$$

As the slope of the tangent line is 0 at this point, we expect either a maximum or a minimum to occur at the point

> [4,f(4)];

$$[4, -7]$$

We can easily decide which in this case (and also confirm that there is either a peak or a valley at $a = 4$) by graphing f. The following graph shows both f and $D(f)$. Can you decide which curve belongs to which function?

> plot({f,D(f)},0..10);

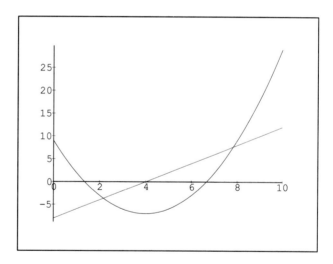

We can also show the tangent line to f at the *critical point* where the slope of f is 0.

> showtangent(f(x), x=4 ,x=0..10);

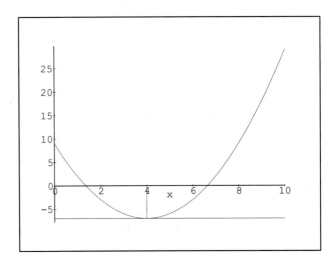

Motion

If the function f represents the position of a particle as a function of time t, then the slope of the curve at $t = a$ is the instantaneous rate of change with respect to time (velocity).

Example 3.4 If the position of a particle is given by the function

```
> f := (t) -> t^3 - 6*t^2 + 9*t ;
```
$$t \mapsto t^3 - 6t^2 + 9t$$

where t is measured in seconds and $s = f(t)$ is measured in meters, find the velocity of the particle when $t = 2$.

Solution The velocity at time a is the slope of the tangent line for the graph of f as a function of time. The slope is

```
> Limit( (f(a+h)-f(a))/h,h=0 ) = D(f)(a);
```
$$\lim_{h \to 0} \frac{(a+h)^3 - 6(a+h)^2 + 9h - a^3 + 6a^2}{h} = 3a^2 - 12a + 9$$

At $t = 2$ this is just

```
> D(f)(2);
```
$$-3$$

meters per second.

REMARK: A negative velocity usually indicates that the direction of travel is downward or perhaps backward.

REMARK: When we ignore the direction of travel, we refer to velocity as *speed*. The speed of this particle at $t = 2$ seconds is

```
> abs( D(f)(2) );
```

$$3$$

meters per second.

Example 3.5 Find the exact time when the particle in Example 3.4 is at rest.

Solution The position of the particle at time $t = a$ is given by

```
> a := 'a':
> f(a);
```

$$a^3 - 6a^2 + 9a$$

The velocity at time $t = a$ is given by

```
> D(f)(a);
```

$$3a^2 - 12a + 9$$

The particle will be at rest when the tangent line to the graph of f is horizontal. Thus, we want to find a value of a for which

```
> D(f)(a) = 0;
```

$$3a^2 - 12a + 9 = 0$$

The values of a for which this occurs are

```
> solve(D(f)(a)=0,a);
```

$$3, 1$$

We have used **solve( )** here instead of **isolate( )** because we want to consider all solutions of the quadratic equation. The following graph clarifies how to interpret these times relative to the motion of the particle.

```
> plot({f,D(f)},0..5);
```

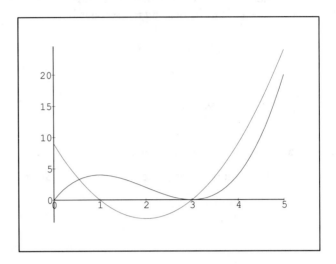

The particle moves away from the origin at a fairly high velocity but slows down and stops at $t = 1$. It returns to the origin, arriving at $t = 3$, and then departs again continuing to increase velocity. The graph of $D(f)$ crosses the t-axis where the tangent line to the graph of f is horizontal. ∎

Growth of Populations

Let $n = f(t)$ be the number of individuals in a population at time t so that $\Delta n = f(t_1) - f(t_2)$ is the change in population between time $t = t_1$ and time $t = t_2$. The average rate of growth from time t_1 to time t_2 is the slope of the secant line defined by the two points $[t_1, f(t_1)]$ and $[t_2, f(t_2)]$.

$$\text{average rate of growth} = \frac{\Delta n}{\Delta t} = \frac{f(t_2) - f(t_1)}{t_2 - t_1}$$

The instantaneous rate of growth is the slope of the tangent line to the graph of f at t_1, which is

```
> D(f)(t1);
```

$$D(f)(t1)$$

It is only an approximation to the rate of growth because to compute the limit represented by this expression, we must assume that f is continuous. Strictly speaking, f is not continuous since it not defined when the population is, for example, 123.5. For large populations this is a reasonable distortion of the truth.

Example 3.6 Find the instantaneous population growth rate in persons per minute at time $t = 10.1$ minutes if the population growth is approximated by the function

```
> p := (t) -> 3*exp(t) + 1000;
```

$$t \mapsto 3e^t + 1000$$

Solution The function exp defined by $(t) \to e^t$ is known as the exponential function and arises in situations where, for example, populations are doubling every ten years. The value of $\exp(t)$ can be computed as a power e^t where the constant[3] e is approximately

```
> evalf(E,40);
```

$$2.718281828459045235360287471352662497757$$

The function exp is noted for its very rapid growth. This can be seen by evaluating it at a few large values and by examining the following graph.

```
> plot( {t^3,exp(t)},t=1..6);
```

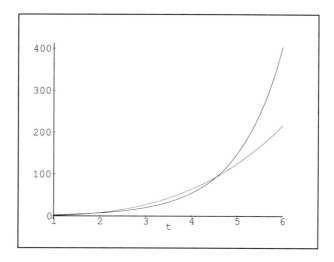

This graph includes a plot of t^3 for comparison. Actually, as t grows large, the value of $\exp(t)$ grows more rapidly than any power of t — in particular a cubic.

The instantaneous rate of growth of a function at $t = a$ is just the slope of the tangent line at $t = a$. The slope[4] of the curve p at $t = a$ is given by

```
> D(p)(a);
```

$$3e^a$$

Note that this formula for the slope also involves the expression $\exp(a) = e^a$. It is quite close in value to $3 p(a)$. At $t = 10.1$ minutes, the population size is

```
> p(10.1);
```

$$74029.02826$$

and the slope of the tangent line is

```
> D(p)(10.1);
```

$$73029.02826$$

---

[3] In *Maple*, the constant e is represented by **E = exp(1)**.
[4] The exponential function and its derivative will be investigated in more detail later in this chapter.

A growth rate that is directly proportional[5] to the function value is characteristic of the exponential function. ∎

Fluid Flow

The velocity of a fluid flowing through a cylindrical tube of radius R and length L is faster near the center of the tube and almost stopped near the walls. This behavior is modeled by the function defined by

```
> V := (r) -> P/(4*eta*L)*(R^2 - r^2);
```

$$r \mapsto \frac{1}{4} \frac{P\left(R^2 - r^2\right)}{\eta L}$$

where r measures the distance from the the center, and the constants R, η, L, and P are respectively the radius of the tube, the viscosity of the fluid, the length of the tube, and the difference in pressure as measured at the two ends.

Consider the case of blood flow through a small blood vessel. Assuming the correct units of measure[6] are used, typical values for the constants are

```
> R := .009:  eta := 0.027:  L := 2.5:  P := 3900:
```

so that at a distance of r from the center of the tube the velocity in cm/s is

```
> V(r);
```

$$1.170000000 - 14444.44444\, r^2$$

Example 3.7 Find a formula for the instantaneous change in the velocity of the blood as we move from the center toward the wall of the blood vessel.

Solution At the center of the blood vessel the blood is flowing at a rate of

```
> V(0);
```

$$1.170000000$$

cm per second. The function

```
> D(V);
```

$$r \mapsto -28888.88888\, r$$

gives the instantaneous change in the fluid velocity as a function of the distance from the center of the tube. The following graph of V shows the actual fluid velocity as a function of the distance from the center of the blood vessel.

---

[5] A value $v_1(t)$ is said to be directly proportional to another value $v_2(t)$ if they are related by an equation of the form $v_1(t) = k v_2(t)$ for some constant k as t varies.
[6] The distances r, R, and L are in centimeters, and P is in dynes/cm^2.

```
> plot( V , 0..R);
```

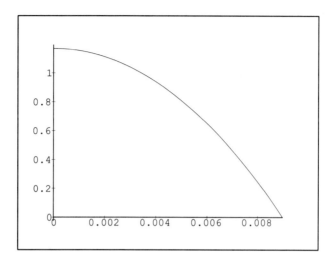

Note the values

```
> D(V)(0), D(V)(R);
```

$$0, -259.9999999$$

These values and the graph both suggest that as we move closer to the outside wall the velocity decreases. ∎

Rates of Change

Example 3.8 A spherical snowball is melting in such a way that its radius is decreasing at a rate of $\frac{1}{10}$ cm/min. At what rate (change per unit of time) is the volume changing when the radius is 5 cm?

Solution The volume of the snowball, expressed in terms of the radius, is computed by the function

```
> V := (r)-> 4/3*Pi*r^3;
```

$$r \mapsto \frac{4}{3}\pi r^3$$

Because V is a function of the radius and not of time, we cannot use it as it stands. However, if we can compute the radius r as a function of time, say $r = r(t)$, then the volume at time t is

```
> V(r(t));
```

$$\frac{4}{3}\pi r(t)^3$$

This new formula can be used to define a function for volume as a function of time.[7]

```
> V2 := makeproc( " , t );
```

$$t \mapsto \frac{4}{3}\pi r(t)^3$$

A graph of the function $V2$ would show how volume changes with respect to time. The tangent lines in this graph would correspond to the instantaneous *rate of change*.

The function

```
> D(V2);
```

$$t \mapsto 4\pi r(t)^2 D(r)(t)$$

computes the slope of these tangent lines. At time t, the snowball is decreasing in size at a rate of

```
> D(V2)(t);
```

$$4\pi r(t)^2 D(r)(t)$$

cubic cm per minute.

We do not know when the radius will reach 5 cm, but we do know that at that instant, $r(t) = 5$ and $D(r)(t) = -1/10$. This is enough information to compute the answer. The rate is

```
> subs( {r(t)=5, D(r)(t) = -1/10} , " );
```

$$-10\pi$$

which is approximately

```
> evalf(");
```

$$-31.41592654$$

cubic cm per minute. ∎

3.1.2 Leibnitz Notation

Typically, functions are defined in terms of an algebraic expression involving a variable x. Sometimes it is convenient to work directly with that algebraic expression.

When the *function* f is not yet defined, the *expression*

```
> f(x);
```

$$f(x)$$

is an *algebraic expression in* x, and the algebraic expression in x corresponding to the function $D(f)$ evaluated at x is

```
> D(f)(x);
```

---

[7] The command **unapply( )** can be used here instead of **makeproc( )**, which is a command specific to the student package and which has additional capabilities related to points and lines.

$$D(f)(x)$$

The Leibnitz notation for derivatives works directly with expressions in *x* instead of just the function *f*. There is no equivalent to $D(f)$ and the expression in *x*, $D(f)(x)$, can be written as

```
> convert(",Diff);
```

$$\frac{d}{dx} f(x)$$

Similarly, we might write

```
> Diff(x^2 + 3*x ,x );
```

$$\frac{d}{dx}\left(x^2 + 3x\right)$$

The result is also *an expression in x*, The symbol $\frac{d}{dx}$ indicates that the derivative of the expression is to be taken with respect to *x* to produce a new expression in *x*.

If *f* is already defined by, for example,

```
> f := (x) -> (x^2 + 2*x)^3;
```

$$x \mapsto \left(x^2 + 2x\right)^3$$

then we can still write

```
> Diff(f(x),x);
```

$$\frac{d}{dx}\left(\left(x^2 + 2x\right)^3\right)$$

The actual computation of the derivative of this expression can be completed by the command

```
> value(");
```

$$3\left(x^2 + 2x\right)^2 (2x+2)$$

which yields the same result as $D(f)(x)$. This result can also be computed immediately from the expression $f(x)$ by the command

```
> diff(f(x),x);
```

$$3\left(x^2 + 2x\right)^2 (2x+2)$$

Remember that **D( )** constructs functions from functions, whereas **Diff( )** and **diff( )** construct expressions in *x* from expressions in *x*. This crucial distinction between functions and expressions will be discussed further in Section 3.4.

3.2 Differentials

One of the most important uses of tangent lines is to construct approximations. If the graph of a function is sufficiently smooth, we can estimate values of the function near a specific point on the curve by assuming the curve is identical to the tangent line at that point. The resulting formula is usually much easier to compute and, providing that we remain close to the point where the tangent line actually meets the curve, the results can be quite accurate.

The slope of a line tangent at $x = x0$ to the curve defined by f is

```
> D(f)(x0);
```

$$D(f)(x0)$$

and so the equation for the line tangent at the point $[x0, f(x0)]$ is

```
> eq1 := y-f(x0) = D(f)(x0) * (x-x0);
```

$$y - f(x0) = D(f)(x0)(x - x0)$$

which can also be rearranged as

```
> eq2 := isolate(",y);
```

$$y = D(f)(x0)(x - x0) + f(x0)$$

The quantity $y - f(x0)$ in $eq1$ is an estimate of the change in the function value generated by moving from $x0$ to x and is usually denoted by dy. The exact change in the x value is denoted by $dx = x - x0$ so that $eq1$ can be written as

$$dy = D(f)(x0)dx$$

REMARK: Remember that dy is only an estimate of the change in function value and that $f(x0) + dy$ is only an estimate of the value $y = f(x)$.

Example 3.9 Approximate $f(x)$ near $x = 1$ by using the tangent line at $x = 1$, for

```
> f := (x) -> x^3 + 2*x^2;
```

$$x \mapsto x^3 + 2x^2$$

Solution The slope of the tangent line at $x = 1$ is

```
> D(f)(1);
```

$$7$$

The tangent line

```
> y - f(1) = D(f)(1)*(x-1);
```

$$y - 3 = 7x - 7$$

approximates the function f near $x = 1$ so that, for example,

```
> isolate(",y);
```

$$y = 7x - 4$$

is an approximation to $f(x)$. The function f and its tangent line appear in the graph

```
> showtangent(f(x),x=1,x=0..2);
```

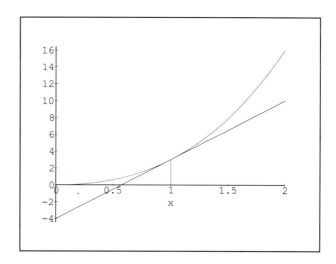

The change in the function value caused by moving from $x = 1$ to $x = a$ is approximated by

```
> D(f)(1)*(a-1);
```

$$7a - 7$$

∎

Example 3.10 Find the error incurred by using the tangent line at $x = 2$ to estimate $f(2.5)$ if the function f is defined by

```
> f := (x) -> x^3 + 3*x^2 - 2*x + 1;
```

$$x \mapsto x^3 + 3x^2 - 2x + 1$$

Solution The tangent line at $x = 2$ is given by the equation

```
> y - f(2) = D(f)(2)*(x-2);
```

$$y - 17 = 22x - 44$$

and so by the function

```
> isolate(",y);
```

$$y = 22x - 27$$

```
> g := makeproc(rhs("),x);
```

$$x \mapsto 22x - 27$$

The error in using g to estimate f is the vertical distance between the two curves at x = 2.5, as shown by the graph

```
> showtangent(f(x),x=2,x=1..3);
```

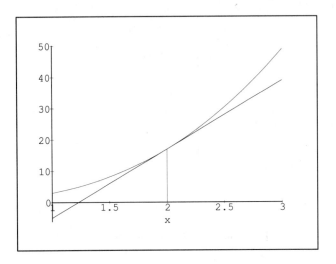

This error is

```
> f(2.5) - g(2.5);
```

$$2.375$$

∎

3.3 Differentiation Formulas

The formal mathematical definition of a derivative makes use of limits. In principle, we could continue to compute all derivatives by setting up and evaluating such limits, but if we did so the process would remain unnecessarily complex.

Mechanisms that allow us to avoid complexity in a computation are very important, whether we intend to carry out the computation by hand or by machine. In the previous sections, once we defined the concept of a derivative, we relied heavily on *Maple* to compute them. Although *Maple* could evaluate the associated limits, *Maple* actually computed these derivatives directly from the structure of the formulas. This more efficient approach was possible because of identifiable patterns that occur in the the algebraic results. These systematic rules for the computation of derivatives are some of the best known examples of this mathematical phenomenon.

The object of this section is to discover some of the useful algebraic patterns that occur when computing derivatives. Once we have identified and confirmed these patterns, they remain available to help us streamline all of our computations – both by hand and by machine.

3.3.1 Some Fundamental Examples

We begin by closely examining the derivatives of a few basic functions. The resulting formulas, together with some rules for combining them through sums and products and compositions, allow us to compute directly all the derivatives that arise in this course.

During our search for formulas we will frequently need to refer to the quotient

```
> (f(x+h)-f(x))/h;
```

$$\frac{f(x+h) - f(x)}{h}$$

since the formal definition of derivatives is in terms of the limit of such expressions as h approaches 0.

This quotient is called the *Newton quotient of f at x*. The following *Maple* procedure can be used to generate such a quotient for a given function f.

```
> nq := makeproc(",f);
```

$$f \mapsto \frac{f(x+h) - f(x)}{h}$$

Here are some examples of its use.

```
> nq(f);
```

$$\frac{f(x+h) - f(x)}{h}$$

```
> nq(g);
```

$$\frac{g(x+h) - g(x)}{h}$$

```
> nq(f+g);
```

$$\frac{f(x+h) + g(x+h) - f(x) - g(x)}{h}$$

Our goal is to discover patterns. To this end, we begin by examining the derivatives of functions defined by very simple formulas or combinations of simple formulas. If we were computing by hand, we would proceed in much the same way but would be forced to spend more time on each calculation.

3.3.2 Constants

A typical *constant function* is

```
> f := (x) -> 3;
```
$$3$$

The Newton quotient of f at x is

```
> nq(f);
```
$$0$$

This value will always be 0 for constant functions because both $f(x)$ and $f(x+h)$ are the same.

```
> f(x), f(x+h);
```
$$3, 3$$

Of course, the limit $\lim_{h \to 0} \text{nq}(f)$ is also 0.

This remains true even when the constant is unspecified. Any formula that does not involve x will remain constant as x changes. Consider the function

```
> f := (x) -> c;
```
$$x \mapsto c$$

We have the same result as before.

```
> f(x), f(x+h);
```
$$c, c$$

```
> nq(f);
```
$$0$$

```
> limit(",h=0);
```
$$0$$

We summarize these facts in Theorem 3.1.

THEOREM 3.1 If f is a constant function so that $f(x) = c$ for some constant c, then $f' = D(f)$ is defined by $D(f)(x) = 0$ for all x.

REMARK: The analogous result using the Leibnitz notation and expressions is

```
> Diff(c,x);
```
$$\frac{d}{dx} c$$

```
> value(");
```
$$0$$

When the value of $\frac{d}{dx} c$ is 0, c remains constant as x changes.

3.3.3 Pure Powers

What happens when the function f is defined in terms of a polynomial that is a pure power, such as x^2 or x^3? Consider

```
> f := (x) -> x^3;
```

$$x \mapsto x^3$$

The Newton quotient at x is

```
> nq(f);
```

$$\frac{(x+h)^3 - x^3}{h}$$

which simplifies to

```
> simplify(");
```

$$3x^2 + 3xh + h^2$$

so the value of $D(f)(x) = f'(x)$ (the derivative of f evaluated at x) is

```
> Limit(nq(f),h=0);
```

$$\lim_{h \to 0} \frac{(x+h)^3 - x^3}{h}$$

```
> value(");
```

$$3x^2$$

A general pattern becomes apparent after examining a few more examples.

```
> g := (x) -> x^4: D(g);
```

$$x \mapsto 4x^3$$

```
> g := (x) -> x^5: D(g);
```

$$x \mapsto 5x^4$$

```
> g := (x) -> x^6: D(g);
```

$$x \mapsto 6x^5$$

The general outcome for this pattern is summarized in Theorem 3.2.

THEOREM 3.2 The derivative of the function f defined by

```
> f := (x) -> x^n;
```

$$x \mapsto x^n$$

where $n \neq 0$ is a constant, is the function

```
> D(f);
```

$$x \mapsto \frac{x^n n}{x}$$

REMARK: The analogous formula for expressions in x is

> Diff(x^n,x);

$$\frac{\partial}{\partial x}(x^n)$$

> value(");

$$\frac{x^n n}{x}$$

REMARK: When $n = 0$, the function is a constant function and the derivative evaluates to 0, by Theorem 3.1.

Proof Here we will consider only the case of positive integer powers. Examples of more general cases follow.

The validity of this formula for the derivative relies on the following limit computation.

> Limit(nq(f),h=0);

$$\lim_{h \to 0} \frac{(x+h)^n - x^n}{h}$$

> value(");

$$\frac{x^n n}{x}$$

Rather than rely solely on *Maple*, we choose to look at this computation in greater detail. Our standard approach to computing such a limit is to simplify the Newton quotient. The numerator is

> numer(nq(f));

$$(x+h)^n - x^n$$

If we can find a common factor, h, then it can be used to cancel the denominator. The binomial theorem allows us to rewrite

> f(x+h);

$$(x+h)^n$$

as

> s := Sum(binomial(n,i)*h^i*x^(n-i) , i=0..n);

$$\sum_{i=0}^{n} \binom{n}{i} h^i x^{n-i}$$

3.3 Differentiation Formulas

For example, at $n = 6$ this becomes

```
> value( subs(n=6,s) );
```

$$x^6 + 6hx^5 + 15h^2x^4 + 20h^3x^3 + 15h^4x^2 + 6h^5x + h^6$$

which is the same as

```
> factor(");
```

$$(x+h)^6$$

The term in s corresponding to $i = 0$ is $g(x)$. Every other term of the expanded sum s, and hence of $g(x+h) - g(x) = s - g(x)$, has a factor of h. The term in $s - g(x)$ corresponding to $i = 1$, nhx^{n-1}, is the only term with exactly one factor of h and so it is the only term, after division by h, that does not tend to 0 as $h \to 0$.

```
> limit(nq(f),h=0);
```

$$\frac{x^n n}{x}$$

□

Example 3.11 Use Newton quotients and limits to verify that for the function f defined by

```
> f := (x) -> x^12;
```

$$x \mapsto x^{12}$$

the derivative is the function $D(f)$ defined by

```
> D(f);
```

$$x \mapsto 12x^{11}$$

Solution The Newton quotient is

```
> nq(f);
```

$$\frac{(x+h)^{12} - x^{12}}{h}$$

This simplifies to

```
> expand(");
```

$$12x^{11} + 66hx^{10} + 220h^2x^9 + 495h^3x^8 + 792h^4x^7 + 924h^5x^6 + 792h^6x^5 + 495h^7x^4 + 220h^8x^3 + 66h^9x^2 + 12h^{10}x + h^{11}$$

All the terms have a factor of h except $12x^{11}$, so the limit evaluates to

Chapter 3 Derivatives

```
> limit(",h=0);
```

$$12x^{11}$$

The rule for derivatives of powers also works for negative powers. The following limit evaluation confirms this without relying on the binomial theorem to expand $(x+h)^n$.

Example 3.12 Use Newton quotients and limits to verify that for the function defined by

```
> f := (x) -> x^(-2);
```

$$x \mapsto x^{-2}$$

the derivative is the function $D(f)$ defined by

```
> D(f);
```

$$x \mapsto -\frac{2}{x^3}$$

Solution The Newton quotient is

```
> nq(f);
```

$$\left((x+h)^{-2} - x^{-2}\right) h^{-1}$$

This simplifies to

```
> normal(");
```

$$-\frac{2x+h}{(x+h)^2 x^2}$$

Because neither the denominator nor the numerator tend to zero[8] as h tends to zero, this can be evaluated directly. The result is

```
> limit(",h=0);
```

$$-\frac{2}{x^3}$$

∎

Even derivatives of fractional powers behave according to the power rule.

Example 3.13 Use Newton quotients and limits to verify that for the function defined by

```
> f := (x) -> x^(3/2);
```

$$x \mapsto x^{\frac{3}{2}}$$

the derivative is the function $D(f)$ defined by

```
> D(f);
```

[8]This assumes that $x \neq 0$.

3.3 Differentiation Formulas

$$x \mapsto \frac{3}{2}\sqrt{x}$$

Solution The Newton quotient is

> nq(f);

$$\left((x+h)^{\frac{3}{2}} - x^{\frac{3}{2}}\right) h^{-1}$$

This time, finding an algebraic simplification is more difficult. It can still be accomplished by introducing a common factor

> b := numer(") + 2*f(x);

$$(x+h)^{\frac{3}{2}} + x^{\frac{3}{2}}$$

We rewrite the numerator as

> numer(nq(f))*b;

$$\left((x+h)^{\frac{3}{2}} - x^{\frac{3}{2}}\right)\left((x+h)^{\frac{3}{2}} + x^{\frac{3}{2}}\right)$$

which expands to

> e1 := expand(");

$$3x^2 h + 3xh^2 + h^3$$

and the denominator as

> e2 := b*h;

$$\left((x+h)^{\frac{3}{2}} + x^{\frac{3}{2}}\right) h$$

Thus, the Newton quotient can be rewritten as

> normal(e1/e2);

$$\left(3x^2 + 3xh + h^2\right)\left((x+h)^{\frac{3}{2}} + x^{\frac{3}{2}}\right)^{-1}$$

As neither the numerator nor denominator are zero,[9] the limit

> Limit(",h=0);

$$\lim_{h \to 0} \left(3x^2 + 3xh + h^2\right)\left((x+h)^{\frac{3}{2}} + x^{\frac{3}{2}}\right)^{-1}$$

can be evaluated directly as

> value(");

$$\frac{3}{2}\sqrt{x}$$

∎

---

[9] Again, we require that $x \neq 0$ for this statement to hold.

3.3.4 Laws for Addition and Multiplication

Specific patterns like those shown in the preceding section already provide numerous computational shortcuts. They become especially useful, however, when they can be combined to build derivatives of expressions that are much more complicated.

We validate the following rules in much the same way that we established the validity of the specific formulas in the previous section. In each case we need to set up the Newton quotient and carefully simplify it to compute the corresponding limit.

THEOREM 3.3 Derivatives of combinations of expressions can be computed by making use of the following basic rules.

```
> constants := constants , c:
> D(c*f);
```
$$c D(f)$$

```
> D(f+g);
```
$$D(f) + D(g)$$

```
> D(f*g);
```
$$D(f)g + f D(g)$$

REMARK: The equivalent rules for expressions in x are

```
> diff(c*f(x),x);
```
$$c \frac{d}{dx} f(x)$$

```
> diff(f(x)+g(x),x);
```
$$\frac{d}{dx} f(x) + \frac{d}{dx} g(x)$$

```
> diff(f(x)*g(x),x);
```
$$\frac{d}{dx} f(x) g(x) + f(x) \frac{d}{dx} g(x)$$

The Derivative of $c * f(x)$

Proof Consider the function g defined as follows.

```
> f:= 'f': g := (x) -> c*f(x);
```
$$x \mapsto c f(x)$$

The Newton quotient for g is

```
> nq(g);
```

$$\frac{cf(x+h) - cf(x)}{h}$$

By factoring this Newton quotient, we see that the constant c plays no real role in computing the resulting limit.

> `factor(");`

$$\frac{c(f(x+h) - f(x))}{h}$$

The limit defining this derivative is

> `limit(",h=0);`

$$c D(f)(x)$$

which can also be written as

> `convert(",diff);`

$$c\frac{d}{dx}f(x)$$

□

The Derivative of a Sum

Proof Consider a typical sum of functions

> `G := f+g;`

$$f + g$$

The derivative is defined in terms of

> `e1 := nq(G);`

$$\frac{f(x+h) + g(x+h) - f(x) - g(x)}{h}$$

But this is just a simplified version of

> `e2 := nq(f) + nq(g);`

$$\frac{f(x+h) - f(x)}{h} + \frac{g(x+h) - g(x)}{h}$$

as verified by the simplification

> `normal(");`

$$\frac{f(x+h) + g(x+h) - f(x) - g(x)}{h}$$

The actual derivative is computed as

> `map( Limit,e2,h=0);`

$$\lim_{h \to 0} \frac{f(x+h) - f(x)}{h} + \lim_{h \to 0} \frac{g(x+h) - g(x)}{h}$$

> `value(");`

$$D(f)(x) + D(g)(x)$$

□

The Derivative of a Product

Proof Consider a typical product of two functions.

```
> H := f*g;
```

$$fg$$

The Newton quotient

```
> e1 := nq(H);
```

$$\frac{f(x+h)g(x+h) - f(x)g(x)}{h}$$

can be written as

```
> e2 := f(x+h)*nq(g)  + g(x)*nq(f);
```

$$\frac{f(x+h)(g(x+h) - g(x))}{h} + \frac{g(x)(f(x+h) - f(x))}{h}$$

(To see that these two expressions are equivalent, expand the equation $e1 = e2$.)

```
> expand(e1=e2);
```

$$\frac{f(x+h)g(x+h)}{h} - \frac{f(x)g(x)}{h} = \frac{f(x+h)g(x+h)}{h} - \frac{f(x)g(x)}{h}$$

Now, as $h \to 0$, $f(x+h) \to f(x)$ and $g(x+h) \to g(x)$ so the limit of $e2$ can be evaluated as

```
> map(Limit,e2,h=0);
```

$$\lim_{h \to 0} \frac{f(x+h)(g(x+h) - g(x))}{h} + \lim_{h \to 0} \frac{g(x)(f(x+h) - f(x))}{h}$$

which evaluates to the indicated formula.

```
> value(");
```

$$f(x)D(g)(x) + g(x)D(f)(x)$$

□

These new rules allow us to extend our collection of special formulas for derivatives to include, for example, all functions defined by polynomials.

Example 3.14 Use the rule for sums to express the derivative of the function f defined by a polynomial in terms of the derivatives of the terms of that polynomial.

Solution To make our discussion more specific, we use a random polynomial, say of degree at most 12.

```
> p := randpoly(x,degree=12);
```

$$56x^5 + 49x^2 + 63x^{11} + 57x^{10} - 59x^9 + 45x^8$$

3.3 Differentiation Formulas

The function defined by this polynomial is

```
> f := makeproc(",x);
```
$$x \mapsto 56x^5 + 49x^2 + 63x^{11} + 57x^{10} - 59x^9 + 45x^8$$

The derivative is computed by differentiating each of these terms separately. The terms are

```
> convert(f(x),list);
```
$$[56x^5, 49x^2, 63x^{11}, 57x^{10}, -59x^9, 45x^8]$$

The derivatives of these terms (as expressions in x) are

```
> map(diff,",x);
```
$$[280x^4, 98x, 693x^{10}, 570x^9, -531x^8, 360x^7]$$

The resulting derivative of $f(x)$ is

```
> convert(",`+`);
```
$$280x^4 + 98x + 693x^{10} + 570x^9 - 531x^8 + 360x^7$$

which can be used to define the function

```
> D(f);
```
$$x \mapsto 280x^4 + 98x + 693x^{10} + 570x^9 - 531x^8 + 360x^7$$

■

Example 3.15 Compare the derivative computed by using the product rule with the derivative produced by the power rule for the function H defined by $H = fg$ with

```
> f := (x)-> x^2:    g := (x)-> x^3:
```

Solution The function

```
> H := f*g;
```
$$fg$$

can be defined directly using the formula

```
> H(x);
```
$$x^5$$

The corresponding formula for the derivative is

```
> diff(H(x),x);
```
$$5x^4$$

resulting in the function definition

```
> DH := makeproc(",x);
```
$$x \mapsto 5x^4$$

for $H' = D(H)$. If we use the product rule directly on H we obtain the derivative[10]

```
> D(H);
```

$$(x \mapsto 2x) g + f (x \mapsto 3x^2)$$

These two apparently quite different functions yield the same result when evaluated at x.

```
> D(H)(x), DH(x);
```

$$5x^4,\ 5x^4$$

■

3.3.5 Derivatives of $\dfrac{1}{g(x)}$

We need a formula for computing derivatives of expressions such as $1/g(x)$. Such a formula would allow us to extend the product rule to general quotients because

$$\frac{f}{g} = f \, \frac{1}{g}$$

Such a rule is constructed as follows. Let the function v be defined by

```
> v := 1/g;
```

$$g^{-1}$$

This function evaluates to, for example

```
> v(x), v(x+h);
```

$$g(x)^{-1},\ g(x+h)^{-1}$$

To find the derivative $D(v) = v'$ at x, we again examine the Newton quotient

```
> nq(v);
```

$$\left(g(x+h)^{-1} - g(x)^{-1}\right) h^{-1}$$

This simplifies to

```
> normal(");
```

$$-\frac{-g(x) + g(x+h)}{g(x+h)g(x)h}$$

which can be regarded as the product of

```
> 1/g(x)/g(x+h);
```

$$\frac{1}{g(x)g(x+h)}$$

---

[10] This strange looking result is still a function even though it is defined as a sum of products of other functions.

3.3 Differentiation Formulas

and the Newton quotient of g. As $h \to 0$, the Newton quotient tends to $D(g)(x)$ and the preceding expression tends to

```
> limit(",h=0);
```

$$g(x)^{-2}$$

The combined result is

```
> diff(v(x),x);
```

$$-\frac{\frac{d}{dx}g(x)}{g(x)^2}$$

This general rule is summarized in Theorem 3.4.

THEOREM 3.4   The derivative of the function v defined by $v = 1/g$ is

```
> v := 1/g:
> D(v);
```

$$-\frac{D(g)}{g^2}$$

When this is evaluated at x, we obtain the expression

```
> D(v)(x);
```

$$-\frac{D(g)(x)}{g(x)^2}$$

which can also be written as

```
> diff(1/g(x),x);
```

$$-\frac{\frac{d}{dx}g(x)}{g(x)^2}$$

By using the product rule on f/g, we obtain the following corollary to this theorem.

COROLLARY 3.1   The derivative of the function v defined by $v = f/g$ is

```
> v := f/g:
> D(v);
```

$$\frac{D(f)}{g} - \frac{f D(g)}{g^2}$$

When this function is evaluated at x, we obtain the expression

```
> D(v)(x);
```

$$\frac{D(f)(x)}{g(x)} - \frac{f(x) D(g)(x)}{g(x)^2}$$

which can also be written as

```
> diff(v(x),x);
```

$$\frac{\frac{d}{dx}f(x)}{g(x)} - \frac{f(x)\frac{d}{dx}g(x)}{g(x)^2}$$

REMARK: This last rule

```
> diff(f(x)/g(x),x);
```

$$\frac{\frac{d}{dx}f(x)}{g(x)} - \frac{f(x)\frac{d}{dx}g(x)}{g(x)^2}$$

for computing derivatives of a quotient is often given in the form

```
> simplify(");
```

$$\frac{\frac{d}{dx}f(x)g(x) - f(x)\frac{d}{dx}g(x)}{g(x)^2}$$

REMARK: When g is defined by $g(x) = x$, the formula for $1/g(x)$ agrees with the power formula

```
> diff(x^n,x);
```

$$\frac{x^n n}{x}$$

The result is

```
> diff(x^(-1),x);
```

$$-x^{-2}$$

3.3.6 Summary

We now have a significant collection of explicit formulas for derivatives. The rules for sums and products allow us to combine these to produce formulas for an even wider class of derivatives. For example, each of the following derivatives can now be computed without going back to the definition in terms of limits:

```
> f := (x) -> x^2*sqrt(x - 7): f(x),D(f)(x);
```

$$x^2\sqrt{x-7},\ 2x\sqrt{x-7} + \frac{1}{2}x^2\frac{1}{\sqrt{x-7}}$$

```
> f := (x) -> 3*x^2: f(x), D(f)(x);
```

$$3x^2,\ 6x$$

```
> f := (x) -> sqrt(x-7): f(x), D(f)(x);
```

$$\sqrt{x-7},\ \frac{1}{2}\frac{1}{\sqrt{x-7}}$$

```
> f := (x) -> sqrt(x-7)/x^2: f(x), D(f)(x);
```

$$\sqrt{x-7}x^{-2}, \; \frac{1}{2}\frac{1}{\sqrt{x-7}}x^{-2} - 2\sqrt{x-7}x^{-3}$$

We can also compute the derivatives of quotients involving any of the examples already computed in this chapter without going back to the formal definition in terms of limits.

Example 3.16 Compute the derivative of the quotient $p(x)/q(x)$ with respect to x when $a = p(x)$ and $b = q(x)$ are polynomials given by

```
> randpoly(x);
```

$$-85x^5 - 55x^4 - 37x^3 - 35x^2 + 97x + 50$$

```
> a := randpoly(x,degree=3);
```

$$79x^3 + 56x^2 + 49x + 63$$

```
> b := randpoly(x,degree=4);
```

$$57x^4 - 59x^3 + 45x^2 - 8x - 93$$

Solution The quotient of polyomials is

```
> a/b;
```

$$\frac{79x^3 + 56x^2 + 49x + 63}{57x^4 - 59x^3 + 45x^2 - 8x - 93}$$

By the product rule, the derivative is

```
> diff(a,x)/b + a*diff(1/b,x);
```

$$\frac{237x^2 + 112x + 49}{57x^4 - 59x^3 + 45x^2 - 8x - 93} - \frac{\left(79x^3 + 56x^2 + 49x + 63\right)\left(228x^3 - 177x^2 + 90x - 8\right)}{\left(57x^4 - 59x^3 + 45x^2 - 8x - 93\right)^2}$$

which is the same as

```
> diff(a/b,x);
```

$$\frac{237x^2 + 112x + 49}{57x^4 - 59x^3 + 45x^2 - 8x - 93} - \frac{\left(79x^3 + 56x^2 + 49x + 63\right)\left(228x^3 - 177x^2 + 90x - 8\right)}{\left(57x^4 - 59x^3 + 45x^2 - 8x - 93\right)^2}$$

The more traditional formula for quotients gives the result in the form

```
> normal(");
```

$$-\frac{4503x^6 + 6384x^5 + 1520x^4 + 9846x^3 + 13543x^2 + 16086x + 4053}{\left(57x^4 - 59x^3 + 45x^2 - 8x - 93\right)^2}$$

■

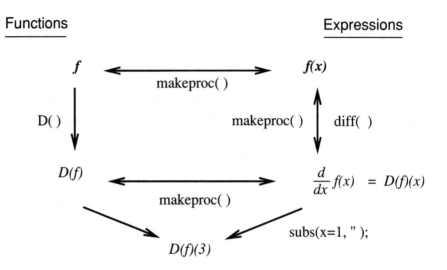

Figure 3.1 Derivatives: An exercise in evaluation

3.4 Expressions versus Functions

Throughout this chapter, we have been referring primarily to the derivatives of functions. The functions have always been defined in terms of a formula that describes what happens to a typical argument. The diagram in Figure 3.1 summarizes the distinction between formulas and functions and, in particular, the distinction between **D(f)** and **diff(f(x),x)**.

Whether we apply the function f to x to construct a formula in x, then use **diff( )** to obtain a formula for the derivative, and finally use this formula to construct the new function $D(f)$ or whether we construct the function $D(f)$ directly, the outcome is the same function. The two methods correspond to the two paths from f to $D(f)$ in Figure 3.1.

If our only goal is to construct the expression $D(f)(x)$, it makes little difference whether we refer to it as $D(f)(x)$ or $\frac{d}{dx}f(x)$. In fact, we can easily convert from one notation to the other.

```
> convert( Diff(f(x),x) , D );
```
$$D(f)(x)$$

```
> convert(",Diff);
```
$$\frac{d}{dx}f(x)$$

We discover an important difference in these representations if we ever need to evaluate them at specific values of x, such as $x = 1$. The substitution

```
> subs(x=1,Diff(f(x),x));
```

$$\frac{d}{d1}f(1)$$

leads to an improper use of the Leibnitz notation because we cannot differentiate with respect to a number. In fact, we get an error when we attempt to evaluate such an expression, as in

```
> value(");
Error, (in g) wrong number (or type) of parameters
in function diff;
```

However, all the intended evaluations are properly represented by

```
> D(f)(1);
```

$$D(f)(1)$$

(and also, **subs(x=1,D(f)(x))**). The expression remains unevaluated until such time as the function $D(f)$ is defined and can in turn be evaluated at 1.

There are extensions of the expression-based Leibnitz notation that do capture this delayed evaluation concept. For example, we may write

$$\left.\frac{d}{dx}f(x)\right|_{x=1}$$

to mean "construct this derivative and then evaluate it at 1." In this course, however, we will rely primarily on the notation $D(f)(1)$.

3.5 Trigonometric Functions

Elegant formulas for the computation of derivatives exist for other functions besides polynomials and their quotients. Recall that in Chapter 2 on limits we introduced the squeeze theorem as a way of computing limits that could not be evaluated directly. This theorem can be used to derive formulas for derivatives involving trigonometric functions.

3.5.1 Limits of Trigonometric Functions

To compute the limits that arise in the definition of the derivative of the sine function,[11] we must be able to compute limits involving the sine function. Figure 3.2 is used to derive some important relationships involving the sine function.

From the diagram it is clear that for any particular value of $x < \pi/2$, $\sin(x) <$ Length(AB). As the angle goes to 0, so does the length of the arc AB, and both numbers remain nonnegative. Thus, the squeeze theorem shows that

```
> Limit(sin(x),x=0,right);
```

$$\lim_{x \to 0^+} \sin(x)$$

---

[11] The sine function is the function **sin( )**.

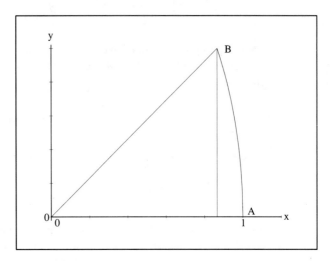

Figure 3.2 Limits for the sine function

evaluates to

> value(");

$$0$$

In a similar way, we can show that

> Limit(cos(x),x=1,right);

$$\lim_{x \to 1^+} \cos(x)$$

evaluates to

> value(");

$$\cos(1)$$

It is not enough to know that $\sin(x)$ goes to 0 as $x \to 0$. The actual *rate* at which $\sin(x)$ approaches 0 is just as important. We investigate such rates by comparing the size of $\sin(x)$ to that of a simpler expression. If the limit of this ratio becomes constant, then the growth rates are essentially the same. If the limit of the ratio goes to 0, then the denominator is growing faster than the numerator. Finally, if the limit of the ratio goes to infinity, then the numerator is growing faster than the denominator.

Consider the following graph.

> plot(sin(x)/x, x=-0.5..0.5);

3.5 Trigonometric Functions

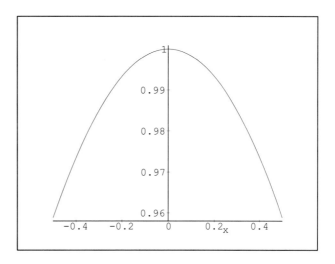

We see that the ratio $\frac{\sin(x)}{x}$ is about 1 near $x = 0$. This confirms the limit value

```
> Limit(sin(x)/x,x=0);
```

$$\lim_{x \to 0} \frac{\sin(x)}{x}$$

```
> value(");
```

$$1$$

A similar result holds for the ratio

```
> (cos(x)-1)/x;
```

$$\frac{\cos(x) - 1}{x}$$

The graph

```
> plot(" , x=-0.5..0.5);
```

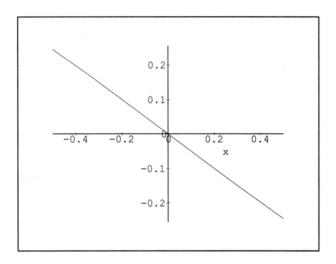

confirms

> Limit((cos(x)-1)/x , x = 0);

$$\lim_{x \to 0} \frac{\cos(x) - 1}{x}$$

> value(");

$$0$$

Both of these limits arise in the computation of the derivative of sin(x).

Example 3.17 Compute

> e1 := Limit(sin(7*x)/(4*x) , x = 0);

$$\lim_{x \to 0} \frac{1}{4} \frac{\sin(7x)}{x}$$

Solution The secret here is to rewrite this in terms of sin(u)/u for a suitable choice of u. The proper choice is determined by examining the argument to sin( ). The change of variables defined by $7x = u$, as in

> with(student):
> 'changevar(7*x = u , e1 , u);

$$\lim_{u \to 0} \frac{7}{4} \frac{\sin(u)}{u}$$

transforms this limit into the more familiar form, which evaluates to

3.5 Trigonometric Functions

```
> value(");
```

$$\frac{7}{4}$$

Example 3.18 Compute

```
> e2 := Limit( cot(x)*x , x = 0 );
```

$$\lim_{x \to 0} \cot(x) x$$

Solution   Again, we must reformulate this limit in terms of known facts. We begin by reexpressing **tan( )** in terms of **sin( )** and **cos( )**. This is accomplished by the substitution

```
> with(student):
> subs(cot=cos/sin,e2);
```

$$\lim_{x \to 0} \frac{\cos(x) x}{\sin(x)}$$

Because $\sin(x)/x$ goes to 1 as $x \to 1$, so does its reciprocal. Thus, it remains to evaluate

```
> powsubs( x/sin(x)=1,");
```

$$\lim_{x \to 0} \cos(x)$$

There are no 0s in the denominator of this expression, so the answer is

```
> value(");
```

$$1$$

3.5.2 Derivatives of Trigonometric Functions

To find formulas for the derivatives of **sin( )** and **cos( )** we again use the Newton quotient.

```
> x := 'x': h := 'h':
> nq := (f) -> (f(x+h)-f(x))/h;
```

$$f \mapsto \frac{f(x+h) - f(x)}{h}$$

In the case of **sin( )** we must compute the limit of

```
> nq(sin);
```

$$\frac{\sin(x+h) - \sin(x)}{h}$$

as $h \to 0$. This expression can be simplified by expanding $\sin(x+h)$.

```
> expand(");
```

$$\frac{\sin(x)\cos(h)}{h} + \frac{\cos(x)\sin(h)}{h} - \frac{\sin(x)}{h}$$

We will need to make use of our special results for limits involving $\frac{\sin(h)}{h}$ and for $\frac{\cos(h)-1}{h}$. To do this, we reformulate the Newton quotient so that $\frac{\sin(h)}{h}$ and $\frac{\cos(h)-1}{h}$ are recognizable. This involves forming two groups of terms: those with $\sin(x)$ and those with $\sin(h)$.

```
> collect(",[sin(x),sin(h)]);
```

$$\left(\frac{\cos(h)}{h} - h^{-1}\right)\sin(x) + \frac{\cos(x)\sin(h)}{h}$$

We can use the command **normal( )** to put each of these groups over a common denominator.

```
> map( normal , " );
```

$$\frac{(\cos(h) - 1)\sin(x)}{h} + \frac{\cos(x)\sin(h)}{h}$$

The limit of this expression is computed as

```
> map(Limit,",h=0);
```

$$\lim_{h \to 0} \frac{(\cos(h) - 1)\sin(x)}{h} + \lim_{h \to 0} \frac{\cos(x)\sin(h)}{h}$$

which evaluates to

```
> value(");
```

$$\cos(x)$$

In a similar manner, we can show

```
> Diff(cos(x),x) = diff(cos(x),x );
```

$$\frac{d}{dx}\cos(x) = -\sin(x)$$

When we use these formulas with the product rule, we can carry out computations such as

```
> Diff(x^2*sin(x),x);
```

$$\frac{d}{dx}\left(x^2 \sin(x)\right)$$

```
> value(");
```

$$2x\sin(x) + x^2\cos(x)$$

or

```
> Diff(tan(x),x);
```

$$\frac{d}{dx}\tan(x)$$

```
> subs(tan=sin/cos,");
```
$$\frac{d}{dx}\left(\frac{\sin(x)}{\cos(x)}\right)$$

```
> value(");
```
$$1 + \frac{\sin(x)^2}{\cos(x)^2}$$

3.6 The Chain Rule

In the preceding sections, we established convenient patterns for the derivatives of several classes of functions, and we worked out how to combine these simple patterns to find formulas for derivatives of functions defined in terms of the addition, subtraction, multiplication, and division of these simple expressions.

Another important way of constructing new functions from old is through functional composition. Specifically, we need to be able to deal with expressions of the form $f(g(x))$.

Example 3.19 For
```
> f := (x) -> sin(x):
```
and
```
> g := (x) -> 4*x^2 + x:
```
the function $k = f \circ g$ evaluates as
```
> k := f@g:
> k(x);
```
$$\sin(4x^2 + x)$$

∎

The derivative of k is
```
> D(k);
```
$$\cos \circ g \, (x \mapsto 8x + 1)$$

When this is evaluated at a point $x = a$, we obtain
```
> D(k)(a);
```
$$\cos(4a^2 + a)(8a + 1)$$

In fact, this expression can always be computed in terms of the known formulas for $D(\sin)$ and $D(g)$. We have
```
> D(f)(t);
```
$$\cos(t)$$

and

> D(g)(x);

$$8x + 1$$

so that with $t = g(a)$ we obtain

> D(f)(g(a))*D(g)(a);

$$\cos(4a^2 + a)(8a + 1)$$

The general rule for compositions is stated in its functional form as follows.

THEOREM 3.5 (*the chain rule*) If f and g are differentiable functions with domains defined in such a way that $v = f \circ g$ is defined on a domain S, then

> D(v) = D(f@g);

$$D(v) = D(f) \circ g\, D(g)$$

The result of applying the function $D(v)$ to x is

> D(v)(x) = D(f@g)(x);

$$D(v)(x) = D(f)(g(x))\, D(g)(x)$$

REMARK: The chain rule, like all our other rules for derivatives, can be stated entirely in terms of how the underlying expressions are transformed. The correct expression is

> Diff(f(g(x)),x) = Diff(f(u),u)*Diff(g(x),x);

$$\frac{d}{dx} f(g(x)) = \frac{d}{du} f(u) \frac{d}{dx} g(x)$$

evaluated at $u = g(x)$.

REMARK: In the formulation of the chain rule in terms of expressions, part of the computation involves first computing

> Diff(f(u),u);

$$\frac{d}{du} f(u)$$

and then evaluating the resulting formula at $u = g(x)$, as in

> subs(u=g(x) ,");

$$\frac{d}{dg(x)} f(g(x))$$

However, if we attempt to do the substitution too early, we encounter a problem with the notation.

```
> value(");
Error, (in g) wrong number (or type) of parameters
in function diff;
```

We must always have just a name in the denominator of the expression $\frac{d}{dx}$. This problem will not arise if the derivative is fully calculated (so that $\frac{d}{du}$ does not appear in the resulting expression) *prior* to the substitution.

REMARK: Again, no notational problem occurs if we use the functional form of the notation. The substitution becomes

```
> subs(u=g(x),D(f)(u));
```

$$D(f)(g(x))$$

In the following examples the emphasis is on recognizing when the chain rule may be used to good effect.

Example 3.20 Find the derivative of $\sqrt{x^2 + 3}$.

Solution This is the composition of the functions g and h given by

```
> g := sqrt:
> h := (x) -> x^2 + 3:
> (g@h)(x);
```

$$\sqrt{x^2 + 3}$$

The derivative of $g \circ h$ evaluated at x is

```
> D(g@h)(x);
```

$$x \frac{1}{\sqrt{x^2 + 3}}$$

This was evaluated as the product of

```
> D(g)(u);
```

$$\frac{1}{2} \frac{1}{\sqrt{u}}$$

with $u = h(x)$ and

```
> D(h)(x);
```

$$2x$$

which can also be computed as

```
> D(g)(h(x))*D(h)(x);
```

$$x \frac{1}{\sqrt{x^2+3}}$$

∎

Example 3.21 Find $\frac{dy}{dx}$ when $y = u^3 + 3*u^2 + 3*u + 1$ and $u = x^2 + 3$. What is its value at $x = 3$?

Solution We can write $y = f(u)$, where

```
> f := (u) -> u^3 + 3*u^2 + 3*u + 1;
```

$$u \mapsto u^3 + 3u^2 + 3u + 1$$

Similarly, we can write $u = g(x)$, where

```
> g := (x) -> x^2 + 3;
```

$$x \mapsto x^2 + 3$$

Then $y = (f \circ g)(x)$, as in

```
> y = (f@g)(x);
```

$$y = (x^2+3)^3 + 3(x^2+3)^2 + 3x^2 + 10$$

The rate of change of y as a function of x is

```
> D(f@g)(x);
```

$$2\left(3(x^2+3)^2 + 6x^2 + 21\right)x$$

At $x = 3$, this becomes

```
> D(f@g)(3);
```

$$3042$$

∎

REMARK: Because we can construct an exact formula for y in terms of x, this computation can been done entirely in terms of our rules for polynomials (i.e., without using the chain rule). The expression

```
> y = (f@g)(x);
```

$$y = (x^2+3)^3 + 3(x^2+3)^2 + 3x^2 + 10$$

can be expanded to the polynomial

```
> expand(");
```

$$y = x^6 + 12x^4 + 48x^2 + 64$$

which can be easily differentiated. We obtain

```
> Diff(y,x) = diff(rhs("),x);
```

$$\frac{d}{dx}y = 6x^5 + 48x^3 + 96x$$

At $x = 3$, this evaluates to

```
> subs(x=3,rhs("));
```

$$3042$$

REMARK: This result can also be constructed using the chain rule for expressions. The result is

```
> Diff( f(u),u)*Diff(g(x),x);
```

evaluated at $u = g(x)$ and then $x = 3$. We obtain

```
> value(");
```

$$2\left(3u^2 + 6u + 3\right)x$$

```
> subs(u=g(x),");
```

$$2\left(3\left(x^2+3\right)^2 + 6x^2 + 21\right)x$$

and finally

```
> subs(x=3,");
```

$$3042$$

REMARK: The graphs of $f(u)$ as a function of u and $f(u)$ as a function of x when $u = g(x)$ are very different graphs. These graphs are shown by the plot command[12]

```
> plot( {f(x),f(g(x))} , x=-1..1);
```

---

[12]It does not matter whether we plot $f(x)$ as a function of x, or $f(u)$ as a function of u, and by using x we can include both graphs in the same plot command.

142   Chapter 3 Derivatives

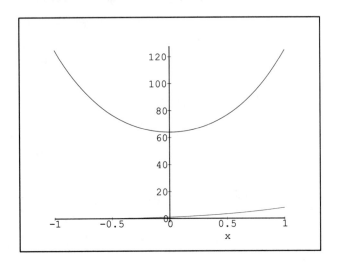

Thus, it is only natural that the graphs of the slopes of the tangent lines for the respective graphs also be very different.

The first slope is given by the function

> D(f);

$$u \mapsto 3u^2 + 6u + 3$$

and the second is given by the function

> D(f@g);

$$u \mapsto 3u^2 + 6u + 3 \circ g \, (x \mapsto 2x)$$

It is always important to compare the relative merits of two computational approaches.

Example 3.22   Differentiate

> y = (x^4 - x^2 + 1)^4;

$$y = \left(x^4 - x^2 + 1\right)^4$$

with respect to x.

Solution   Again we have two choices. We might regard this example as computing the derivative of $v = f \circ g$ with

> f := (x) -> x^4;

$$x \mapsto x^4$$

> g := (x) -> x^4 - x^2 + 1;

$$x \mapsto x^4 - x^2 + 1$$

so that by the chain rule we obtain the expression

```
> D(f)(u)*D(g)(x);
```

$$4u^3\left(4x^3-2x\right)$$

with $u = g(x)$, as in

```
> D(f@g)(x);
```

$$4\left(x^4-x^2+1\right)^3\left(4x^3-2x\right)$$

Alternatively, we might construct a formula for $f(g(x))$ and then differentiate this formula. We obtain

```
> (f@g)(x);
```

$$\left(x^4-x^2+1\right)^4$$

which expands to

```
> expand(");
```

$$x^{16}-4x^{14}+10x^{12}-16x^{10}+19x^8-16x^6+10x^4-4x^2+1$$

so that by differentiating the polynomial we obtain

```
> diff(",x);
```

$$16x^{15}-56x^{13}+120x^{11}-160x^9+152x^7-96x^5+40x^3-8x$$

Factoring this formula, as in

```
> factor(");
```

$$8x\left(-1+2x^2\right)\left(x^4-x^2+1\right)^3$$

helps us see that it is really the same as the first result. ∎

Example 3.23 Differentiate

```
> y = (x^4 - x^2 + 1)^100;
```

$$y = \left(x^4-x^2+1\right)^{100}$$

with respect to x.

Solution This time we have only one practical choice. The chain rule allows us to avoid expanding the expression for y, which in this case would result in a polynomial of degree 400. The expression can be viewed as the result of the composition

```
> f := (u) -> u^100: g := (x) -> x^4 - x^2 + 1:
> (f@g)(x);
```

$$\left(x^4-x^2+1\right)^{100}$$

By the chain rule, the derivative of this expression is found to be

```
> D(f@g)(x);
```

$$100\left(x^4 - x^2 + 1\right)^{99}\left(4x^3 - 2x\right)$$

(Compare the complexity of this answer with that of the result obtained by first expanding $(f \circ g)(x)$!)

∎

Example 3.24 Find $D(f)$ where $f(x) = \frac{1}{(x^3-3)^{1/3}}$.

Solution We cannot expand the given formula because of the fractional exponent. Thus, we must use the chain rule. In order to use the chain rule, we introduce the functions

```
> g := (u) -> u^(-1/3);
```

$$u \mapsto \frac{1}{\sqrt[3]{u}}$$

```
> h := (x) -> x^3 -3 ;
```

$$x \mapsto x^3 - 3$$

Their composition yields the expression

```
> (g@h)(x);
```

$$\frac{1}{\sqrt[3]{x^3 - 3}}$$

so, by the chain rule, the derivative of this expression is

```
> D(g)(h(x)) * D(h)(x) ;
```

$$-x^2\left(x^3 - 3\right)^{-\frac{4}{3}}$$

∎

Example 3.25 Find $D(g)$ where $g(t) = \left(\frac{t^2 - 1}{t^3 + 1}\right)^{19}$.

Solution Again, it is possible in principle to compute this derivative without using the chain rule, but it would be silly to do so. Why work with a quotient of two polynomials of large degree when, with the chain rule, we get

```
> f1 := (t) -> (t^2 - 1 ) / (t^3 + 1);
```

$$t \mapsto \frac{t^2 - 1}{t^3 + 1}$$

3.6 The Chain Rule

```
> f2 := (u) -> u^19 ;
```

$$u \mapsto u^{19}$$

The expression is

```
> (f2@f1)(t);
```

$$\frac{(t^2-1)^{19}}{(t^3+1)^{19}}$$

and the resulting derivative is

```
> D(f2)(f1(t)) * D(f1)(t);
```

$$19\left(t^2-1\right)^{18}\left(\frac{2t}{t^3+1} - \frac{3\left(t^2-1\right)t^2}{\left(t^3+1\right)^2}\right)\left(t^3+1\right)^{-18}$$

∎

REMARK: The order of composition really does make a difference. For example, given

```
> f := (x) -> x^3;
```

$$x \mapsto x^3$$

we obtain two very different expressions and two very different derivatives depending on which order we do the composition. They are

```
> (f@sin)(x);
```

$$\sin(x)^3$$

```
> D(f)(sin(x)) * D(sin)(x);
```

$$3\sin(x)^2\cos(x)$$

and

```
> (sin@f)(x);
```

$$\sin(x^3)$$

```
> D(sin)(f(x)) * D(f)(x);
```

$$3\cos(x^3)x^2$$

3.6.1 Compositions Involving More Than Two Functions

If you are given a composition involving more than two functions, proceed by composing all but one of the functions into a single new function. This reduces the problem of computing such a derivative to the same process outlined in the previous section for compositions of two functions.

Example 3.26 Compute the derivative of the composite function f where f is defined by

```
> h := (x) -> x^2 + 3;
```

$$x \mapsto x^2 + 3$$

```
> f := h@cos@sin;
```

$$h \circ \cos \circ \sin$$

Solution The function f is defined in terms of the expression $f(x)$ given by

```
> (h@cos@sin)(x);
```

$$\cos(\sin(x))^2 + 3$$

To be able to use the chain rule, simply think of function f as $h \circ g$ with $g = \cos \circ \sin$, so that we have the expression

```
> g := 'g':
> (h@g)(x);
```

$$g(x)^2 + 3$$

The derivative of f is

```
> D(f@g);
```

$$x \mapsto 2x \circ \cos \circ \sin \circ g - \sin \circ \sin \circ g \cos \circ g\, D(g)$$

which simplifies further as

```
> value( subs(g=cos@sin,") );
```

$$x \mapsto 2x \circ \cos \circ \sin \circ \cos \circ \sin - \sin \circ \sin \circ \cos \circ \sin \cos \circ \cos \circ \sin - \sin \circ \sin \cos$$

The result of applying this derivative to x is the expression

```
> D(h@cos@sin)(x);
```

$$-2\cos(\sin(x))\sin(\sin(x))\cos(x)$$

Note that we used the chain rule a second time to compute

```
> D(cos@sin);
```

$$-\sin \circ \sin \cos$$

■

3.6.2 Summary

The chain rule is an extremely powerful tool. Its use often results in much simpler formulations of answers. The main challenge is to be able to recognize the compositions. After that, computing individual derivatives using the rules for sum and product and the chain rule is straightforward. Some of the types of computations this rule facilitates include:

```
> f := (x) -> x^(3/2);
```

$$x \mapsto x^{\frac{3}{2}}$$

```
> g := (x) -> x + sin(h(x));
```

$$x \mapsto x + \sin(h(x))$$

```
> k := (x) -> x^5 + sqrt(g(x));
```

$$x \mapsto x^5 + \sqrt{x + \sin(h(x))}$$

```
> v := (f@k);
```

$$f \circ k$$

Can you detect how the chain rule was applied in computing each of the following?

```
> diff(f(x),x);
```

$$\frac{3}{2}\sqrt{x}$$

```
> diff(g(x),x);
```

$$1 + \cos(h(x))\frac{d}{dx}h(x)$$

```
> diff(k(x),x);
```

$$5x^4 + \frac{1}{2}\left(1 + \cos(h(x))\frac{d}{dx}h(x)\right)\frac{1}{\sqrt{x + \sin(h(x))}}$$

```
> diff(v(x),x);
```

$$\frac{3}{2}\sqrt{x^5 + \sqrt{x + \sin(h(x))}}\left(5x^4 + \frac{1}{2}\left(1 + \cos(h(x))\frac{d}{dx}h(x)\right)\frac{1}{\sqrt{x + \sin(h(x))}}\right)$$

3.7 Derivatives of Exponentials and Logarithms

Consider exponential functions. These are functions of the form

```
> f := (x) -> a^x;
```

$$x \mapsto a^x$$

where a is a nonnegative constant. We encountered them in Example 3.6. In particular, if $a > 1$, the value of such functions grows faster than any fixed power of x as $x \to \infty$.

The formula for their derivatives is particularly elegant. To derive it we start with the Newton quotient of f at x.

```
> e1 := (f(x+h)-f(x))/h;
```

$$\frac{a^{x+h} - a^x}{h}$$

This can be rewritten as

```
> e2 := normal(expand("));
```

$$\frac{a^x \left(a^h - 1\right)}{h}$$

and since $a^0 = 1$, this can be interpreted as a^x multiplied by the Newton quotient at $x = 0$. In the limit, the remaining Newton quotient must be $f'(0)$. However, *Maple* reports this as

```
> Limit(",h=0);
```

$$\lim_{h \to 0} \frac{a^x \left(a^h - 1\right)}{h}$$

```
> value(");
```

$$a^x \ln(a)$$

suggesting that $f'(0) = \ln(a)$. This limit can in principle be computed for any nonnegative value of a, so we could define the function ln in this way. The graph of **ln( )** has the general shape

```
> plot(ln(a),a=1/2..5);
```

3.7 Derivatives of Exponentials and Logarithms

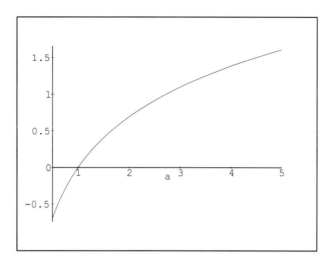

The derivative is just the function f itself when the base a is chosen so that $f'(0) = 1$. Let e be that a. The function f corresponding to e is called **exp( )** and it corresponds to $\exp(1) = e^1$. We have already seen that the value[13] of e is approximately

```
> evalf(exp(1),50);
```

$$2.7182818284590452353602874713526624977572470937000$$

THEOREM 3.6 The derivative of the function **exp( )** is **exp( )**. The derivative of the general exponential function defined by

```
> eval(f);
```

$$x \mapsto a^x$$

is

```
> D(f);
```

$$x \mapsto a^x \ln(a)$$

The equivalent expressions in terms of x are

```
> Diff(exp(x),x) = diff(exp(x),x);
```

$$\frac{d}{dx} e^x = e^x$$

and

```
> Diff(f(x),x) = diff(f(x),x);
```

$$\frac{\partial}{\partial x}(a^x) = a^x \ln(a)$$

[13] We use the notation **exp(1)** instead of e to avoid any confusion that might occur if the variable e is used in other contexts (such as a name for an equation).

These rules, together with the chain rule for compositions, enable us to obtain formulas directly for the following kinds of examples.

Example 3.27 Find a formula for the derivative of the expression

```
> e1 := exp(sin(x));
```
$$e^{\sin(x)}$$

with respect to x.

Solution Note that it is much easier to recognize a functional composition when it is written as exp(sin(x)) instead of $e^{\sin(x)}$. Either way, the desired formula is obtained by using the chain rule. We obtain the result

```
> D(exp)(sin(x))*D(sin)(x);
```
$$e^{\sin(x)} \cos(x)$$

∎

3.7.1 Logarithms as Inverse Exponential Functions

The function **exp( )** is strictly increasing because it is its own derivative, and it is strictly positive. Thus, it has an inverse. This inverse function is also of special interest in connection with derivatives.

Let us call the inverse function g. Because the functions exp and g are inverses we know that

```
> e2 := (exp@g)(x) = x;
```
$$e^{g(x)} = x$$

This equation can be differentiated on both sides, using the chain rule to yield

```
> map(diff,",x);
```
$$\frac{d}{dx}g(x)e^{g(x)} = 1$$

Isolating $\frac{d}{dx}g(x)$ in the above, we obtain the equation

```
> isolate(",diff(g(x),x));
```
$$\frac{d}{dx}g(x) = e^{g(x)-1}$$

and so by *e2* we have

```
> e3 := subs(e2,");
```
$$\frac{d}{dx}g(x) = x^{-1}$$

This approach also gives us a formula for the derivative of the inverse of our general exponential function f. This time, for practice, we do the simplifications by working directly with the functions. Let u be such that the composition of f and u is the identity function, as in

3.7 Derivatives of Exponentials and Logarithms

```
> f@u = ((x)-> x);
```

$$f \circ u = x \mapsto x$$

This can be differentiated on both sides to obtain

```
> map(D,");
```

$$x \mapsto a^x \ln(a) \circ u \, D(u) = 1$$

which leads to the formula

```
> isolate(",D(u));
```

$$D(u) = x \mapsto a^x \ln(a) \circ u^{-1}$$

On evaluating this at x we obtain the equivalent relationship as expressions in x.

```
> "(x);
```

$$D(u)(x) = \frac{1}{a^{u(x)} \ln(a)}$$

The inverse functions g and u are usually called $\ln(\ )$ and $\log_a(\ )$. In *Maple* these are written as **ln( )** and **log[a]( )**. The latter evaluates to

```
> log[a](x);
```

$$\frac{\ln(x)}{\ln(a)}$$

THEOREM 3.7 The derivative of the inverse function of **exp( )**, **ln( )**, is

```
> D(ln);
```

$$a \mapsto a^{-1}$$

The equivalent rule for expressions in x is

```
> diff(ln(x),x);
```

$$x^{-1}$$

The derivative of the inverse of the general exponential function evaluated at x is

```
> Diff(log[a](x),x) = diff(log[a](x),x);
```

$$\frac{\partial}{\partial x} \left(\frac{\ln(x)}{\ln(a)} \right) = \frac{1}{x \ln(a)}$$

Example 3.28 Use the chain rule to evaluate

```
> e1 := Diff(ln( sin(x)),x);
```

$$\frac{d}{dx} \ln(\sin(x))$$

Solution This is evaluated as

```
> D(ln)(sin(x))*D(sin)(x);
```

$$\frac{\cos(x)}{\sin(x)}$$

The chain rule in conjunction with logarithms gives an extremely powerful way of computing derivatives of complicated products.

Example 3.29 Find the derivative of

> e2 := product((1+i*x^(2*i+1))/(1+i*x^(2*i)),i=1..5);

$$\frac{\left(1+x^{3}\right)\left(1+2x^{5}\right)\left(1+3x^{7}\right)\left(1+4x^{9}\right)\left(1+5x^{11}\right)}{\left(1+x^{2}\right)\left(1+2x^{4}\right)\left(1+3x^{6}\right)\left(1+4x^{8}\right)\left(1+5x^{10}\right)}$$

Solution We could multiply this out in full and then use the product rule. Instead, we make use of the fact that for an arbitrary function g

> Diff(ln(g(x)),x) = diff(ln(g(x)),x);

$$\frac{d}{dx}\ln(g(x)) = \frac{\frac{d}{dx}g(x)}{g(x)}$$

For the problem at hand we can obtain the derivative by computing

> e3 := expand(ln(e2));

$$\ln(1+x^{3}) - \ln(1+x^{2}) + \ln(1+2x^{5}) - \ln(1+2x^{4}) + \ln(1+3x^{7}) - \ln(1+3x^{6}) + \ln(1+4x^{9}) - \ln(1+4x^{8}) + \ln(1+5x^{11}) - \ln(1+5x^{10})$$

> map(diff,",x);

$$\frac{3x^{2}}{1+x^{3}} - \frac{2x}{1+x^{2}} + \frac{10x^{4}}{1+2x^{5}} - \frac{8x^{3}}{1+2x^{4}} + \frac{21x^{6}}{1+3x^{7}} - \frac{18x^{5}}{1+3x^{6}} + \frac{36x^{8}}{1+4x^{9}} - \frac{32x^{7}}{1+4x^{8}} + \frac{55x^{10}}{1+5x^{11}} - \frac{50x^{9}}{1+5x^{10}}$$

and multiplying this by *e2*.

3.8 Implicit Derivatives

Until now, all the techniques we have used for finding derivatives have relied on our being able to find $f(x)$ explicitly in terms of x. However, this may not always be practical. In fact, the following example shows why, even when we can find an explicit solution, we may not want to.

Example 3.30 Find an expression for $\frac{dy}{dx}$ in the following equation:

> e1 := 3*x^2 - x^2*y^3 + 4*y = 12;

$$3x^{2} - x^{2}y^{3} + 4y = 12$$

An Explicit Solution

Solution Because the equation is cubic in y, there are potentially three solutions.[14]

```
> A := [ solve(e1,y) ]:
```

For brevity we consider only the first solution, though the technique can be used for all three solutions. Even so, the expressions are quite large.

```
> simplify(A[1]);
```

$$\frac{1}{18}\sqrt[3]{2}\,3^{\frac{2}{3}}\left(\sqrt[3]{2}\,3^{\frac{2}{3}}\,\text{`%2`}^{\frac{2}{3}} + 24\right)x^{-1}\frac{1}{\sqrt[3]{\text{`%2`}}}$$

$$\frac{1}{18}\sqrt[3]{2}\,3^{\frac{2}{3}}\left(\sqrt[3]{2}\,3^{\frac{2}{3}}\,\text{`%2`}^{\frac{2}{3}} + 24\right)x^{-1}\frac{1}{\sqrt[3]{\text{`%2`}}}$$

$$\text{`%1`} = -256 + 243x^6 - 1944x^4 + 3888x^2$$

$$\text{`%2`} = 27x^3 - 108x + \sqrt{3}\sqrt{\text{`%1`}}$$

The value of dy/dx for the first solution (expanded to facilitate line-breaking of this relatively large expression) is

```
> diff(",x):
> expand(");
```

$$\frac{9}{2}2^{\frac{2}{3}}\sqrt[3]{3}\,x\,\text{`%2`}^{-\frac{2}{3}} - 62^{\frac{2}{3}}\sqrt[3]{3}\,\text{`%2`}^{-\frac{2}{3}}x^{-1} + \frac{81}{2}2^{\frac{2}{3}}3^{\frac{5}{6}}x^4\,\text{`%2`}^{-\frac{2}{3}}\frac{1}{\sqrt{\text{`%1`}}} -$$

$$2162^{\frac{2}{3}}3^{\frac{5}{6}}x^2\,\text{`%2`}^{-\frac{2}{3}}\frac{1}{\sqrt{\text{`%1`}}} + 2162^{\frac{2}{3}}3^{\frac{5}{6}}\,\text{`%2`}^{-\frac{2}{3}}\frac{1}{\sqrt{\text{`%1`}}} - \frac{1}{6}2^{\frac{2}{3}}\sqrt[3]{3}\,\sqrt[3]{\text{`%2`}}\,x^{-2} -$$

$$\frac{4}{3}\sqrt[3]{2}\,3^{\frac{2}{3}}x^{-2}\frac{1}{\sqrt[3]{\text{`%2`}}} - 36\sqrt[3]{2}\,3^{\frac{2}{3}}x\,\text{`%2`}^{-\frac{4}{3}} + 48\sqrt[3]{2}\,3^{\frac{2}{3}}x^{-1}\,\text{`%2`}^{-\frac{4}{3}} -$$

$$972\sqrt[3]{2}\,\sqrt[6]{3}\,x^4\,\text{`%2`}^{-\frac{4}{3}}\frac{1}{\sqrt{\text{`%1`}}} + 5184\sqrt[3]{2}\,\sqrt[6]{3}\,x^2\,\text{`%2`}^{-\frac{4}{3}}\frac{1}{\sqrt{\text{`%1`}}} -$$

$$5184\sqrt[3]{2}\,\sqrt[6]{3}\,\text{`%2`}^{-\frac{4}{3}}\frac{1}{\sqrt{\text{`%1`}}}$$

$$\text{`%1`} = -256 + 243x^6 - 1944x^4 + 3888x^2$$

$$\text{`%2`} = 27x^3 - 108x + \sqrt{3}\sqrt{\text{`%1`}}$$

∎

---

[14] Do not worry if you cannot remember how to solve a general cubic equation. This example is included primarily to motivate the simpler solution that follows.

An Implicit Solution

Solution Given equality between two algebraic expressions in x, the derivative of each side with respect to x must be equal. (If the two sides were changing at different rates, they would soon have to differ for some value of x as we changed x.) This observation allows us to approach the problem by constructing an equation for the desired derivative.

As we change our point of view, some caution is in order. We have begun with an equation relating x and y, but the very fact that we want to *differentiate with respect to x* forces us to think of y as the result of applying a function of x. We can make our thinking explicit by writing $y(x)$ everywhere we see y.

```
> e2 := subs(y=y(x),e1);
```

$$3x^2 - x^2 y(x)^3 + 4y(x) = 12$$

By doing this immediately, we avoid any possible misinterpretation.

Now we are ready to differentiate both sides of the equation. We obtain

```
> e3 := map(diff,e2,x);
```

$$6x - 2xy(x)^3 - 3x^2 y(x)^2 \frac{d}{dx} y(x) + 4 \frac{d}{dx} y(x) = 0$$

Now if we recall that

```
> diff(y(x),x);
```

$$\frac{d}{dx} y(x)$$

is just a fancy name for the result of carrying out the indicated computation, we can isolate this term in $e3$.

```
> solve(e3 , diff(y(x),x));
```

$$-\frac{6x - 2xy(x)^3}{-3x^2 y(x)^2 + 4}$$

Note that the final answer involves $y^2(x)$ and $y^3(x)$ but does not involve $y(x)$ explicitly. Thus, in some circumstances, we will have completely avoided computing the cube root of the original equation. ∎

Example 3.31 Find the equation of the line tangent to the point $[x, y] = [6, 8]$ given that

```
> eq := x^2 + y^2 = 100;
```

$$x^2 + y^2 = 100$$

must be satisfied.

Solution The commands

```
> s := [solve(eq,y)];
> map(diff,",x);
> subs(x=6,");
> subs(x=6,eq);
```

will find the slope of the tangent line by explicitly solving the quadratic for y and then differentiating and evaluating the resulting expressions at $x = 6$. There is a solution for

3.8 Implicit Derivatives

each of the top and bottom halves of the circle so we still need to figure out which is which.

The approach based on implicit derivatives bypasses the need to determine which half of the circle the point is on. It proceeds as follows.

```
> eq2 := subs(y=y(x),eq);
```

$$x^2 + y(x)^2 = 100$$

```
> map(diff,",x);
```

$$2x + 2y(x)\frac{d}{dx}y(x) = 0$$

```
> isolate(",diff(y(x),x));
```

$$\frac{d}{dx}y(x) = -\frac{x}{y(x)}$$

The slope of the tangent line at [6, 8] is simply

```
> m := subs( {y(x)=8,x=6},rhs(") );
```

$$-\frac{3}{4}$$

(Note that we still have not defined the function y!) and the equation for the tangent line is

```
> L := y - 8 = m*(x-6);
```

$$y - 8 = -\frac{3}{4}x + \frac{9}{2}$$

We conclude this example by plotting the tangent line and actual curve.[15]

```
> solve(L,y);
```

$$\frac{25}{2} - \frac{3}{4}x$$

```
> plot( { " , solve(eq,y) } , x=-15..15 );
```

---

[15]We generate all of the curve defined by *eq* by solving for *y*; note, however, that we did not need this information to get the equation of the tangent line.

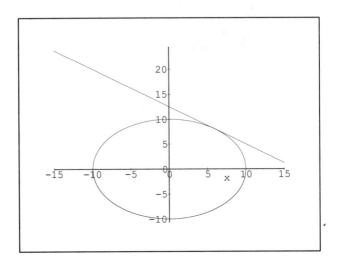

3.8.1 Treating x and y as Functions of t

In each of the preceding cases, we began our discussion with an equation relating x and y and then arbitrarily decided that we wished to treat y as a function of x. It is equally convenient to think of both y and x as functions of yet a third variable — say t. The advantage of this point of view is that we can now think of both x and y as functions.

Example 3.32 Find an expression for dy/dx at $x = 3$ in the following equation:

```
> e1 := 3*x^2 - x^2*y^3 + 4*y = 12;
```
$$3x^2 - x^2 y^3 + 4y = 12$$

if $y = 1$ when $x = 2$.

Solution Although this problem could be solved by either of the earlier methods, we will approach it this time by thinking of both x and y as functions of t.

The advantage of thinking in this way is that we become free to work with an equation relating functions. Among other things, this avoids the first step of converting everything to expressions, and it also simplifies the notation.

```
> e2 := map(D,e1);
```
$$6 D(x)x - 2 D(x)xy^3 - 3x^2 D(y)y^2 + 4 D(y) = 0$$

Note that we can reinterpret this equation as an equation for expressions in t at any time simply by doing

```
> lhs(e2)(t) = rhs(e2)(t);
```
$$6 D(x)(t)x(t) - 2 D(x)(t)x(t)y(t)^3 - 3x(t)^2 D(y)(t)y(t)^2 + 4 D(y)(t) = 0$$

(equivalently, **(e2)(t);**). This linear equation relating the functions can now be solved for $D(y)$.

```
> e3 := isolate(e2,D(y));
```

$$D(y) = \frac{-6D(x)x + 2D(x)xy^3}{-3x^2y^2 + 4}$$

All that remains is for us to interpret $D(x)$ on the right-hand side of e3. We are free to do this in any manner appropriate to the problem at hand. For example, if we know that $x(t) = t$, then we know that $D(x) = 1$, or we may be told directly that $D(x)$ is a constant function, say $(x) \to 12$, so that $D(x) = 12$.

Because both x and y are functions of the same quantity t, the rate of change of y with respect to x will be

```
> e3/D(x);
```

$$\frac{D(y)}{D(x)} = \frac{-6D(x)x + 2D(x)xy^3}{D(x)\left(-3x^2y^2 + 4\right)}$$

which simplifies to

```
> normal(");
```

$$\frac{D(y)}{D(x)} = -\frac{2x\left(-3 + y^3\right)}{3x^2y^2 - 4}$$

For the current problem, we know that $y = 1$ when $x = 2$, so without knowing anything more about how the functions x and y are defined, or even a value for t, we obtain the rate of change of y with respect to x at $x = 2$ as

```
> subs({y=1,x=2},rhs("));
```

$$1$$

This can be expressed as

```
> Diff(y(x),x) = ";
```

$$\frac{d}{dx}y(x) = 1$$

at $x = 2$. ∎

REMARK: The method of implicit differentiation constructs a "linear equation" for $D(y)$. Thus, obtaining an expression for $D(y)$ or $D(y)/D(x)$ in terms of the functions x and y is easy and there is only one solution. This is a tremendous advantage if, for example, we are interested only in a rate of change at precomputed pairs $[x, y]$.

3.8.2 Related Rates

Now that we have introduced the notion of implicit differentiation, we are ready to turn to some applications. In the examples that follow we also use *Maple* to keep track of the units of computation. It is helpful to specify some of the special properties of unit names. For example, units are generally regarded as positive. This can be specified as

```
> assume( meters > 0): assume( second > 0 ):
> assume( cm > 0 ): assume( minutes > 0 ):
```

We indicate that these units are constants by adding them to the list of known system constants

```
> constants;
```

$$\text{false, } \gamma, \infty, \text{ true, Catalan, } e, \pi$$

by the command

```
> constants := constants,meters,second,cm,mn:
```

Now compare the solution to Example 3.33 with that obtained for Example 3.8, which is essentially the same problem.

Example 3.33 A spherical snowball is melting in such a way that its radius is decreasing at a rate of $\frac{1}{2}$ cm/min. At what rate is the volume changing when the radius is 5 cm? At what radius is the volume changing at a rate of 150 cm³/min?

Solution Regard both V and r as functions of t. An equation relating these two functions is

```
> V = 4/3*Pi*r^3;
```

$$V = \frac{4}{3}\pi r^3$$

The corresponding equation relating their derivatives as functions is

```
> eq := map(D,");
```

$$D(V) = 4\pi\, D(r) r^2$$

and the actual function values at time t are related by the equation

```
> eq2 :- subs({D(V)=D(V)(t),r=r(t),D(r)=D(r)(t)},eq);
```

$$D(V)(t) = 4\pi\, D(r)(t) r(t)^2$$

At the specific time that the radius is 5 cm, several of these function values are known. They can be substituted directly into this equation as

```
> subs({D(r)(t)=1/2*cm/minute ,r(t)=5*cm}, " );
```

$$D(V)(t) = \frac{50\pi\,\text{cm}^3}{\text{minute}}$$

and the value of $D(V)(t)$ becomes approximately

3.8 Implicit Derivatives

```
> evalf(");
```

$$D(V)(t) = \frac{157.0796327 \, cm^3}{minute}$$

To find the value of $r(t)$ when the volume is changing at a rate of 150 cm³/min, we isolate r in *eq*. The only unknown in the resulting equation is the value of the radius $r(t)$.

```
> subs({D(r)(t) = 1/2*cm/minute, D(V)(t)=150*cm^3/minute},
> eq2 );
```

$$\frac{150 cm^3}{minute} = \frac{2\pi \, cm \, r(t)^2}{minute}$$

This equation can be solved for $r(t)$.

```
> isolate(",r(t));
```

$$r(t) = \frac{\sqrt{75} \, cm}{\sqrt{\pi}}$$

When the volume is changing at a rate of 150cm³/min, the value of $r(t)$ is approximately

```
> evalf(");
```

$$r(t) = 4.886025119 \, cm$$

■

Example 3.34 A 5-meter ladder is standing on level ground and resting against a vertical wall. If the top of the ladder starts sliding down the wall at a rate of 1/3 meter per second, how fast is the base of the ladder moving away from the wall when the top has fallen 1 meter?

Solution We first set up equations relating the various functions by using Pythagoras's theorem.

```
> eq1 := x^2 + y^2 = L^2;
```

$$x^2 + y^2 = L^2$$

For this problem to make sense, we make the following assumptions.

```
> assume(y>=0):
> constants := constants , L:
```

The rates of change of these quantities, regarded as functions of time, are related by

```
> eq2 := map(D,eq1);
```

$$2 D(x)x + 2 D(y)y = 0$$

We can express the function x in terms of y by

```
> eq3 := isolate(eq1,x);
```

$$x = \sqrt{|L^2 - y^2|}$$

This simplifies further if we make the assumption that $L^2 - y^2$ is nonnegative.[16]

```
> eq3 := subs(abs( L^2 - y^2)= L^2-y^2,eq3 ): eq3;
```

$$x = \sqrt{L^2 - y^2}$$

At time t we have $y(t) = 4$ meters and $L = 5$ meters, so at time t the value of $x(t)$ is[17]

```
> subs({y=4 *meters,L=5*meters},rhs(")):
> xval := simplify(");
```

$$3\text{meters}$$

Next we rearrange *eq2* to isolate $D(x)$

```
> isolate(eq2,D(x)):
```

and substitute in the values of the remaining unknowns.

```
> subs({y=4*meters,
>      D(y) = -1/3*meters/second, x = xval } , rhs(")):
```

This must be interpreted as the value of $D(x)(t)$ at the instant the top of the ladder has dropped by 1 meter. It is

```
> value(");
```

$$\frac{4}{9}\frac{\text{meters}}{\text{second}}$$

∎

Example 3.35 Water is leaking out of an inverted conical tank at a rate of 1 meter³ per minute. The tank is 6 meters high and 6 meters in diameter at the top. When will the surface of the water be falling at a rate of 5 meters per minute?

Solution The functions volume and height are related by the following equation:

```
> eq1 := V = 1/3*Pi*r^2*h;
```

$$V = \frac{1}{3}\pi r^2 h$$

Here, r can be eliminated, as we are told that

```
> r/h = 1/2;
```

$$\frac{r}{h} = \frac{1}{2}$$

yielding

---

[16] This is an equation between functions. Because of the presence of the nested function calls in **sqrt(L^2-y^2)**, this equation should really use compositional notation, as in **convert(",'@')**. This matters only when we actually apply the functions to a value.

[17] We bypass the construction of an equation for $x(t)$ and $y(t)$ by substituting directly for y.

```
> isolate(",r);
```

$$r = \frac{1}{2}h$$

The simplified equation *eq1* is

```
> assign("):
> eq1;
```

$$V = \frac{1}{12}\pi h^3$$

The derivatives of these functions are related by the equation:

```
> eq2 := map(D,");
```

$$D(V) = \frac{1}{4}\pi \, D(h) h^2$$

For the problem at hand, we have $D(h) = 5$ meters per second and $D(V) = 1$ meter^3 per second so

```
> eq3 := subs({ D(h) = 5 *meters / seconds,
> D(V) = 1*meters^3 / seconds}, eq2 );
```

$$\frac{\text{meters}^3}{\text{seconds}} = \frac{5\,\pi\,\text{meters}\,h^2}{4\,\text{seconds}}$$

The height of the surface at this time is

```
> isolate(eq3,h);
```

$$h = \frac{1}{5}\frac{\sqrt{4}\sqrt{5}\,\text{meters}}{\sqrt{\pi}}$$

which is approximately

```
> evalf(");
```

$$h = 0.5046265044 \text{ meters}$$

■

Example 3.36 At 3:00 P.M., the ship *Endeavor* is 110 km due west of the ship *Enterprise* and proceeding north at 22 knots. The ship *Enterprise* continues east at 17 knots. How fast is the distance between the two ships changing at 5:00 P.M.?

Solution Let the function s measure the distance between the two ships, and let the functions x and y measure the position in nautical miles of the ships *Enterprise* and *Endeavor* relative to the original position of the *Endeavor*, all as functions of time. All rates of changes will be in terms of knots, and the various functions are related by the equation

```
> eq1 := (s^2 = x^2 + y^2);
```

$$s^2 = x^2 + y^2$$

Their derivations are related by

```
> eq2 := map(D,eq1);
```
$$2D(s)s = 2D(x)x + 2D(y)y$$

which can be rearranged as

```
> eq3 := isolate(",D(s));
```
$$D(s) = \frac{1}{2}\frac{2D(x)x + 2D(y)y}{s}$$

```
> y := (t) -> 22*t;
```
$$t \mapsto 22t$$

The function $D(s)$ computes the rate of change of the distance s as a function of time, and it has been expressed entirely in terms of the functions $D(x)$ and $D(y)$. If east is the positive direction, then the position of the *Enterprise* is given by the function

```
> x := (t) -> 110 + 17*t;
```
$$t \mapsto 110 + 17t$$

If north is the other positive direction, then the position of the *Endeavor* is given by the function

```
> y := (t) -> 22*t;
```
$$t \mapsto 22t$$

With these definitions for x and y, *eq3* becomes

```
> eq3;
```
$$D(s) = \frac{1}{2}\frac{34x + 44y}{s}$$

At $t = 2$ this evaluates to

```
> "(2);
```
$$D(s)(2) = \frac{3416}{s(2)}$$

We can compute the value of $s(2)$ from *eq1*. It is

```
> eq1(2);
```
$$s(2)^2 = 22672$$

```
> map(sqrt,");
```
$$s(2) = 4\sqrt{1417}$$

yielding a rate of change of

```
> subs(",eq3(2));
```
$$D(s)(2) = \frac{854}{1417}\sqrt{1417}$$

```
> evalf(");
```
$$D(s)(2) = 22.68678449$$

The fact that this is positive indicates that the ships are moving apart. ■

3.9 A Derivation of Newton's Formula

Since $f(x_0 + h) \approx f(x_0) + hf'(x_0)$, we can find h (and hence $x_0 + h$) such that $f(x_0 + h) \approx 0$. We simply solve the equation

```
> 0 = f(x0) + D(f)(x0)*h;
```
$$0 = f(x0) + D(f)(x0)h$$

as

```
> solve(",h);
```
$$-\frac{f(x0)}{D(f)(x0)}$$

The new x value is $x1$ given by

```
> x1 := x0 + ";
```
$$x0 - \frac{f(x0)}{D(f)(x0)}$$

and it is generally a better guess at the solution to $f(x) = 0$ than $x0$.

By using this formula over and over again with better and better guesses for $x0$ we should be able to make as accurate a guess as we want to the value of the root of the equation.

The following *Maple* procedure takes a function **f** and an initial guess **startvalue** and generates new guesses until the value of the function **f** at the current guess is sufficiently small.

```
> newton := proc(f,startvalue) local x0,x1;
>     x0 := startvalue;
>     while abs(f(x0) ) > .001 do
>         x1 := x0 - f(x0)/D(f)(x0);
>         x0 := x1;
>         print('current approximation: ', x0);
>     od;
>     RETURN(x0);
> end:
```

Example 3.37 Find a solution to $x^2 - 2 = 0$.

Solution The function under consideration is

```
> g := (x) -> x^2-2;
```
$$x \mapsto x^2 - 2$$

If we begin with an initial guess of 1.0, we find the approximate root

```
> s1 := newton(g,1.0);
      current approximation: , 1.500000000

      current approximation: , 1.416666667

      current approximation: , 1.414215686
```

1.414215686

The guess 1.5 corresponds to the zero of the line tangent to g at x = 1, as in

> with(student): with(plots):

> showtangent(g(x),x=1,x=.9..1.6);

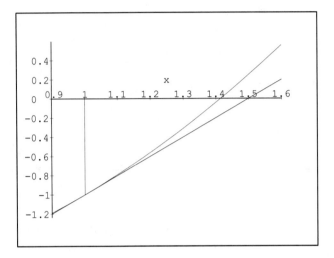

The second number, 1.416666667, corresponds to the zero of the line tangent to g at x = 1.5, as in

> showtangent(g(x),x=1.5,x=.9..1.6);

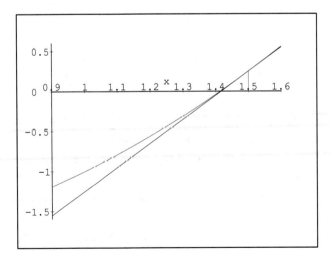

and is already quite close. Compare this with the solution produced by the *Maple* command **fsolve( )**

```
> s2 := fsolve( g(x)=0 ,x ,1..2);
```
$$1.414213562$$

```
> abs(s1 - s2);
```
$$0.000002124$$

∎

Different starting values may yield different roots. For example, if we use the initial guess of -1.5 for a root of g, we find the approximate root

```
> newton(g,-1.5);
      current approximation: , -1.416666667

      current approximation: , -1.414215686
```
$$-1.414215686$$

Some initial guesses never lead to a root. For example, if we start with an initial guess of 0.0 in the preceding root-finding problem, we obtain the message

```
>   newton(g,0.0);
Error, (in newton) division by zero
```

What went wrong? The reason this failed is apparent if we look at a graph of the tangent line.

```
> showtangent( g(x),x=0,x=-2..2);
```

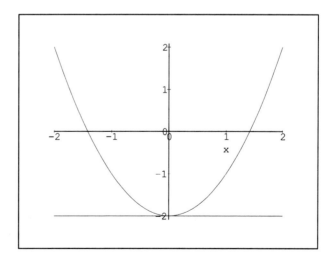

The tangent line is parallel to the x-axis so the two lines can never intersect. The exact values for the two roots we have found are

```
>    solve( g(x)=0,{x});
```
$$\{x = \sqrt{2}\}, \{x = -\sqrt{2}\}$$

Example 3.38 Find an approximate solution to $\sin(x^2 - 3) = 0$.

Solution The function of interest is

```
>    g := (x)-> sin(x^2-3);
```
$$x \mapsto \sin(x^2 - 3)$$

Depending on where we start, we obtain different roots. Some sample roots are

```
>    newton(g,1.0);
```
$$-1.732051981$$

```
>    newton(g,5.0);
```
$$4.999114834$$

```
>    newton(g,10.0);
```
$$10.01944959$$

These are consistent with the roots shown by

```
>    plot(g,0..10);
```

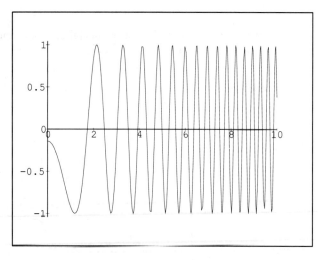

(Try tracing tangent lines for the given start values.)

An exact solution to this problem, corresponding to sin(0), is

```
>    solve( g(x)=0,{x});
```
$$\{x = -\sqrt{3}\}, \{x = \sqrt{3}\}$$

∎

EXERCISE SET 3

The following exercises review the topics that were encountered in this chapter. The new Maple commands that were introduced appear in Table 3.1. Feel free to use these commands and others to investigate the properties of the various mathematical functions that arise.

Procedure	Description
additionally(L>=y);	Add to assumptions about a symbol
assume(meters > 0);	Make new assumptions about a symbol
binomial(n,i);	Compute the binomial coefficient $\binom{n}{i}$
changevar(h=b-a,m,h);	Change limit or differentiation variable
collect(",[sin(x),sin(h)])	Regroup terms of an expression
constants;	All the constants that are known to *Maple*
convert(Diff(f(x),x),D);	Change to the **D** notation
convert(D(f)(x),Diff);	Change to the **Diff** notation
convert(D(f)(x),diff);	Change to the **diff** notation
diff(f(x),x);	Compute the derivative of $f(x)$ with respect to x
Diff(f(x),x);	The inert form of **diff**
D(f);	Differentiate a function
D(f)(x);	Evaluate $D(f)$ at x
isolate(eq,a);	Solve the equation *eq* for *a*
newton(g,10.0);	Newton's algorithm (see page 163)
nq(f);	The Newton quotient of f (see page 115)
normal(e);	Cancel common factors from a quotient
product(i,i=1..3);	Multiply a sequence of terms together
randpoly(x);	Generate a random polynomial

Table 3.1 New Commands Used in Chapter 3

1. Consider the function f defined by

    ```
    > f := (x) -> 2*x^2 - 16*x + 9 ;
    ```
    $$x \mapsto 2x^2 - 16x + 9$$

    a. Find all x values where the slope of f is 2 and compute the function values for those points.
    b. Draw a graph of f and its tangent line at each of these points.
    c. Find an equation for the tangent line to f at $x = 3$. Confirm that it is a tangent line by plotting both $f(x)$ and your line over a suitable range.

2. Let f be defined by:

    ```
    > f := (x) -> (x-1/2)*(x-2/3)*(x);
    ```
    $$x \mapsto \left(x - \frac{1}{2}\right)\left(x - \frac{2}{3}\right)x$$

Approximate the slope of f at $x = 3/4$, by using pairs of points to define a secant line. How close do the x values of the two points need to be to result in an estimate for the slope that is accurate to within 0.001 of the correct slope?

3. Define the function f by

```
> f := (x) -> (x-5)*(x-2)*(x-3)*(x-4);
```
$$x \mapsto (x-5)(x-2)(x-3)(x-4)$$

Use plotting to investigate the relationship between f and its derivative. Use f' to identify the regions where
- f is increasing
- f is decreasing
- f peaks or bottoms out.

4. Use plotting to find where the function f defined by

```
> f := (x) -> -(x-3)^2 + 7*x - 4 ;
```
$$x \mapsto -(x-3)^2 + 7x - 4$$

is maximized. Use f' to locate this maximum precisely.

5. Let the position of a particle be given by the function

```
> f := (t) -> t^3 - 7*t^2 + 8*t ;
```
$$t \mapsto t^3 - 7t^2 + 8t$$

where t is measured in seconds and $s = f(t)$ is measured in meters. Plot the position of the object as a function of time for the first six seconds. Find the velocity of the particle at $t = 3$ seconds. How does this differ from the speed of the object at $t = 3$? At what time does the object halt to return to the origin for the second time?

6. It has been observed that the population of a certain country is growing at a rate of 2.7% per year. The population of that country in 1980 was 23 million, so that the population in millions after t years is given by

```
> p := (t) -> 23*(1.027)^t;
```
$$t \mapsto 23 \cdot 1.027^t$$

Graph the population growth over 30 years. What population does this model predict for the year 2010? Compare the instantaneous rate of population growth (increase / total population) at time $t = 0$ and at time $t = 30$.

7. If f is the focal length of a convex lens and an object is at a distance of p from the lens, then its image will be at a distance of q from the lens. If f, p, and q are related by the *lens equation*

$$\frac{1}{f} = \frac{1}{p} + \frac{1}{q}$$

find the rate of change of p with respect to q.

8. During a hail storm, small droplets of water freeze into hail stones and gradually grow in size as additional moisture condenses on their surface. If the radius of a hail stone is increasing at a rate of $\frac{1}{10}$ cm/min, at what rate (change per unit of time) is the volume changing when the radius is 1 cm?

9. Use Newton quotients and limits to verify that for the function f defined by

> f := (x) -> x^2 + 3*x^3;
$$x \mapsto x^2 + 3x^3$$

the derivative is the function $D(f)$ defined by

> D(f);
$$x \mapsto 2x + 9x^2$$

10. Use Newton quotients and limits to verify that for the function f defined by

> f := (x) -> x^(-3);
$$x \mapsto x^{-3}$$

the derivative is the function $D(f)$ defined by

> D(f);
$$x \mapsto -\frac{3}{x^4}$$

11. Use Newton quotients and limits to verify that for the function f defined by

> f := (x) -> 3*x^(5/2);
$$x \mapsto 3x^{\frac{5}{2}}$$

the derivative is the function $D(f)$ defined by

> D(f);
$$x \mapsto \frac{15}{2} x^{\frac{3}{2}}$$

12. Use the rule for sums to express the derivative of the function f defined by

> f := unapply(randpoly(x) , x);
$$x \mapsto -85x^5 - 55x^4 - 37x^3 - 35x^2 + 97x + 50$$

in terms of the derivatives of the terms of that polynomial.

13. Use Newton quotients to reconstruct the proof of the product rule for the specific function H defined by $H = fg$ for the specific functions

> f := (x) -> x^4;
$$x \mapsto x^4$$

> g := (x) -> x^7;
$$x \mapsto x^7$$

14. Show the derivatives that it would be necessary to compute in order to use the product rule to compute the derivative of a/b with respect to x when

> a := randpoly(x);
$$-58x^5 - 90x^4 + 53x^3 - x^2 + 94x + 83$$

```
> b := randpoly(x);
```
$$-86x^5 + 23x^4 - 84x^3 + 19x^2 - 50x + 88$$

Use them to construct the derivative $\frac{d}{dx}\frac{a}{b}$.

15. Given that $f(x) = 2x^3+3$, and g is defined by $g(f(x)) = x$ find $\frac{d}{dx}g(x)$ by differentiating both sides of this equation.

16. Use plotting to estimate the value of the limit

```
> Limit( sin(9*x)/(3*x) , x = 0 );
```
$$\lim_{x \to 0} \frac{1}{3} \frac{\sin(9x)}{x}$$

and then compute its true value by transforming it into one of the standard forms already computed in this chapter.

17. Use plotting to estimate the value of the limit

```
> Limit( cot(2*x)*x , x = 0 );
```
$$\lim_{x \to 0} \cot(2x)x$$

and then compute its true value by transforming it into a combination of some of the standard forms already computed in this chapter.

18. Review the rules for differentiations of sums, products, and compositions by applying D to each of $f + g$, fg and $f \circ g$. Convert each of the results to the Leibnitz notation.

19. Given

```
> f := (x) -> tan(x):
```
and
```
> g := (x) -> 7*x^3 + x^2:
```
indicate how to use $D(f)$ and $D(g)$ to compute $D(f \circ g)$.

20. Which of the special rules for computing derivatives introduced in this chapter are used to find the derivative of $\sqrt{x^3 + 3x}$ without computing limits?

21. Find $\frac{dy}{dx}$ when $y = u^5 + 3u^3 + 2u + 1$ and $u = x^3 + 7$. What is its value at $x = 2$?

22. Compare the result of expanding and then differentiating

```
> y = (x^4 - x^2 + 1)^5;
```
$$y = (x^4 - x^2 + 1)^5$$

with respect to x to the result of using the chain rule on $f \circ g$ with

```
> f := (u) -> u^5:
```
and
```
> g := (x) -> x^4 -x^2 + 1;
```
$$x \mapsto x^4 - x^2 + 1$$

Show that the two results are really the same.

23. Differentiate

```
> (x + a)^100;
```

$$(95x - 58x^5 - 90x^4 + 53x^3 - x^2 + 83)^{100}$$

with respect to x, without using the chain rule. What would you need to do to this result to reconstruct the answer given by the chain rule?

24. Find $D(f)$ where $f(x) = \frac{1}{x^5 - 3^{1/5}}$.
25. Find $D(g)$ where $g(t) = \frac{t^2-1}{t^3+1}^{19}$.
26. Compute the derivative of the composite function f where f is defined by

```
> h := (x) -> x^3 + 5;
```

$$x \mapsto x^3 + 5$$

```
> f := sin@sin@h;
```

$$\sin^{(2)} \circ h$$

27. The expression

```
> b^(ln(a)*c);
```

$$(-86x^5 + 23x^4 - 84x^3 + 19x^2 - 50x + 88)^{\ln(-58x^5 - 90x^4 + 53x^3 - x^2 + 94x + 83)c}$$

can be rewritten as

```
> convert(",exp);
```

$$e^{\ln(-86x^5 + 23x^4 - 84x^3 + 19x^2 - 50x + 88)\ln(-58x^5 - 90x^4 + 53x^3 - x^2 + 94x + 83)c}$$

Use this to show that $b^{c\ln(a)} = a^{c\ln(b)}$.

28. Find $\frac{dy}{dx}$ if $e^{x^2+xy+y^2} = e^{x^2 y^2}$.
29. Use *Maple* to compute $\frac{dy}{dx} \ln(f(x))$ for an arbitrary function f and solve the resulting equation for $\frac{dy}{dx} f(x)$. This leads to a general approach to computing the derivative of expressions such as

```
> e1 := (x-1)^(3/2)*(x-2)^(5/2)*(x-3)^(7/2)*(x-4)^(-3/2);
```

$$(x-1)^{\frac{3}{2}} (x-2)^{\frac{5}{2}} (x-3)^{\frac{7}{2}} (x-4)^{-\frac{3}{2}}$$

Simply compute **diff(expand(ln(e1)),x)** and multiply by $e1$. Compare the result of computing the derivative in this way with the result you get by doing **diff(e1,x)**. What can you conclude about how *Maple* approaches this problem?

30. Find an expression for $\frac{dy}{dx}$ in the following equation.

```
> e1 := 3*x^2 - 5*x^2*y^3 + 4*x*y = 12;
```

$$3x^2 - 5x^2 y^3 + 4xy = 12$$

31. Find the equation of the tangent line given to the point $[6, 8]$ given that

```
> eq := 3*x^2 + 5*y^2 = 100;
```

$$3x^2 + 5y^2 = 100$$

must be satisfied.

32. Find an expression for $\frac{dy}{dx}$ at $x = \sqrt{3}$ in the following equation

```
> e1 := 5*x^2 - x^2*y^3 + 4*y = 16;
```

$$5x^2 - x^2y^3 + 4y = 16$$

if $y = 1$ when $x = \sqrt{3}$.

33. A 4-meter ladder is standing on level ground and resting against a vertical wall. If the top of the ladder is being pushed up the wall at a rate of 1/4 meter per second, how fast is the base of the ladder moving toward the wall when the top is at 3 meters?

34. Water is being pumped into an inverted conical tank at a rate of 2 meter^3 per minute. The tank is 10 meters high and 5 meters in diameter at the top. When will the surface of the water be rising at a rate of 5 meters per minute?

35. At 4:00 P.M., the ship *Nostalgia* is 200 km due east of ship *No Regrets* and proceeding south at 20 knots. The ship *No Regrets* continues west at 16 knots. How fast is the distance between the two ships changing at 6:00 P.M.?

36. Use the tangent line at $x = 3$ to approximate the change dy as an expression in x if

```
> f := (x) -> sin(3*x^2)/x;
```

$$x \mapsto \frac{\sin(3x^2)}{x}$$

Use this to estimate $f(3.5)$.

37. Find the exact error incurred by using the tangent line at $x = 2$ to estimate $f(2.01)$ if the function f is defined by

```
> f := (x) -> x^3 + 4*x^2 - 6*x + 2;
```

$$x \mapsto x^3 + 4x^2 - 6x + 2$$

38. Use Newton's algorithm to find a solution to $x^3 - 2 = 0$. Compare your result with that obtained by *Maple*'s **fsolve( )**.

39. Use Newton's algorithm to find at least three solutions to $\cos(3x^2 - 5) = 0$. Compare your results with that obtained by *Maple*'s **fsolve( )**. You can use the **plot( )** command to help identify appropriate intervals and starting points.

40. We can construct polynomials that pass through particular points by beginning with undetermined coefficients and then evaluating the polynomial at certain points.
    a. Find and graph a polynomial of degree at most 3 that passes through the points $[0,0]$, $[1,2]$, $[2,1]$, and $[3,3]$.
    b. Find and graph a polynomial of degree at most 3 that has horizontal tangent lines at $x = 0$ and at $x = 3$, and which passes through the points $[0,0]$ and $[3,5]$.

41. An aircraft is flying level at an altitude of h above a level runway that is at a distance of R miles straight ahead. If the landing approach follows a cubic curve exactly, and performs a perfectly smooth landing, find the polynomial that describes this smooth approach.

42. Find the shortest distance between the two curves

$$f : (x) \to x^2 + 2x + 5$$

and

$$g : (x) \to -x^2 + 6x - 1$$

4 Optimal Solutions and Extreme Values

4.1 Maximums and Minimums

In many practical situations, solving a problem involves maximizing or minimizing some quantity. This *optimization* requires that we find a value of x that maximizes or minimizes the value of a particular function $f(x)$. We call these solutions *extreme points*.

DEFINITION 4.1 A function has an *absolute maximum value* at c if $f(c) \geq f(x)$ for all x in the domain of f. A function has an *absolute minimum value* at c if $f(c) \leq f(x)$ for all x in the domain of f.

Where possible, we use plotting to help understand the behavior of the function and to estimate where such extrema occur.

Example 4.1 Find the maximum value and the minimum value of $\cos(x)$.

Solution The following plot suggests that there might not be a unique solution for these values.

```
> plot(cos(x),x=-5*Pi...5*Pi,y=-2..2);
```

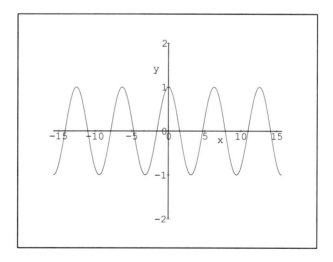

The maximum value of this function is 1, but it occurs at regularly spaced values of x along the real axis. The cosine function is defined in terms of a ratio of two sides of a right angle triangle in a manner that prevents it from ever being bigger than 1. One solution (i.e., a value of x that maximizes $\cos(x)$) is given by the command

```
> solve(cos(x)=1,{x});
```
$$\{x = 0\}$$

Near this maximum the graph is concave downward.

```
> plot(cos(x),x=-1/10..+1/10);
```

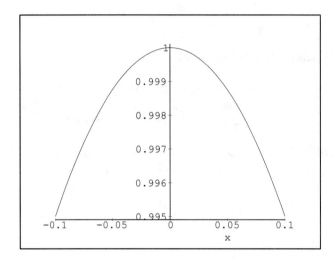

A minimum function value, -1, also occurs at regularly spaced points. One such point is given by

```
> solve( cos(x)=-1,{x});
```
$$\{x = \pi\}$$

The following graph

```
> plot( cos(x),x=Pi-1/10..Pi+1/10);
```

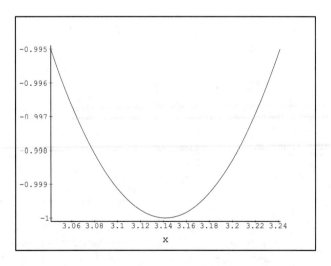

shows that, near this minimum, the graph is concave upward.

Approximate locations of other extreme points on this curve can be found by commands such as

```
> fsolve( cos(x)=1,x,6..7);
```

$$6.283185307$$

and

```
> fsolve( cos(x)=-1,x,9..10);
```

$$9.424777961$$

The exact solutions are multiples of π. ∎

Not all functions have a global maximum and a global minimum.

Example 4.2 Minimize and maximize the function f defined by

```
> f := (x)-> x^2 + 2;
```

$$x \mapsto x^2 + 2$$

Solution A plot of this function provides some insight.

```
> plot(f,-10..10);
```

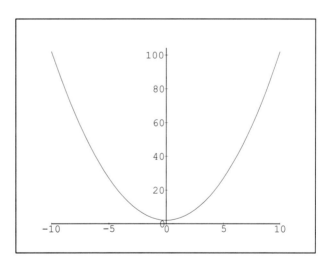

A minimum value of 2 appears to occur at $x = 0$. The location of this minimum, and its value, can be identified more exactly by using a narrower domain for the plot.

```
> plot(f,-1..1,1..3);
```

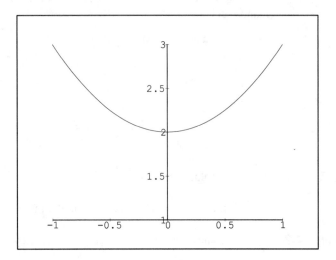

As helpful as these graphs are in locating the minimum, they do not guarantee that we have found a global extremum.

In this particular case, because x^2 is always nonnegative, we can show that we really do have a minimum. There is no maximum because the value of x^2 can be made as large as we like simply by making x very large. In fact, the limit

```
> Limit(f(x),x=infinity);
```

$$\lim_{x \to \infty} (x^2 + 2)$$

has the value

```
> value(");
```

$$\infty$$

∎

Global extrema need not exist at all. Functions defined by odd-degree polynomials over the real numbers have neither a global maximum nor a global minimum.

Example 4.3 Show that following function has neither an absolute maximum nor an absolute minimum.

```
> f := (x) -> x^3 - x^2 + 4*x + 4 ;
```

$$x \mapsto x^3 - x^2 + 4x + 4$$

Solution The reason is readily apparent from the plot

```
> plot(f,-6..6);
```

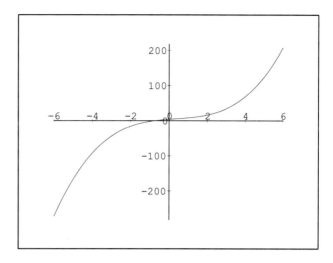

and from the fact that the values of the function can be made as large or small as we like just by taking extremely large or small values of *x*. The maximum is

```
> limit(f(x),x=infinity);
```

$$\infty$$

and the minimum is

```
> limit(f(x),x=-infinity);
```

$$-\infty$$

so we say the function is unbounded. ∎

Plotting can help us approximate the solution to any particular optimization problem. The main difficulty lies in finding an interval in which a maximum actually occurs. Some use can be made of "infinity" plots. For

```
> f(x);
```

$$x^3 - x^2 + 4x + 4$$

we have

```
> plot( f,-infinity..infinity);
```

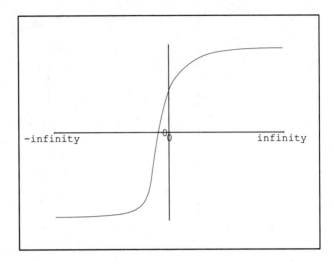

This sort of plot drastically distorts the true shape of the curve in order to collapse an infinitely long *x*-axis into a finite picture. However, it can provide useful hints on where to start looking for maximums, minimums, and zeros.

To see the distortion, consider the graph of a function you know well, such as

```
> g := (x)-> x^2;
```

$$x \mapsto x^2$$

```
> plot(g,-infinity..infinity);
```

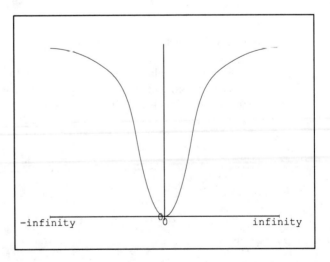

The true picture — the one with constant scaling — is concave upward and not, as shown here, in the shape of an upside down bell.

4.1.1 Optimizations in Maple

We can often use the algebraic properties of functions to discover their maximums or minimums. These properties are used by the *Maple* commands **maximize( )** and **minimize( )**. For example, the results for **cos( )** in Example 4.1 correspond to

```
> readlib(minimize):
> minimize(cos(x),{x});
```

$$-1$$

```
> maximize(cos(x),{x});
```

$$1$$

and the results for f in Example 4.3 correspond to

```
> minimize(f(x),{x});
```

$$-\infty$$

```
> maximize(f(x),{x});
```

$$\infty$$

The characterizations of maximum or minimum that are used to confirm these extrema algebraically have a much more fundamental use. They allow us to reverse the modeling process by first specifying the desired features and then building functions that model these features. Consider the following motivating example.

> *Build a function that behaves like a quadratic polynomial and has a maximum at $x = 4$, and passes through $[0, 2]$ and $[4, 0]$.*

How do we force the graph of a polynomial to go through a certain point? How do we force it to have a maximum or a minimum at a specified location? A primary goal of this chapter is to help you develop the mathematical tools and concepts necessary to characterize these maximums and minimums algebraically and to combine them to solve problems such as this.

4.1.2 Local Maximums and Minimums

A first step in achieving an algebraic characterization of extrema (i.e., maximums or minimums) is to examine how the function behaves near such points.[1] The endpoints of the domain always require special attention. For example, consider

```
> f(x);
```

$$x^3 - x^2 + 4x + 4$$

---
[1] In most practical problems we are interested only in extrema on some finite domain such as x in $[0, 5]$.

Chapter 4 Optimal Solutions and Extreme Values

On the domain [0, 5] its graph is

```
> plot(f(x),x=0..5);
```

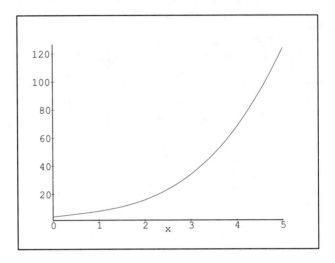

A global maximum (on this restricted domain) clearly occurs at $x = 5$.

REMARK: Always check the endpoints of the domain for possible solutions to an optimization problem.

With the exception of maximums at the endpoints of an interval, when we look near a maximum, we always find a graph that is concave downward.[2] For these "internal" maximums, it is clear that the maximum is the point that is higher than all its closest neighbors. This leads to the following technical definition.

DEFINITION 4.2 A function has a *local*[3] *maximum value* at c if we can find a δ such that $f(c) \geq f(x)$ for all x in the domain $[x - \delta, x + \delta]$. A function has a *local minimum value* at c if we can find a δ such that $f(c) \leq f(x)$ for all x in the domain $[x - \delta, x + \delta]$.

In practice, the crucial observation is that a tangent line, if one exists, is horizontal at a local extremum.

Example 4.4 Show that the function f defined by,

```
> f := (x) -> 7*x^3+14*x^2-7*x-14;
```

$$x \mapsto 7x^3 + 14x^2 - 7x - 14$$

has a local maximum and a local minimum on the domain $[-2, 2]$.

---
[2] In an effort to focus on what we would like to have happen, we have deliberately left out any discussion of nasty examples.
[3] These types of extrema are also called *relative* extrema.

Solution The graph

> plot(f(x),x=-2..2);

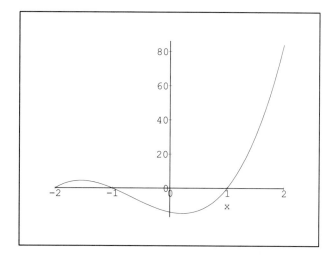

shows two locations where the tangent line is horizontal, thus confirming the presence of two local extrema.

We can use the tools developed in the previous chapter to find these points. Because differentiation can be used to find the slope of a curve at a point, all we really need are the values of x for which $f'(x) = 0$. The equation specifying that these slopes be 0 is

> D(f)(x) = 0;

$$21x^2 + 28x - 7 = 0$$

By solving this quadratic equation for x, we obtain two possible x values

> sol := solve(",x);

$$-\frac{2}{3} + \frac{1}{3}\sqrt{7}, \; -\frac{2}{3} - \frac{1}{3}\sqrt{7}$$

In order of size, they are approximately

> sol := sort(evalf(["]));

$$[-1.548583770, 0.2152504369]$$

The tangent line for the first of these is

> with(student):

> showtangent(f(x),x=sol[1],x=-2..2);

Chapter 4 Optimal Solutions and Extreme Values

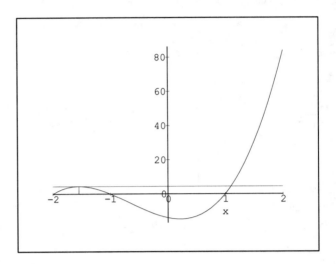

which clearly corresponds to the local maximum. ■

Example 4.5 Find the local maximum and minimum of the function

```
> f := (x) -> 3*x^4 - 16*x^3 + 18*x^2;
```

$$x \mapsto 3x^4 - 16x^3 + 18x^2$$

on the interval $x \in [-1, 4]$.

Solution On this interval the graph of the function is

```
> plot(f(x),x=-1..4);
```

4.1 Maximums and Minimums

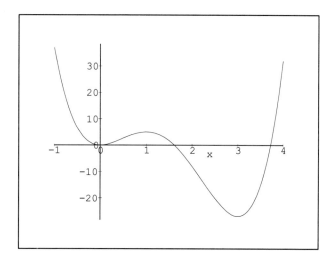

For this domain, the global maximum would appear to be at the left endpoint. To confirm this, compare the function values at the two endpoints.[4]

```
> f(-1),f(4);
```

$$37,\ 32$$

There are also three local extrema: two minimums and a maximum. Each occurs at a point where the tangent line is horizontal and their locations can be found by solving the equation

```
> D(f)(x) = 0;
```

$$12x^3 - 48x^2 + 36x = 0$$

The three exact solutions to this equation are

```
> solve(",x);
```

$$0,\ 3,\ 1$$

The local maximum occurs at $x = 1$.

```
> showtangent(f(x),x=1,x=-1..4);
```

---

[4] A local maximum is potentially a global maximum, but in this instance the graph shows that its value is less than the function values at the endpoints.

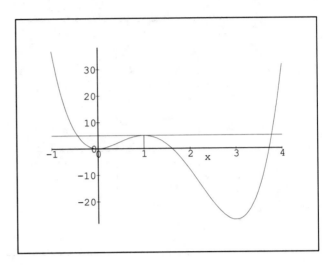

Armed with the knowledge that local maximums occur at the the same place as horizontal tangent lines, we can even tackle our earlier challenge (see page 179) of constructing a function with specific characteristics.

Example 4.6 Build a function that behaves like a quadratic polynomial and has a maximum at $x = 4$, and passes through $[0, 2]$ and $[4, 0]$.

Solution The desired function has the form

```
> a := 'a': b := 'b': c := 'c':
> f := (x)-> a*x^2 + b*x + c;
```

$$x \mapsto ax^2 + bx + c$$

We collect additional information about the function by evaluating it, or its derivatives, at selected points. For example, because a maximum is to be at $x = 4$, the tangent line at $x = 4$ should be horizontal. We obtain the equation

```
> e1 := D(f)(4) = 0;
```

$$8a + b = 0$$

We are also given two other function evaluations.

```
> e2 := f(0)=2, f(4)=0;
```

$$c = 2, \quad 16a + 4b + c = 0$$

This gives us a total of three equations that can be solved together to find values for the three unknowns $\{a, b, c\}$. We obtain the result

```
> solve( {e1,e2} );
```

$$\left\{a = \frac{1}{8}, c = 2, b = -1\right\}$$

A graph of the corresponding polynomial is given by

```
> assign("):
```

```
> plot(f(x),x=-1..5);
```

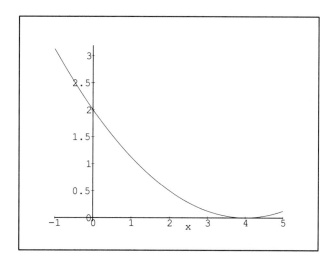

■

We finish this section with an example where our differentiation technique fails to provide useful information. Any comprehensive solution strategy will need to allow for this type of situation.

Example 4.7 Find a relative minimum for the function defined by

```
> f := (x) -> abs(x):
```

on the interval $[-2, 2]$.

Solution The following graph shows that a relative minimum exists at $x = 0$.

```
> plot( f,-2..2);
```

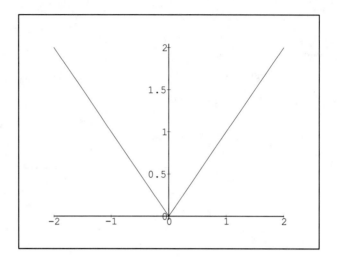

However, this time the slope of the tangent line is not defined at 0. The derivative of f is

```
> D(f);
```

$$a \mapsto \frac{|a|}{a}$$

and when we try to evaluate this formula at $x = 0$, we end up dividing by 0.

```
> D(f)(0);
Error, (in unknown) division by zero
```

■

REMARK: For locating local extrema, the places where the derivative of a function becomes undefined are just as important as the places where it evaluates to 0.

Whether the derivative becomes 0 or is undefined, it is a warning that the function values may be changing from increasing to decreasing or vice versa. All these locations are good places to look for extrema, and so they are given a special name.

DEFINITION 4.3 A *critical number* of a function f is a number c in the domain of f where either $f'(c) = 0$ or $f'(c)$ does not exist.

Although it may not seem like much, the fact that we have reduced our search zone for extrema from "everywhere" to just a few points is very important.

4.1.3 The Extreme Value Theorem

For which functions will this differentiation technique work? A partial answer is given by the following theorem.

THEOREM 4.1 If a function f is defined on a closed interval $[a, b]$ and is continuous, then f always attains its absolute maximum for some $x = c$ in that interval.

To see that the condition of continuity cited in this theorem is important, consider the following example.

Example 4.8 The piecewise-defined function, defined by using

```
> f := (x) -> if x >= 0 and x < 1 then x^2+1
> else undefined  fi:
```

if $0 \leq x < 1$, and

```
> g := (x) -> if x >= 1 and x <= 2 then -x^2+3/2
> else undefined fi:
```

if $1 \leq x \leq 2$, is not continuous. It also does not attain a maximum on the interval $[0, 2]$.

Solution In spite of the appearance of the graph

```
> plot({f,g},-1..3);
```

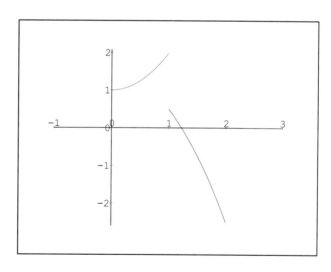

this function has no maximum. The apparent candidate for a maximum is

```
> Limit(x^2+1,x=1,left);
```

$$\lim_{x \to 1^-} \left(x^2 + 1\right)$$

which evaluates to

```
> value(");
```

$$2$$

However, this is deceptive because f is not actually defined at $x = 1$. By choosing x close to 1, we can make $f(x)$ arbitrarily close to 2, but we can never actually get there. ∎

REMARK: The value 2 is an example of an *upper bound*. This upper bound is not a maximum because there is no x for which $f(x) = 2$.

4.1.4 Summary

Our approach to finding local extrema is to examine the points where the derivative of the function is 0 or undefined. This approach is justified by the following theorem.

THEOREM 4.2 (*Fermat's theorem*) If f has a local extremum (maximum or minimum) at c, and $f'(c)$ exists, then $f'(c) = 0$.

For global extrema, we need to compare the value of f at these points with the endpoints of the domain.

If a local maximum occurs at $x = c$ and $f'(c)$ is defined, then the slope of the tangent line must make a smooth transition from increasing to decreasing as x moves from left to right in this differentiable region that includes c. The result is a smooth curve that is concave downward. Similarly, a local minimum at a differentiable point c must correspond to a smooth curve that is concave upward. The curve f is smooth near such points because differentiability is defined at points in the interior of a region rather than at the boundary. When f is differentiable at c it must be differentiable on an entire region that includes c, and this prevents the slope of the tangent line from changing abruptly in this neighborhood.

4.2 The Mean Value Theorem

Smooth continuous functions play an important role in our attempt to characterize maximums and minimums. They are precisely the kinds of functions for which Fermat's theorem is most useful. A first step toward finding such extrema is the following theorem.

THEOREM 4.3 (*Rolle's theorem*) If f satisfies the following conditions

1. f is continuous on $[a, b]$,
2. f is differentiable on (a, b),
3. $f(a) = f(b)$,

then there is a number c in (a, b) such that $D(f)(c) = 0$.

This theorem states that, for a smooth continuous function f, if $f(a) = f(b)$, there is at least one horizontal tangent line to be found somewhere between a and b. At that special location, c, the curve f is concave downward or upward. This can help find maximums and minimums as follows.

Example 4.9 Use Rolle's theorem to show that there is a local maximum or minimum in the interval $[2, 6]$ for the function s defined by

```
> s := (t)-> 8*t+288-t^2;
```

$$t \mapsto 8t + 288 - t^2$$

Find that position if possible.

Solution The function s is continuous on $[2, 6]$ and differentiable on $(2, 6)$, and

```
> s(2)=s(6);
```

$$300 = 300$$

From the graph of s on the interval $[2, 6]$ we see that there is an extremum, as predicted by Rolle's theorem.

```
> plot(s,2..6);
```

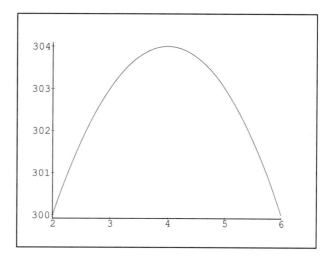

Rolle's theorem guarantees that there is a c such that

```
> D(s)(c) = 0;
```

$$8 - 2c = 0$$

To find c, we solve this equation.

```
> solve(",{c});
```

$$\{c = 4\}$$

∎

REMARK: By simply knowing that $f(a) = f(b)$, we are guaranteed that an extremum must occur in $[a, b]$. This is especially important if we cannot solve $D(f)(c) = 0$.

This theorem can also be used to draw conclusions about the number of roots of smooth continuous functions. If there are two distinct roots of f, then we have two x values for which $f(x) = 0$. By Rolle's theorem we must also have a horizontal tangent line somewhere between them. If it is impossible to have a horizontal tangent line, then there cannot possibly be two distinct roots.

Example 4.10 Show that the equation $x^3 + x - 1 = 0$ has exactly one real root.

Solution The function of interest is

```
> f := (x) -> x^3 + x - 1;
```

$$x \mapsto x^3 + x - 1$$

This smooth continuous function changes sign between $x = 0$ and $x = 1$, so by the intermediate value theorem (see Section 2.8) there is at least one real root.

```
> f(0), f(1);
```

$$-1, 1$$

If there were two roots, then by Rolle's theorem there would be a solution to

```
> D(f)(c)=0;
```

$$3c^2 + 1 = 0$$

Such a solution for c cannot exist, however, because two positive numbers can never sum to 0. ∎

Even when the equation $f'(c) = 0$ has solutions for c, it may be very hard to find them. Most of the examples we encounter will involve finding roots of polynomials. We can also resort to approximations such as those produced via Newton's algorithm or **fsolve( )**.

In *Maple* we rely on tools such as **solve( )** and **fsolve( )**, but remember that the failure of such commands to find a root is only partial evidence that such roots do not exist. An actual proof that a solution does not exist requires additional evidence such as provided in Example 4.10.

4.2.1 The Average Slope

One interpretation of Rolle's theorem is that for smooth continuous functions with $f(a) = f(b)$, an *average slope* of 0 on that particular portion of the curve is actually "realized" as an instantaneous slope somewhere in that interval. The following theorem states that the average slope across an interval (total rise over change in x) is realized in that interval even when the average slope is nonzero.

4.2 The Mean Value Theorem

THEOREM 4.4 *(The Mean Value Theorem)* If f satisfies the following conditions:

1. f is continuous on $[a, b]$,
2. f is differentiable on (a, b),

then there is a number c in (a, b) such that $D(f)(c) = \dfrac{f(b) - f(a)}{b - a}$.

Proof This proof is a simple application of Rolle's theorem. We define an appropriate function, apply Rolle's theorem to this new function to show that c exists, and then rearrange the resulting equation involving c to obtain a formula for the slope of the new curve at the special point c.

To help focus on the key ideas just outlined, we delegate all the computation and manipulation to *Maple*.

A Specific Instance

Our first task is to define an appropriate function. To learn how this is done in general, we consider a specific example in detail. Consider the function f

```
> f := (x) -> 3*x^4 - 16*x^3 + 18*x^2:
```

on the interval $[-1, 3]$. The two endpoints of the curve over this interval are

```
> A := [-1,f(-1)]:   B := [3,f(3)]:
> A,B;
```

$$[-1, 37], [3, -27]$$

The line defined by these two points is

```
> with(student):
> g := makeproc(A,B);
```

$$x \mapsto -16x + 21$$

The following is a simultaneous plot of the two functions, f and g.

```
> plot( {f,g},-1..3);
```

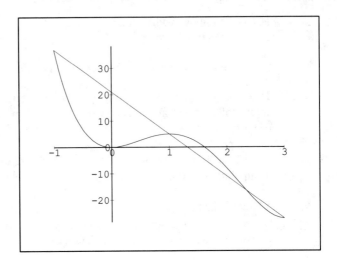

This graph shows that there is at least one point in the interval $[-1, 3]$ where the tangent line is parallel to the straight line g. The average slope

```
> slope(A,B);
```

$$-16$$

is realized at that point.

The key observation that enables us to use Rolle's theorem to guarantee the existence of a suitable point c is that the difference between the two curves f and g will be 0 at the endpoints of the interval. Let h be the function that measures this vertical distance between the curves f and g.

```
> h := f-g;
```

$$f - g$$

```
> plot(h,-1..3);
```

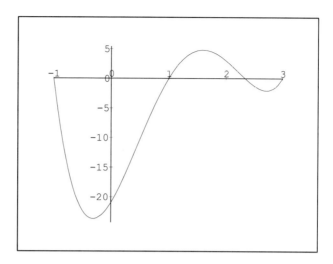

The new function $h = f - g$ will be smooth and continuous if f and g are, so it has all the properties required to enable us to use Rolle's theorem.

The theorem guarantees that there is a point c in $[-1, 3]$ such that

```
> D(h)(c) = 0;
```
$$12c^3 - 48c^2 + 36c + 16 = 0$$

In this example, we can solve[5] the equation explicitly for c. A numeric approximation to the candidate values for c is

```
> sols := [fsolve(",c)];
```
$$[-0.3081146640, 1.594978011, 2.713136653]$$

Any of these three values will do, because they are all in the interval $[-1, 3]$. They also have the property that $D(h)(c) \approx 0$, as verified by the command

```
> map(D(h),sols);
```
$$[0, -0.00000003, 0.00000001]$$

and they satisfy $D(f)(c) \approx \text{slope}(A, B)$, as verified by the command

```
> map(D(f),sols);
```
$$[-16.0, -16.00000003, -15.99999999]$$

The General Case

The proof of the theorem proceeds almost identically to the preceding example, using even the same computations, with only one minor variation at the end.

We restart *Maple* to eliminate references to specific functions or quantities.

```
> restart: with(student):
```

Then the points on the curve, using the same notation as before, are

---

[5] For an exact solution try **solve(",c)**.

```
> A := [a,f(a)];
```
$$[a, f(a)]$$

```
> B := [b,f(b)];
```
$$[b, f(b)]$$

The line corresponding to the average slope is defined by the function

```
> g := makeproc( A, B );
```
$$x \mapsto \frac{f(a)x}{a-b} - \frac{f(a)a}{a-b} - \frac{f(b)x}{a-b} + \frac{f(b)a}{a-b} + f(a)$$

and the function to which we apply Rolle's theorem is

```
> h := f-g;
```
$$f - g$$

We must always check that the conditions required to use Rolle's theorem are satisfied. The function h is continuous and differentiable whenever f and g are, and it evaluates to 0 at the beginning and endpoints of the interval. To see this, note that for the first endpoint we have

```
> h(a);
```
$$0$$

and for the second, we have

```
> h(b);
```
$$f(b) - \frac{f(a)b}{a-b} + \frac{f(a)a}{a-b} + \frac{f(b)b}{a-b} - \frac{f(b)a}{a-b} - f(a)$$

which simplifies to

```
> normal(");
```
$$0$$

Just as before, Rolle's theorem guarantees that the equation

```
> D(h)(c) = 0;
```
$$D(f)(c) - \frac{f(a)}{a-b} + \frac{f(b)}{a-b} = 0$$

has a solution.

If this were a specific example, we could finish the computation and the proof by solving this equation for c. However, we cannot do that here because f has not been specified. A moment of reflection shows that this does not matter. What we really want is a formula for $D(f)(c)$. The preceding equation is linear in $D(f)(c)$ and easy to solve. We obtain the solution

```
> isolate(",D(f)(c));
```

$$D(f)(c) = \frac{f(a)}{a-b} - \frac{f(b)}{a-b}$$

which simplifies to

```
> normal(");
```

$$D(f)(c) = \frac{-f(a)+f(b)}{-a+b}$$

This proves the theorem, because the instantaneous slope at c is the same as the average slope, as given by

```
> slope(A,B);
```

$$\frac{f(a)-f(b)}{a-b}$$

□

Example 4.11 Find a point c, in (0, 2), that satisfies

```
> D(f)(c) = slope([0,f(0)],[2,f(2)]);
```

$$D(f)(c) = -\frac{1}{2}f(0) + \frac{1}{2}f(2)$$

for f defined by

```
> f := (x) -> x^3 - x;
```

$$x \mapsto x^3 - x$$

Solution Let

```
> A := [0,f(0)]:  B := [2,f(2)]:
```

By the mean value theorem, there is a c in (0, 2) such that

```
> D(f)(c) = slope(A,B);
```

$$3c^2 - 1 = 3$$

Two approximate solutions to this equation are

```
> sols := sort( [fsolve(",c)] );
```

$$[-1.154700538, 1.154700538]$$

Only one of these is in the interval [0, 2]. It is

```
> c := sols[2]:
```

The tangent line at this point has the slope of the line from A to B given by

```
> g := makeproc( A , B );
```

$$x \mapsto 3x$$

We can display these two curves and the tangent line at $x = c$ simultaneously, as follows.

```
> p1 := plot( {f , g} , 0..2 ):
> p2 := showtangent( f(x) , x=c,x=0..2,y=-1..6):
> plots[display]([p1,p2]);
```

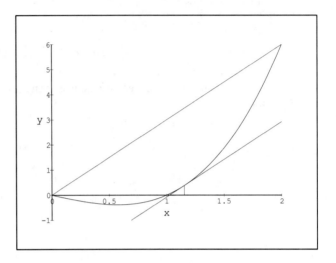

The graph confirms that the tangent line and the line depicting the average slope are parallel. ∎

We can use this strong relationship between average slope and instantaneous slope to reach a number of useful conclusions.

THEOREM 4.5 If $D(f)(x) = 0$ for all x in an interval (a, b), then f is constant on (a, b).

Proof If there are any two points $x1$ and $x2$ in (a, b) for which $f(x1) \neq f(x2)$, then the slope of the line going through the two points

```
> A := [x1,f(x1)]: B := [x2,f(x2)]:
```

namely

```
> m := slope(A,B);
```

$$\frac{f(x1) - f(x2)}{x1 - x2}$$

is nonzero. Furthermore, the mean value theorem for f on the interval $(x1, x2)$ guarantees that there is a c such that $D(f)(c) = m \neq 0$. But this would contradict the assumption that $D(f)(x) = 0$ for all x in the interval. □

COROLLARY 4.1 If $D(f)(x) = D(g)(x)$ in (a, b), then $f - g$ is constant on (a, b).

Proof Apply the previous theorem to $h = f - g$. □

4.3 Monotonic Functions

DEFINITION 4.4 A function is *increasing* on an interval I if $f(x_1) < f(x_2)$ whenever $x_1 < x_2$ in I.
A function is *decreasing* on an interval I if $f(x_1) > f(x_2)$ whenever $x_1 < x_2$ in I.
A function is *monotonic* on I if it is *increasing* on I or *decreasing* on I, but not both.

THEOREM 4.6 Suppose f is continuous on $[a, b]$ and differentiable on (a, b). If $D(f)(x) > 0$ for all x in (a, b), then f is increasing on $[a, b]$. If $D(f)(x) < 0$ for all x in (a, b), then f is decreasing on $[a, b]$.

Proof For any choice of $x_1 \leq x_2$ in (a, b) we have — by the mean value theorem — that $f(x_2) - f(x_1) = D(f)(c)(x_2 - x_1)$ for some c in (a, b). The sign of the right-hand side depends entirely on the sign of $D(f)(c)$. □

This theorem justifies using $f' = D(f)$ to break up the domain of a function f into increasing and decreasing segments.

Example 4.12 For f defined by

```
> f := (x) -> 3*x^4 - 4*x^3 - 12*x^2 + 5:
```

find where f is increasing and decreasing.

Solution The function f is increasing when $D(f)$ is positive and decreasing when $D(f)$ is negative. By experimentation, we have found that all the interesting behavior for this function occurs in the interval $[-2, 3]$. The following graph of both f and $D(f)$ reveals the appropriate regions.

```
> plot( {f,D(f)}, -2..3);
```

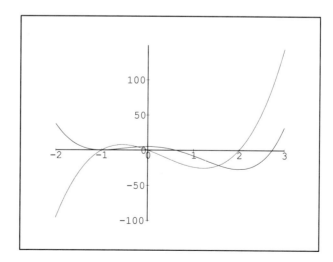

The function f is monotonic between the adjacent extreme points which, in this example, are found by solving the equation

Chapter 4 Optimal Solutions and Extreme Values

```
> D(f)(x) = 0;
```
$$12x^3 - 12x^2 - 24x = 0$$

The solutions[6] are

```
> solve( " );
```
$$0, 2, -1$$

Between these "boundaries" the sign of $D(f)(x)$ remains the same for the entire interval. The slope of each interval can be determined by testing a single point in that interval. Given the values

```
> D(f)(-2.1), D(f)(-.5) , D(f)(.9), D(f)(3);
```
$$-113.652, 7.500, -22.572, 144$$

we know that f is decreasing, increasing, decreasing, and increasing again as it moves from region to region left to right. ■

4.3.1 The First Derivative Test

DEFINITION 4.5 The points on the curve that correspond to the critical numbers are called *critical points*.

You can determine if a critical point is a maximum or a minimum (or neither) by examining the sign of $D(f)(x)$ near the corresponding critical number. If $D(f)(x)$ changes sign from negative to positive as x moves (increases) past the critical number c, then you must have a minimum. If $D(f)(x)$ changes sign from positive to negative, you must have a maximum. If $D(f)(x)$ does not change sign, then you do not have an extremum.

Example 4.13 Find the local extrema of f.

```
> f := (x) -> x*abs(1-x)^(2/5);
```
$$x \mapsto x \, |1-x|^{\frac{2}{5}}$$

Solution The derivative

```
> D(f)(x);
```
$$|1-x|^{\frac{2}{5}} - \frac{2}{5} x \, |1-x|^{\frac{2}{5}} (1-x)^{-1}$$

simplifies to

```
> e1 := normal(");
```
$$\frac{1}{5} |1-x|^{\frac{2}{5}} (-5 + 7x)(-1 + x)^{-1}$$

---

[6]You can also use commands such as **factor( )** or **fsolve( )**.

The critical numbers are the places where this expression is zero or undefined. The zeros occur when the numerator is zero, as in

```
> solve( numer(e1)=0,{x});
```

$$\left\{x = \frac{5}{7}\right\}, \{x = 1\}$$

and the derivative is undefined when the denominator is zero,[7] as in

```
> solve( denom(e1)=0,{x});
```

$$\{x = 1\}$$

The graph

```
> p1 := plot( D(f),0..1,-10..10):
> p2 := plot( D(f),1..2,-10..10):
> plots[display]([p1,p2]);
```

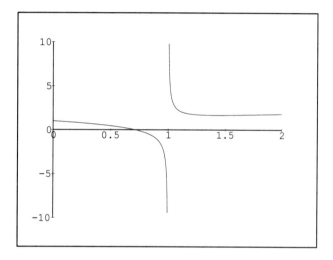

shows that the graph of f has a maximum at $x = 5/7$ and a minimum at $x = 1$.   ∎

Example 4.14 Find the local extrema of the function f defined by

```
> f := (x) -> x^3*(x-2)^2;
```

$$x \mapsto x^3 (x - 2)^2$$

Solution In this case we have a polynomial. Polynomials and their derivatives are defined everywhere so all the critical numbers will correspond to solutions of

```
> D(f)(x)= 0;
```

$$3x^2 (x - 2)^2 + 2x^3 (x - 2) = 0$$

---

[7] Here, *Maple* reports three solutions that are all the same, suggesting that *Maple* was solving a cubic at some intermediate step in the solution process. Can you see how this would arise?

For polynomials, the problem of finding roots may sometimes be simplified by factoring

```
> factor(");
```

$$x^2 (5x - 6)(x - 2) = 0$$

```
> solve(",{x});
```

$$\{x = 0\}, \{x = 0\}, \left\{x = \frac{6}{5}\right\}, \{x = 2\}$$

or we can find approximations to the roots using **fsolve( )**. Note the locations of the extrema in the following plot and that f is monotonic between these critical numbers.

```
> plot({f,D(f)},-1..4,-10..10);
```

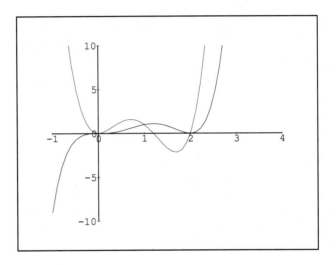

■

We can also use monotonicity to make deductions about inequalities.

Example 4.15 Prove that $(1 + x)^n > 1 + nx$ for $x > 0$ and $n > 1$.

Solution The statement of this theorem can be interpreted as saying that $f(x)$ must be positive if $n > 1, x > 0$, and f is defined by

```
> f := (x) -> (1+x)^n - (1 + n*x);
```

$$x \mapsto (1 + x)^n - 1 - nx$$

To see that this claim makes some sense, we examine a few specific instances such as

```
> subs(n=2,f(x));
```

$$(1 + x)^2 - 1 - 2x$$

```
> subs(n=3,f(x));
```
$$(1+x)^3 - 1 - 3x$$

Note that both
```
> expand( (1+x)^2 );
```
$$1 + 2x + x^2$$

and
```
> expand( (1+x)^3 );
```
$$1 + 3x + 3x^2 + x^3$$

contain terms, that cancel $1 + nx$, leaving a sum of positive terms, as in
```
> expand( subs(n=3,f(x)) );
```
$$3x^2 + x^3$$

We can explain this result as follows.[8] A monotonically increasing function that is already nonnegative when evaluated at $x = 0$ can never become negative as x increases. To determine the monotonicity of f, we examine
```
> D(f)(x);
```
$$\frac{(1+x)^n n}{1+x} - n$$

We must show that this is positive for $n > 1$ and $x > 0$. If we collect the terms with respect to n
```
> collect(",n);
```
$$\left(\frac{(1+x)^n}{1+x} - 1\right)n$$

and simplify the powers that occur, we can rewrite this as
```
> simplify(",power);
```
$$\left((1+x)^{n-1} - 1\right)n$$

No matter what value of $n > 1$ we choose, the cofactor of n in this expression is exactly one less than something that is at least as big as 1. Therefore, the derivative is positive, and $f(0) = 0$ for any $n > 0$, so the function is positive for the indicated domains of n and x. ∎

---

[8]This result can also be proved by using the binomial theorem to show that the negative terms in the expanded sum always cancel with equivalent positive terms.

4.4 Concavity and Points of Inflection

The graph of f is concave upward on an interval if it lies above all its tangent lines. This occurs when $D(f)$ is monotonically increasing. Similarly, the graph of f is concave downward if it lies entirely below all its tangent lines. This occurs when $D(f)$ is monotonically decreasing. This means that the second derivative $D(D(f))$ — the derivative of $D(f)$ — can be used to investigate the concavity of f.

THEOREM 4.7 Let f be twice differentiable on an interval I. If $D(D(f))(x) > 0$ for all x in I, then the graph of f is concave upward. If $D(D(f))(x) < 0$ for all x in I, then the graph of f is concave downward.

Proof Let g denote the function that constructs the tangent line to f at $x = a$. Then

```
> restart: with(student):
> e1 := slope( [a,f(a)] , [x,g(x)] ) = D(f)(a);
```

$$\frac{f(a) - g(x)}{a - x} = D(f)(a)$$

Note that $g(a) = f(a)$.

Also, by the mean value theorem, there is a $c > a$ in I satisfying

```
> e2 := slope([a,f(a)],[x,f(x)]) = D(f)(c);
```

$$\frac{f(a) - f(x)}{a - x} = D(f)(c)$$

Furthermore, because $f'' = D(D(f))$ is positive and $c > a$,

```
> D(f)(a) < D(f)(c);
```

$$D(f)(a) < D(f)(c)$$

we have

```
> lhs(e1) < lhs(e2);
```

$$\frac{f(a) - g(x)}{a - x} < \frac{f(a) - f(x)}{a - x}$$

which can be solved to yield

```
> assume(a<x):
> solve(lhs(e1)<lhs(e2),g(x));
```

$$\{g(x) < f(x)\}$$

□

DEFINITION 4.6 A point P on the curve is a *point of inflection* if the curve changes from concave upward to concave downward at P.

4.4 Concavity and Points of Inflection

Example 4.16 Find the points of inflection of

```
> f := (x) -> x^3 - 3*x + 1;
```

$$x \mapsto x^3 - 3x + 1$$

Solution The only possible points of inflection occur where

```
> D(D(f))(x) = 0;
```

$$6x = 0$$

The solution to this equation is

```
> solve(",x);
```

$$0$$

From the graph

```
> plot(D(D(f)),-2..2);
```

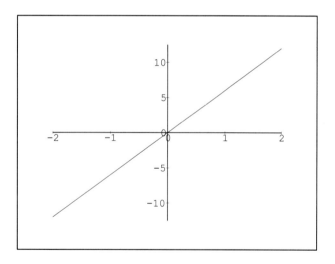

we see that f is concave downward for $x < 0$ and concave upward for $x > 0$ so that the point

```
> P := [0,f(0)];
```

$$[0, 1]$$

is a true inflection point. ∎

REMARK: At an inflection point, the tangent line will cross the curve, as in

```
> showtangent( f(x),x=0,x=-3..3,y=-5..5);
```

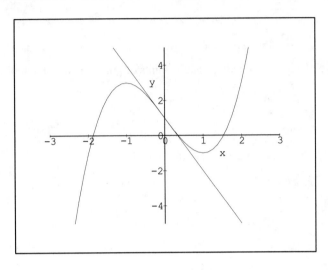

The graph of this region includes two critical points in addition to the point of inflection. They are located at the critical numbers

```
> solve( D(f)(x)=0 , x );
```
$$1, -1$$

and both are local extrema.

Example 4.17 Find the points of inflection and the critical points of

```
> f := (x) -> x^4-8*x^3+18*x^2-16*x+8;
```
$$x \mapsto x^4 - 8x^3 + 18x^2 - 16x + 8$$

and use this to investigate the convexity of f and the nature of any local extrema.

Solution The candidates for inflection points in this graph are derived from solutions to

```
> D(D(f))(x) = 0;
```
$$12x^2 - 48x + 36 = 0$$

which are

```
> solve(",x);
```
$$3, 1$$

In this example, both solutions correspond to genuine inflection points, and they subdivide the domain into three intervals of convexity.

```
> lines := [1,0,1,f(1)], [3,0,3,f(3)];
```
$$[1, 0, 1, 3], [3, 0, 3, -13]$$

4.4 Concavity and Points of Inflection

```
> plot({lines,f(x)},x=0..5 );
```

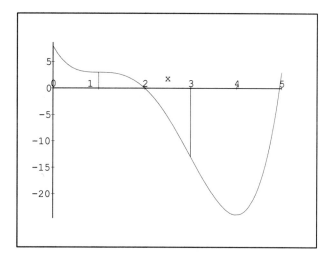

To identify the convexity of f in a particular subinterval, evaluate $f'' = D(D(f))$ at some point in that subinterval. For example, the values

```
> D(D(f))(-1), D(D(f))(2), D(D(f))(4);
```

$$96, -12, 36$$

indicate that the function f is concave upward, concave downward, and concave upward as we work our way across the three regions from left to right.

The candidates for maximums and minimums are the critical points.[9] The critical numbers

```
> solve(D(f)(x)=0 , x );
```

$$4, 1, 1$$

yield the critical points

```
> map( (x) ->([x,f(x)]) , ["] );
```

$$[[4, -24], [1, 3], [1, 3]]$$

At $x = 4$ we have

```
> D(D(f))(4);
```

$$36$$

so the curve f is concave upward and $[4, -24]$ is a minimum. However, at $x = 1$, we have

```
> D(D(f))(1);
```

$$0$$

[9]Note that there are three of them, one for each solution to the cubic equation, even though one root is repeated.

which tells nothing about concavity of f at $[1, f(1)]$ and hence nothing about the nature of this critical point.

By examining $D(f)$ more closely in this region, in this case by plotting $D(f)$, as in

```
> plot(D(f),0..2);
```

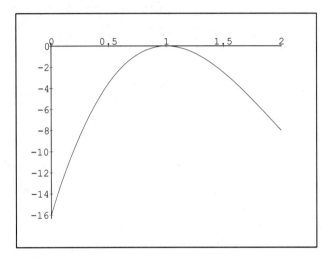

we see that the derivative of f does not change sign in this small region about 1, so the critical point $[1, f(1)]$ is not an extremum. ∎

REMARK: A point such as $[1, f(1)]$ cannot be an extremum and an inflection point at the same time.

The next example involves critical points at x values where the derivative is not defined.

Example 4.18 Find the relative extrema and the inflection points for

```
> f := (x) -> x^(2/3)*(6-x)^(1/3):
```

Solution The graph of this function near $x = 0$ is

```
> plot(f,-10..10);
```

4.4 Concavity and Points of Inflection

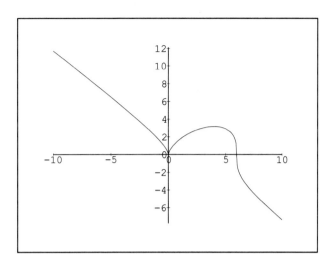

The equation we solve to find the critical numbers is

> e1 := D(f)(x) = 0;

$$\frac{2}{3}\sqrt[3]{6-x}\frac{1}{\sqrt[3]{x}} - \frac{1}{3}x^{\frac{2}{3}}(6-x)^{-\frac{2}{3}} = 0$$

The zeros of $D(f)(x)$ are zeros of the numerator, so we can solve the simpler problem

> numer(lhs(e1)) = 0;

$$12 - 3x = 0$$

to obtain one critical number

> isolate(",x);

$$x = 4$$

There are also two zeros in the denominator:

> denom(lhs(e1));

$$3\sqrt[3]{x}(6-x)^{\frac{2}{3}}$$

Both $x = 0$ and $x = 6$ are zeros of this expression, as confirmed by the graph

> plot(",x=-1..8);

208   Chapter 4   Optimal Solutions and Extreme Values

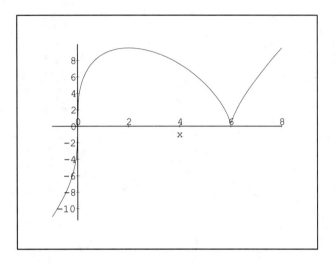

The corresponding critical points are

> `[[4,f(4)],[0,f(0)],[6,f(6)]];`

$$[[4, 4^{\frac{2}{3}} \sqrt[3]{2}], [0, 0], [6, 0]]$$

The first two of these are relative extrema.
   The first candidates for inflection points are the solutions of

> `D(D(f))(x) = 0;`

$$-\frac{2}{9}\sqrt[3]{6-x}\, x^{-\frac{4}{3}} - \frac{4}{9}\frac{1}{\sqrt[3]{x}}(6-x)^{-\frac{2}{3}} - \frac{2}{9}x^{\frac{2}{3}}(6-x)^{-\frac{5}{3}} = 0$$

We can find these by solving the simpler equation

> `numer(lhs(")) = 0;`

$$-2(6-x)^2 - 4x(6-x) - 2x^2 = 0$$

There are no solutions, and hence no such inflection points, because this equation simplifies to

> `expand(");`

$$-72 = 0$$

The remaining candidates for inflection points are at the points where the second derivative of f is undefined. These are zeros $x = 0$ and $x = 6$ of the denominator

> `denom( D(D(f))(x) );`

$$9 x^{\frac{4}{3}} (6-x)^{\frac{5}{3}}$$

The graph

```
> plot(D(D(f))(x),x=-2..8,y=-5..5);
```

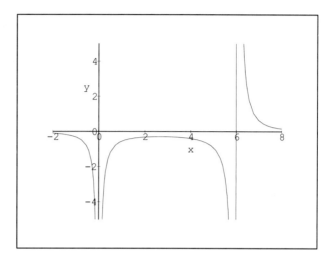

shows that the function f changes convexity only at one of these, $x = 6$ (see the graph of f on page 207), and that it is concave downward to the left of 6, and concave upward to the right of 6. ∎

4.5 Asymptotes

Two other algebraic features of a function help us develop an intuition for its general structure.

The first of these is the behavior of the function as x gets very large or very small. We can deduce this behavior by comparing the rate of growth of the function with that of one of our standard, well-behaved functions, such as powers of x. The second feature is the behavior of the function at points where the function or one of its derivatives (e.g., f' or f'') is undefined.

The first of these characteristics is often referred to as its *asymptotic behavior*. We have already seen simple examples of this when, for instance, we computed a limit as $n \to \infty$.

Recall the following definition.

DEFINITION 4.7 Let f be defined on an interval (a, ∞). Then

$$\lim_{x \to \infty} f(x) = L$$

means that the values of $f(x)$ can be made arbitrarily close to L by taking x sufficiently large.

One interpretation of the existence of such a limit is that the function behaves asymptotically as if it were the straight line g defined by $(x) \to L$.

Example 4.19 Show that, asymptotically, f defined by $(x) \to 1/x + 3$ behaves the same as g defined by $(x) \to 3$.

Solution This behavior is evident from plotting the two graphs.

```
> plot( {1/x + 3, 3} , x = 1..100,y=0..4);
```

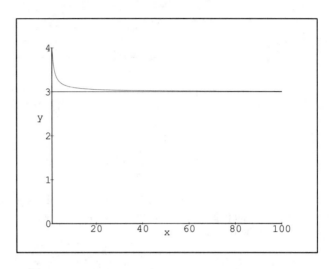

The algebraic justification of this claim is in terms of limits. We have

```
> Limit(1/x+ 3 ,x =infinity):
> " = value(");
```

$$\lim_{x \to \infty} (x^{-1} + 3) = 3$$

The justification works in both directions.

```
> Limit(1/x+3, x=-infinity);
```

$$\lim_{x \to -\infty} (x^{-1} + 3)$$

```
> " = value(");
```

$$\lim_{x \to -\infty} (x^{-1} + 3) = 3$$

The ratio of any two polynomials of the same degree has this same property.

Example 4.20 Find the asymptotic behavior of

```
> f := (x) -> (x^2-1)/(x^2+1);
```

$$x \mapsto \frac{x^2 - 1}{x^2 + 1}$$

4.5 Asymptotes

Solution We simply compute the limits at infinity.

```
> limit(f(x),x=infinity);
```
$$1$$

```
> limit(f(x),x=-infinity);
```
$$1$$

∎

DEFINITION 4.8 If $\lim_{x \to \infty} f(x) = L$, then the line defined by $y = L$ is a *horizontal asymptote*.

Such comparisons can be generalized to comparisons with functions other than constants.

Example 4.21 Show that the function defined by the ratio of the two polynomials

```
> p := 3*x^3 + 5*x + 7;
```
$$3x^3 + 5x + 7$$

and

```
> q := 7*x + 3;
```
$$7x + 3$$

behaves asymptotically the same as a quadratic.

Solution Such a comparison is accomplished by taking a ratio of the two functions. If the ratio is constant, then the two functions must be growing at some rate that is directly proportional.

Our first function is

```
> f := makeproc(p/q,x);
```
$$x \mapsto \frac{3x^3 + 5x + 7}{7x + 3}$$

All we know about the second is that it has the general form

```
> g := (x) -> a*x^2 + b*x + c;
```
$$x \mapsto ax^2 + bx + c$$

Their ratio at x is

```
> f(x)/g(x);
```
$$\frac{3x^3 + 5x + 7}{(7x + 3)(ax^2 + bx + c)}$$

In the limit[10] we obtain

```
> Limit(",x=infinity):
> eq := " = value(");
```

$$\lim_{x \to \infty} \frac{3x^3 + 5x + 7}{(7x + 3)(ax^2 + bx + c)} = \frac{3}{7a}$$

The solution is a constant, so indeed we have similar asymptotic behavior. ∎

Note that in the solution to Example 4.21, b and c did not enter into the final answer. In fact, by choosing

```
> a := 7/3: b := 0: c := 0:
```

we find that the equation eq becomes

```
> eq;
```

$$\lim_{x \to \infty} \frac{3}{7} \frac{3x^3 + 5x + 7}{(7x + 3) x^2} = \frac{9}{49}$$

DEFINITION 4.9   Two functions f and g behave identically asymptotically if

$$\lim_{x \to \infty} \frac{f(x)}{g(x)} = 1$$

We have just shown that the functions defined by

```
> f(x);
```

$$\frac{3x^3 + 5x + 7}{7x + 3}$$

and

```
> g(x);
```

$$\frac{7}{3} x^2$$

behave identically asymptotically.

REMARK: For polynomials, only the highest degree term will be involved in any asymptotic investigation as $x \to \infty$.

The fact that we can often ignore all the detail provided by smaller terms in the polynomial makes asymptotic analysis very useful for categorizing the general shape of the graph of a function. Complicated expressions can be explained by comparing them with simple powers. In particular, asymptotes are useful in providing general descriptions of a graph.

[10] Do you remember how to compute such a limit? If not, then check back to Chapter 2.

4.6 Applied Maximum and Minimum Problems

This section contains applications involving the solution of an extremum problem. As you examine each of them, please note the following:

1. Your primary objective is to understand the problem structure and how the tools that we have developed can be used to tackle such problems.
2. The *Maple* commands can provide a useful summary of the significant steps. Their very presence states our goal clearly at each step of the problem analysis without going into elaborate detail.
3. Just as you should not let the computational details get in the way of analyzing the problem, do not let the details of forming the commands interfere. Here it is enough to know that such a command exists and that it can be used with effect. The important information in commands such as

   ```
   > eq1 := V = 4/3*Pi*r^3;
   > D(f)(x);
   ```

   is "eq1 is an equation relating volume and radius" or "evaluate the derivative of f at x." If you know that the command (technique) exists, then you can always look up details such as spelling, placement of commas, and so on when you start to work through new problems based on these models.
4. The details of the computations that arise have all been covered in the earlier chapters. If you do not know how to approach a particular step in the computation, make note of this fact. These "elusive details" are legitimate material for follow-up in a later lesson or self-study. To have recognized a specific weakness is half the battle in mastering the material.
5. Use the graphical, the numerical, and the algebraic information about the problem and cross-check them for consistency. Generally speaking, the graphical representation provides intuition and approximations while the algebraic approach quantifies the relationships in a way that allows us to reuse them in related problems.

Example 4.22 A farmer has 2400 feet of fencing and wants to fence off a rectangular field bordering a straight river. He needs no fencing along the river. What are the dimensions of the field with the largest possible area?

Solution The following equations specify some known relationships between the area A and the perimeter P.

```
> eqn1 := A = x*y;
```

$$A = xy$$

```
> eqn2 := P = 2*x + y;
```

$$P = 2x + y$$

We use the second of these to obtain an expression for y in terms of x.

```
> sol := solve(eqn2,{y});
```

$$\{y = P - 2x\}$$

This information is used to eliminate *y* from the first equation.

> subs(sol,eqn1);

$$A = x(P - 2x)$$

The result is an expression that can be used to compute the area of the field as a function of *x*.

> f := makeproc(rhs("),x);

$$x \mapsto x(P - 2x)$$

Our goal now is to maximize *f* over a finite domain (our field cannot have a negative length of fencing). The graph of *f* is

> P := 2400:

> plot(f,0..2400/2);

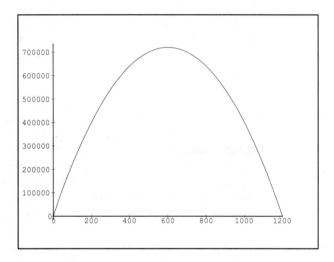

The maximum occurs when

> D(f)(x)=0;

$$2400 - 4x = 0$$

The optimum is at

> sol2 := solve(",{x});

$$\{x = 600\}$$

This solution is unique because the function is concave downward. The area is

> subs(",f(x));

$$720000$$

Example 4.23 A can is being designed to hold one liter of oil. Find the dimensions of the can that will minimize the amount of metal required to make the can.

Solution If we assume we are using sheet metal to make the can, then the amount of metal required will be directly proportional to its surface area.

Equations relating the surface area and volume of the can to its radius and height are

```
> eqn1 :=   A = 2*Pi*r^2 + 2*Pi*r*h;
```

$$A = 2\pi r^2 + 2\pi r h$$

```
> eqn2 :=   V = Pi*r^2*h;
```

$$V = \pi r^2 h$$

Because the volume is constant, we can use the second equation to eliminate either r or h from the first equation. We choose to eliminate h because linear equations are easy to solve.

```
> solve(eqn2,{h});
```

$$\left\{ h = \frac{V}{\pi r^2} \right\}$$

We use this to eliminate h from equation 1.

```
> subs(",eqn1);
```

$$A = 2\pi r^2 + \frac{2V}{r}$$

The result can be used to define area as a function of the radius of the can.

```
> f := makeproc(rhs("),r);
```

$$r \mapsto 2\pi r^2 + \frac{2V}{r}$$

We now need to maximize f. The radius should not be too large or the height will become negative. A plot of f over a fairly restricted range is shown here.

```
> V := 1:
```

```
> plot(f,0..2);
```

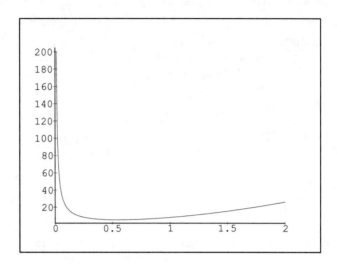

The graph appears to be concave upward with an optimum near $r = 0.5$.

To find the solution algebraically, we look for critical numbers by solving

```
> D(f)(r) = 0;
```

$$4\pi r - \frac{2}{r^2} = 0$$

for r. The result is approximately

```
> fsolve(",r,0..1);
```

$$0.5419260701$$

To prove that this critical number corresponds to a minimum, observe that the function $D(f)$ is monotonically increasing for positive r. Alternatively, note that

```
> D(D(f))(");
```

$$4\pi + 25.13274124$$

is positive. ∎

Often we will want to minimize distances between a point and a curve, or between two lines.

Example 4.24 Find the point on the parabola $y^2 = 2x$ that is closest to [1,4].

Solution Our object is to minimize the quantity distance $d1$ given by

```
> d1 := distance( [1,4],[x,y] );
```

$$\sqrt{(y-4)^2 + (x-1)^2}$$

but subject to the condition that

4.6 Applied Maximum and Minimum Problems

```
> eqn := y^2 = 2*x;
```
$$y^2 = 2x$$

We use *eqn* to reexpress *d1* entirely as a function of one variable. Although we could solve for either *x* or *y*, solving for *x* only requires that we solve a linear equation.

```
> solve(eqn,{x});
```
$$\left\{x = \frac{1}{2}y^2\right\}$$

```
> f := makeproc(subs(",x),y);
```
$$y \mapsto \frac{1}{2}y^2$$

In this way, $[f(y), y]$ denotes a point on the curve defined by *eqn* and

```
> subs(x=f(y),d1):
> d1 := makeproc(",y);
```
$$y \mapsto \sqrt{(y-4)^2 + \left(\frac{1}{2}y^2 - 1\right)^2}$$

redefines *d1* as a function of *y*. The following graph[11] of *d1* over reasonable values of *y* shows that the minimum distance occurs somewhere near $y = 2$.

```
> plot(d1,0..5);
```

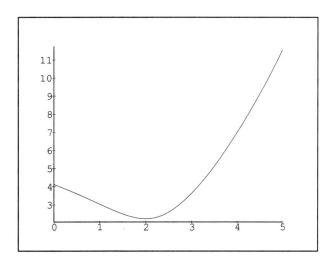

To find the minimum algebraically, we look for the critical points by solving the equation

---

[11] Note that the horizontal axis corresponds to *y* in this plot.

```
> D(d1)(y) = 0;
```

$$\frac{1}{2}\left(2y-8+2\left(\frac{1}{2}y^2-1\right)y\right)\frac{1}{\sqrt{(y-4)^2+\left(\frac{1}{2}y^2-1\right)^2}}=0$$

Although exact solutions exist(try **solve**( )), once again it is sufficient to obtain a numerical approximation.

```
> fsolve(",y);
```

$$2.0$$

The point on the curve closest to [1, 4] is

```
> P := [f("),"];
```

$$[2.0, 2.0]$$

To see that this is reasonable, we plot the graph of the curve and and the line segment joining the point to the curve.[12] To force the x-axis to be horizontal, we resolve the original equation for y, obtaining the two solutions

```
> s := solve( eqn, y );
```

$$\sqrt{2}\sqrt{x},\ -\sqrt{2}\sqrt{x}$$

one for each part of the curve. The graph

```
> pltoptions := x=-1..10,y=-5..5,scaling=constrained:

> plot( { s , PltLine([1,4],P) }, pltoptions );
```

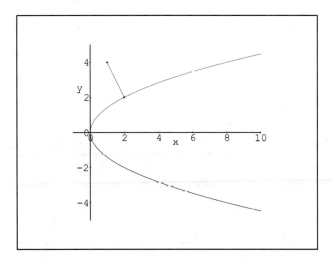

shows the relation of the point P to these two curves. ∎

---

[12] See Appendix A for a definition of the short procedure **PltLine**( ), which constructs a visual representation of line segments.

4.6 Applied Maximum and Minimum Problems 219

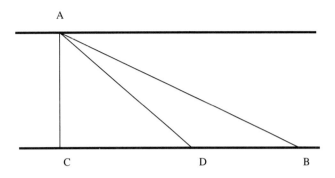

Figure 4.1 Crossing a channel

Example 4.25 A man is at a point A on a bank of a straight channel 3 km wide. What is the fastest route to take to reach a point B, 8 km down the far bank if the man can row at a speed of 6 km/hr and run at a speed of 8 km/hr? Assume that his travel involves some rowing followed by running on the far shore and that there is no current involved.

Solution We can compute the time for each of the two segments of the trip as

$$\text{time} = \frac{\text{distance}}{\text{rate}}$$

Label the point straight across the channel from A as C, and label the actual point of contact on the far shore as D (see Figure 4.1).

If x is the distance from A to B, then the time for each of the segments is

```
> T1 := distance( [3,x],[0,0] ) * 1/6;
```

$$\frac{1}{6}\sqrt{9 + x^2}$$

```
> T2 := distance( [3,8],[3,x] ) * 1/8;
```

$$\frac{1}{8}|8 - x|$$

The total time as a function of x is

```
> T := makeproc( T1 + T2 , x );
```

$$x \mapsto \frac{1}{6}\sqrt{9 + x^2} + \frac{1}{8}|8 - x|$$

To minimize this algebraically, solve for the critical points.

```
> D(T)(x) = 0;
```

$$\frac{1}{6}x\frac{1}{\sqrt{9+x^2}} - \frac{1}{8}\frac{|8-x|}{8-x} = 0$$

```
> s := [ solve(",x)];
```

$$[-\frac{1}{7}\sqrt{81}\sqrt{7}, \frac{1}{7}\sqrt{81}\sqrt{7}]$$

There are two solutions; they can be simplified to

```
> s := simplify(s);
```

$$[-\frac{9}{7}\sqrt{7}, \frac{9}{7}\sqrt{7}]$$

To determine which of these corresponds to a minimum, we simply compare these extrema with what happens at the boundary of the valid domain for x.

```
> T(0), T(s[1]), T(s[2]) , T(8);
```

$$\frac{1}{6}\sqrt{9}+1, \; \frac{1}{42}\sqrt{144}\sqrt{7}+1+\frac{9}{56}\sqrt{7}, \; \frac{1}{42}\sqrt{144}\sqrt{7}+1-\frac{9}{56}\sqrt{7}, \; \frac{1}{6}\sqrt{73}$$

Often such comparisons are easier for approximations.

```
> evalf(["]);
```

$$[1.500000000, 2.181138978, 1.330718914, 1.424000624]$$

■

Example 4.26 Find the area of the largest rectangle that can be inscribed in a semicircle of radius r, as shown in Figure 4.2.

Solution The equation describing the circle is

```
> eqn := x^2 + y^2 = r^2;
```

$$x^2 + y^2 = r^2$$

The rectangle corresponding to a point [x,y] on the circle has area

```
> A := 2*x*y;
```

$$2xy$$

To find A as a function of x, use *eqn* to eliminate y from this formula. (Recall that r is a constant.)

```
> isolate(eqn,y);
```

$$y = \sqrt{|r^2 - x^2|}$$

4.6 Applied Maximum and Minimum Problems

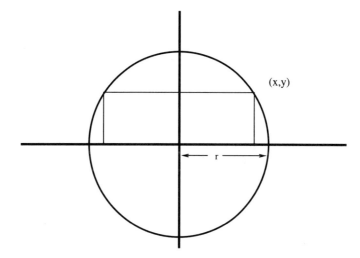

Figure 4.2 An inscribed rectangle

```
> subs(",A);
```
$$2x\sqrt{|r^2 - x^2|}$$

As a function of x, A becomes
```
> A := makeproc(",x);
```
$$x \mapsto 2x\sqrt{|r^2 - x^2|}$$

To maximize A algebraically, we find the critical points.
```
> D(A)(x) = 0;
```
$$2\sqrt{|r^2 - x^2|} - 2x^2\sqrt{|r^2 - x^2|}\left(r^2 - x^2\right)^{-1} = 0$$

```
> solve(",x);
```
$$RootOf(-r^2 + 2\_Z^2), RootOf(-r^2 + 2\_Z^2)$$

The expression **RootOf** is *Maple*'s way of representing the set of all the possible roots of the indicated expression without having to choose one. In this case, both solutions to the **solve( )** command are the same set of roots. A more conventional representation of these solutions is

```
> allvalues("[1]);
```
$$\frac{1}{2}\sqrt{2}\,r, \ -\frac{1}{2}\sqrt{2}\,r$$

only one of which is positive. The corresponding area is
```
> A(r/2*sqrt(2));
```
$$r^2$$

■

222 Chapter 4 Optimal Solutions and Extreme Values

Command	Description
(x) -> if x >= 0 then x^2+1 else x^2 -1 fi;	A piecewise defined function
Limit(x^2+1,x=1,left);	An unevaluated limit
[[4,f(4)],[0,f(0)]];	A list of points
allvalues("[1]);	Converting **RootOf( )**
assign(");	Convert equations to assignments
assume(a<x):	Define assumptions
collect(",n);	Regroup terms of a polynomial
denom(e1);	Bottom part of a quotient
display([p1,p2]);	Redisplay graphics
fsolve(f(x)=-1,x,4..8);	Find a root in an interval
lhs(e1);	Left side of an equation
makeproc(A,B);	A line defined by two points
makeproc(f(x),x);	A curve defined by an expression
maximize(cos(x),{x});	Maximize an expression
minimize(cos(x),{x});	Minimize an expression
numer(e1);	Numerator of a quotient
q := 7*x + 3;	An assignment
readlib(minimize):	Load a library routine
restart: with(student):	Resetting the *Maple* environment
[solve(",x)];	A list of solutions
solve(eqn,y);	Solve an equation
scaling = constrained;	A plot option
showtangent(f(x),x=0,x=-3..3,y=-5..5);	Generate a curve with a tangent line
simplify(s);	Simplify *s*
simplify(",power);	Simplify the exponents
slope(A,B);	Compute the slope a line segment
solve(eqn,y);	Solve an equation
sort(");	Sort a list
subs(sol,eqn1);	Use an equation to replace an object
value(");	Compute the value of an expression
with(student):	Load a package

Table 4.1 Some of the Commands Used in Chapter 4

EXERCISE SET 4

Some of the new commands used in this chapter are shown in Table 4.1. Don't hesitate to use Maple to compute final answers or to plot expressions and generally explore the behavior of a particular function.

The commands that are used summarize what needs to be done to solve selected problems. As you explore the problems, try variations on the problems and verify that the same approach still works, or devise new approaches.

A very important aspect of any computation, by hand or machine, is that the outcome must make sense in light of the other information you know about the problem. Always confirm your algebraic results graphically and numerically. The general patterns represented by the algebra can be discovered or confirmed by checking the outcome for specific instances of the problem.

Throughout, your primary focus should be on recognizing what can be done and what needs to be done to solve a particular problem. After this is mastered, you can reflect on the actual operations represented by those commands and how you would approach the calculations they represent.

1. Find the maximum value and the minimum value of sin(2x). Show that the maximum is achieved at more than one location. Use graphics and **fsolve( )** to find approximate locations for several of these extrema. Is your solution exact or approximate?

2. Given the function definition

   ```
   > f := (x)-> x^2  + 2;
   ```
   $$x \mapsto x^2 + 2$$

   investigate the position and value of the extreme points of the graphs of $f(x + 2)$, $f(x - 2)$, $f(3x)$ and $f(-x)$. How are these related to the extreme point(s) of $f(x)$?

3. Show that following function has neither an absolute maximum nor an absolute minimum.

   ```
   > f := (x)  -> x^3 + 2*x^2 + 5*x + 8 ;
   ```
   $$x \mapsto x^3 + 2x^2 + 5x + 8$$

   Use $D(f)$ to prove that it also has no local extreme points. Does the graph of $f(x+a)$, where a is a constant, have the same properties?

4. Given the function f defined by,

   ```
   > f := (x)  -> 7*x^3-28*x^2+21*x;
   ```
   $$x \mapsto 7x^3 - 28x^2 + 21x$$

   use plotting to locate a local maximum and a local minimum. Use $D(f)$ and **fsolve( )** to locate these extrema more precisely. Experiment with graphs of $f(ax)$, and $f(x+a)$, with the aim of predicting how the value and locations of these extrema are related to those of the function f.

5. Carry out the investigation described in the Exercise 4 on the function g defined by $(x) \to xf(x)$.

6. Given the function f defined by

   ```
   > f := (x)  -> a*x^2 + b*x + c;
   ```

$$x \mapsto ax^2 + bx + c$$

use f and $D(f)$ to construct a system of equations that can be used to find values for a, b, and c so that $f(x)$ has a minimum at $x = 4$ and passes through the points $[0, 3]$ and $[3, 0]$.

7. Find a relative minimum for the function defined by

```
> f := (x) -> abs(x):
```

on the interval $[-2, 2]$.

8. Consider the piecewise-defined function

```
> f := (x) -> if x >= 0 and x < 1 then x^2+3
> elif x > 1 and x <= 2 then -x^2-4
> else undefined  fi:
```

Is f continuous? Does it achieve its maximum on this domain? Generate a plot (perhaps by using **plots[display]( )**) that accurately reflects the continuity of this function on the domain $[-5, 5]$. If the answer to either question asked here is no, modify the definition of f to produce a function for which both answers are yes and generate an appropriate plot.

9. To use Rolle's theorem to guarantee a local maximum or minimum for a function s, we require an a and a b for which $s(a) = s(b)$. Use **solve( )** or **fsolve( )** to find suitable values for a and b for the function s defined by

```
> s := (t)-> 4*t+300-t^2;
```

$$t \mapsto 4t + 300 - t^2$$

10. Generate a plot of the function from $-\infty$ to ∞.

```
> f := (x) -> x^3 - 6*x^2 + 13*x - 11;
```

$$x \mapsto x^3 - 6x^2 + 13x - 11$$

How many solutions to $f(x) = 0$ are there? Locate any roots you find more precisely by graphing f over a more restricted domain. Examine the graph of $D(f)$. Use this information in conjunction with Rolle's theorem to prove that there is only one real solution to $x^3 - 6x^2 + 13x - 11 = 0$.

11. Find a point c in $(0, 2)$ that satisfies

```
> D(f)(c) = slope([1,f(1)],[3,f(3)]);
```

$$3c^2 - 12c + 13 = 2$$

for f defined by

```
> f := (x)-> x^3 - 3*x + 2*x:
```

Let the function g correspond to the line passing through the points $[1, f(1)]$ and $[3, f(3)]$. Find the point x that maximizes $(g - f)(x)$. How does this compare with c?

12. For f defined by

```
> f := (x) -> 3*x^4 - 4*x^3 - 12*x^2 + 5:
```

find where f is increasing and where it is decreasing.

13. Find the local extrema of f.

```
> f := (x) -> x*abs(2-x)^(3/7);
```

$$x \mapsto x|2-x|^{\frac{3}{7}}$$

14. Given that $x > 0$ and $n > 1$, which is bigger: $\dfrac{(2+x)^n}{2^n}$ or $1 + \dfrac{nx}{2}$? To investigate this problem, first generate a graph of the two expressions or their difference.

15. Investigate the convexity of

    ```
    > f := (x) -> x^3 - 3*x + 1;
    ```
    $$x \mapsto x^3 - 3x + 1$$

    by relating the actual convexity to the points of inflection of the graph of f.

16. Find the points of inflection and the critical points of

    ```
    > f := (x) -> x^4-4*x^3+3;
    ```
    $$x \mapsto x^4 - 4x^3 + 3$$

    and use this information to investigate the convexity of f and the nature of any local extrema.

17. Generate a plot of f defined by

    ```
    > f := (x) -> cos(x)^(cos(x));
    ```
    $$x \mapsto \cos(x)^{\cos(x)}$$

    that shows clearly as much of the essential information about this function as possible. Your plot should help you describe the domain and range of the function, as well as the regions where the function is concave upward and downward, the locations of the local extrema, and the symmetry of the graph.

18. Consider the function

    ```
    > f := (x)-> (-55*x^5-37*x^4-35*x^3+97*x^2+50*x+79)
    > /(56*x^2+49*x-37);
    ```
    $$x \mapsto \frac{-55x^5 - 37x^4 - 35x^3 + 97x^2 + 50x + 79}{56x^2 + 49x - 37}$$

    Compare a graph of f over the domain $[-100, 100]$ with that of the cubic $-\frac{55}{56}x^3$. Which grows faster as x tends to ∞? Now compare these two on the domain $[-3, 3]$. Investigate and compare the convexity of the two curves by locating the inflection points.

19. Let f and g be defined by

    ```
    > f := (x) -> (x-3)^2 + 3:
    > g := (x) -> -(x+5)^2 - 5:
    ```

    We can find the minimum distance between two curves by defining two arbitrary points, one on each of the two curves, and then minimizing the distance between these two points.

    ```
    > dist := distance( [x1,f(x1)],[x2,f(x2)] );
    ```

    Use the plot command

    ```
    > plot({f,g});
    ```

    to predict values for $x1$ and $x2$ that will minimize $d1$. If we regard $x1$ as a constant, then $\frac{d}{dx1} dist = 0$ at the minimum. Similarly, if we regard $x2$ as a constant, then

$\frac{d}{dx2}$ *dist* $= 0$ at the minimum. Find suitable values for *x1* and *x2* by solving these two equations simultaneously. Verify that the solution you have found is valid by examining the three dimensional plot of *dist* as a function of *x1* and *x2* near that solution.

20. A rectangular beam is to be cut from a circular log of radius 16 in. If the strength of this beam is given by

    ```
    > strength := (x,y) -> x*y^2;
    ```
    $$(x, y) \mapsto xy^2$$

    where *x* denotes the width and *y* denotes the depth, find the cross-sectional dimensions of the strongest beam that can be cut from this log.

21. Find the point on the parabola $y^2 = 2x$ that is closest to [2,7].

22. Let *x* denote distance down range, and *y* denotes height. If the path of a projectile fired with an initial velocity *v* at an angle of inclination θ above the horizontal can be approximated by the equation

    $$y = tan(\theta)x - \frac{g}{2v^2 \cos^2(\theta)} x^2$$

    where $g = 9.8$ m/sec^2 is the acceleration caused by gravity, find the angle θ that maximizes the range if the terrain is level.

23. A man is at a point *A* on a bank of a straight channel 2 km wide. What is the fastest route to take to reach a point *B*, 10 km down the far bank if the man can row at a speed of 4 km/hr and run at a speed of 8 km/hr? Assume that his travel involves some rowing followed by running on the far shore and that there is no current.

5 Integration

5.1 Introduction

In this chapter we address the problem of finding the areas between curves. Often one or more of the curves will be a straight line, such as the *x*-axis. For example, we may want to find the area bounded by the curve *f* defined by $(x) \to x^2 + 3$, the *x*-axis, and the vertical lines corresponding to $x = 1$ and $x = 5$, as in the graph

```
> f := (x) -> x^2 + 3:

> plot( {f(x),[1,0,1,4],[5,0,5,28]},x=0..6 );
```

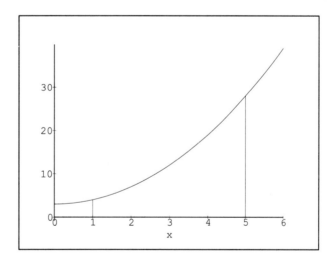

When one of the bounding curves is the *x*-axis, as in this example, we may refer to the area as the area *beneath* the curve *f*.

Because we can compute the area of a rectangle exactly, we can compute the exact area under a curve defined by a constant function. We can also use rectangles to "estimate" the area under a more general curve. For example, given *f* defined by $(x) \to x^2+3$, we can use several equal-width rectangles of varying height, as shown in the following graph.

```
> leftbox( x^2 + 3 , x=1..5 , 4 );
```

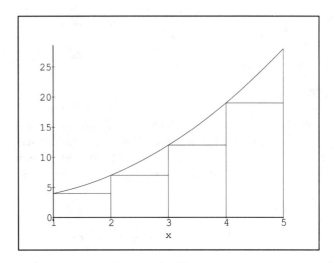

The total area of the four rectangles of width 1 is

> f(1) + f(2) + f(3) + f(4);

$$42$$

The true area in this case is slightly larger than this estimate. Roughly speaking, the more rectangles used, the more accurate will be our estimate of the total area.

5.2 Summations

To be able to work effectively with this method of area approximation, we need to be familiar with the following very powerful notation for summations.

DEFINITION 5.1 Let $x_1, x_2, \cdots, x_k$ be a finite sequence of k x values. Then

$$\sum_{i=1}^{k} f(x_i) = f(x_1) + f(x_2) + \cdots + f(x_k)$$

The Σ indicates our intention to compute a sum. The term $f(x_i)$ is a typical *summand*, and it must assume a value for each value of i. The notation must always specify the following:

1. The variable name i (often called the "summation variable") that is to range over a set of integer values.
2. A lower bound (in this case $i = 1$) for i. This is the first value of i in the sequence of summands.

3. An upper bound (in this case $i = k$) for i. This is the last value of i in the sequence of summands.

In *Maple*, a summation is indicated by the command

```
> Sum( f(x[i]),i=1..k);
```

$$\sum_{i=1}^{k} f(x_i)$$

It too must specify a summand, a summation variable, and a lower and upper bound for the summation variable, i.

We can force *Maple* to attempt to evaluate such unevaluated sums by using the **value( )** command, as in

```
> value(");
```

$$\sum_{i=1}^{k} f(x_i)$$

This particular attempt at evaluation failed, as there was insufficient information available to carry out the computation. Nevertheless, the **Sum( )** command has been transformed into the **sum( )** command — a fact revealed by the command

```
> lprint(");
sum(f(x[i]),i=1..k)
```

REMARK: The **sum( )** command attempts to compute the value of the summation, whereas the inert **Sum( )** simply represents the summation.

The *unevaluated return* involving **sum( )** represents the intended summation in much the same way that **Sum( )** does. Its chief advantage over **Sum( )** is that the computation will proceed further as soon as we have supplied enough additional information and without any additional action on our part. For example, by substituting $k = 4$ into this **sum( )**, as in

```
> subs(k=4,"):
```

the resulting expression, upon being used, simplifies to

```
> ";
```

$$f(x_1) + f(x_2) + f(x_3) + f(x_4)$$

For some simple summands, we can construct a formula for the value of the summation. The formula usually involves the upper and lower bounds on the summation variable but never the summation variable itself. For example,

```
> Sum( i,i=1..k);
```

$$\sum_{i=1}^{k} i$$

evaluates to

```
> value(");
```

$$\frac{1}{2}(k+1)^2 - \frac{1}{2}k - \frac{1}{2}$$

and simplifies to

```
> factor(");
```

$$\frac{1}{2}k(k+1)$$

More generally,

```
> Sum( a*x^2 + b*i + c ,i=1..n);
```

$$\sum_{i=1}^{n} (ax^2 + bi + c)$$

evaluates to

```
> value(");
```

$$ax^2(n+1) + \frac{1}{2}b(n+1)^2 - \frac{1}{2}b(n+1) + c(n+1) - ax^2 - c$$

Such results can always be combined to form solutions to more complicated problems. For example, constants that appear as a factor in the summand can be moved outside the summation[1] so that

```
> restart: with(student):
> Sum(c,i=1..k);
```

$$\sum_{i=1}^{k} c$$

can be rewritten as

```
> expand(");
```

$$c\sum_{i=1}^{k} 1$$

which evaluates to

```
> value(");
```

$$ck$$

Similarly,

```
> Sum( f(i) + g(i),i=1..n)
```

can be written as

[1] We restart *Maple* and load the student package because the special rules for expanding summations are part of the student package.

> expand(");

$$\sum_{i=1}^{n} f(i) + \sum_{i=1}^{n} g(i)$$

The actual techniques for computing such formulas will be developed later. For the moment it suffices to know that such formulas exist.

5.2.1 Summations and Area Under a Curve

Let us reexamine our first attempt at estimating an area under f defined by $(x) \to x^2 + 3$ (see page 227).

Example 5.1 Find the area under the curve

> f := (x) -> x^2 + 3:

from $x = 1$ to $x = 5$.

Solution We originally approximated the area by using a sequence of four equal-width, adjacent rectangles, each one of a height that touched the curve. However, the more rectangles there are, the better the approximation. For example, with 50 rectangles we have

> leftbox(f(x),x=1..5,50);

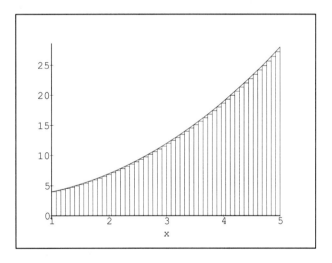

With 50 rectangles, every rectangle is of width

> w := (5-1)/50;

$$\frac{2}{25}$$

and the ith rectangle is of height

> f(1 + i*w);

$$\left(1+\frac{2}{25}i\right)^2+3$$

and has area $wf(1+iw)$. The sum of the areas of all the rectangles is

```
> S := Sum(w*",i=0..49);
```

$$\sum_{i=0}^{49}\left(\frac{2}{25}\left(1+\frac{2}{25}i\right)^2+\frac{6}{25}\right)$$

which evaluates to

```
> value(S);
```

$$\frac{32736}{625}$$

∎

Our next task is to learn how such a computation is carried out.

5.2.2 Rules for Combining Sums

Formulas such as those in the preceding section are constructed from formulas for simpler summations. The following algebraic rules[2] can be used to combine or expand sums. Let $s1$, $s2$, and $s3$ be defined by

```
> s1 := Sum(f(i),i=1..n):
> s2 := Sum(g(i),i=1..n):
```

and

```
> s3 := Sum( f(i) + g(i),i=1..n):
```

We have seen that

```
> s1 + s2 = s3;
```

$$\sum_{i=1}^{n}f(i)+\sum_{i=1}^{n}g(i)=\sum_{i=1}^{n}(f(i)+g(i))$$

We can transform these expressions from one form into the other by using the commands

```
> expand(s3);
```

$$\sum_{i=1}^{n}f(i)+\sum_{i=1}^{n}g(i)$$

---

[2] These rules for **expand( )** are defined only after loading the student package.

or
```
> combine(s1 + s2);
```
$$\sum_{i=1}^{n} (f(i) + g(i))$$

REMARK: When two sums are combined, the range of the summation variable must be identical. Here i goes from 1 to n in both sums.

Similarly, constants can be moved into or out of a sum, as in
```
> combine(a*s1);
```
$$\sum_{i=1}^{n} af(i)$$

and
```
> expand(");
```
$$a \sum_{i=1}^{n} f(i)$$

This leads to the general rule
```
> Sum( a*f(i) + b*g(i) , i = 1..n );
```
$$\sum_{i=1}^{n} (af(i) + bg(i))$$

```
> expand(");
```
$$a \sum_{i=1}^{n} f(i) + b \sum_{i=1}^{n} g(i)$$

and allows us to reexpress the computation of
```
> Sum( (i^2 + 3*i + 2)/n , i=1..n);
```
$$\sum_{i=1}^{n} \frac{i^2 + 3i + 2}{n}$$

in terms of the computation of each of the individual sums
```
> expand(");
```
$$\frac{\sum_{i=1}^{n} i^2}{n} + \frac{3 \sum_{i=1}^{n} i}{n} + \frac{2 \sum_{i=1}^{n} 1}{n}$$

In the next section we investigate how to compute these special sums to obtain
```
> value(");
```
$$\left(\frac{1}{3}(n+1)^3 - \frac{1}{2}(n+1)^2 + \frac{1}{6}n + \frac{1}{6} \right) n^{-1} + 3 \left(\frac{1}{2}(n+1)^2 - \frac{1}{2}n - \frac{1}{2} \right) n^{-1} + 2$$

5.2.3 Formulas for Specific Sums

How would you find the sum of the integers 1 to 10,000 without doing thousands of additions?

```
> sum(i,i=1..10000);
```
$$50005000$$

By using a formula for the more general sum

```
> Sum(i,i=1..n);
```
$$\sum_{i=1}^{n} i$$

we can reduce this computation to a handful of arithmetic operations.[3]

The well-known formula for this sum is

```
> value(");
```
$$\frac{1}{2}(n+1)^2 - \frac{1}{2}n - \frac{1}{2}$$

which, for $n = 10,000$, yields

```
> subs(n=10000,");
```
$$50005000$$

Two important questions that such a result should trigger are "How do we know this is right?" and "How was such a formula obtained?"

Inductive Verification

Verifying a given formula is often easier than deriving the correct formula in the first place. For example, given the formula

```
> Sum(i,i=1..n):
> " = value(");
```
$$\sum_{i=1}^{n} i = \frac{1}{2}(n+1)^2 - \frac{1}{2}n - \frac{1}{2}$$

we are claiming that the value of such a sum as a function of n is

```
> f := makeproc(rhs("),n);
```
$$n \mapsto \frac{1}{2}(n+1)^2 - \frac{1}{2}n - \frac{1}{2}$$

This is certainly correct for small values of n. For example, we have

```
> f(1) = 1;
```
$$1 = 1$$

---

[3] Ask your instructor about Gauss's solution to this problem. The solution we show here is chosen because it provides a systematic approach to such problems.

```
> f(2) = 1 + 2;
```
$$3 = 3$$

```
> f(3) = 1 + 2 + 3;
```
$$6 = 6$$

Now imagine that we have successfuly verified the formula for all values of n up to $n = k$. With very little additional work, we can use this knowledge to verify the formula for $n = k + 1$, without knowing exactly what k is.

Because we have verified that the formula for $f(k)$ is correct, we know that the correct answer for $f(k + 1)$ ought to be

```
> f(k) + (k+1);
```
$$\frac{1}{2}(k+1)^2 + \frac{1}{2}k + \frac{1}{2}$$

How does this compare with the answer predicted by our general formula?

```
> f(k+1);
```
$$\frac{1}{2}(k+2)^2 - \frac{1}{2}k - 1$$

The following command computes the difference of these two solutions and verifies that this difference is 0.

```
> expand(""- ");
```
$$0$$

Thus, our general formula for $f(k + 1)$ is correct, providing that the general formula for $f(k)$ was correct.

The correctness of the formula for $f(k+1)$ has been linked to the correctness of the formula for $f(k)$, even though k has not been specified. We also know that the formula for $f(k)$ was true at least once (i.e., $f(1)$). Thus, we can conclude that the formula is correct for $k \geq 1$. This idea, based on linking the correctness of $f(k + 1)$ to the correctness of $f(k)$, independent of the value of k, is known as *mathematical induction*.

5.2.4 Discovering Formulas

Finding a workable formula for a particular summation can be difficult. One general approach that is used for sums of the type we are dealing with is outlined in this section. It is based on *finite differences*.

Example 5.2 Find a formula for

```
> S := Sum(i,i=1..n);
```
$$\sum_{i=1}^{n} i$$

Solution The key to this solution (as is often the case in problem solving) relies on writing the same expression in two different ways. For the summand i, we start with

```
> (i+1)^2;
```

$$(i+1)^2$$

This can also be written as

```
> expand(");
```

$$i^2 + 2i + 1$$

so that

```
> "" = ";
```

$$(i+1)^2 = i^2 + 2i + 1$$

If we move all terms involving squares to the left-hand side of the equation, then something very special happens as we sum both sides of the equation over the specified range for i.

```
> e1 := " - (i^2=i^2);
```

$$(i+1)^2 - i^2 = 2i + 1$$

```
> e2 := map(Sum,lhs(e1),i=1..n) = Sum(rhs(e1),i=1..n);
```

$$\sum_{i=1}^{n}(i+1)^2 + \sum_{i=1}^{n}-i^2 = \sum_{i=1}^{n}(2i+1)$$

On the left-hand side of this equation, the summands almost cancel. The first sum is just the second sum with the index shifted by one, so all the terms in the overlapping portions of the ranges subtract to leave just the first and the last term. The overall equation simplifies to

```
> value(lhs(e2)) = rhs(e2);
```

$$(n+1)^2 - 1 = \sum_{i=1}^{n}(2i+1)$$

The original unknown summation S is a term in the expansion of the right-hand side and so can be isolated.

```
> expand(");
```

$$n^2 + 2n = 2\sum_{i=1}^{n}i + \sum_{i=1}^{n}1$$

```
> isolate(",S);
```

$$\sum_{i=1}^{n} i = \frac{1}{2}n^2 + n - \frac{1}{2}\sum_{i=1}^{n} 1$$

The remaining sum, on the right-hand side, evaluates to n so our final formula for S is

```
> S = value(rhs("));
```

$$\sum_{i=1}^{n} i = \frac{1}{2}n^2 + \frac{1}{2}n$$

which is usually written in the form

```
> factor(");
```

$$\sum_{i=1}^{n} i = \frac{1}{2}n(n+1)$$

∎

This solution generalizes to any sum of the form

```
> S := Sum( i^k, i=1..n);
```

$$\sum_{i=1}^{n} i^k$$

with almost no additional work.

Example 5.3 Find a formula for

```
> Sum(i^2, i=1..n);
```

$$\sum_{i=1}^{n} i^2$$

Solution We use the same approach as in the previous example. The sum we wish to compute is

```
> k := 2;
```

$$2$$

```
> S;
```

$$\sum_{i=1}^{n} i^2$$

For the summand i^k, we use

```
> (i+1)^(k+1);
```

$$(i+1)^3$$

This can also be written as

> expand(");

$$i^3 + 3i^2 + 3i + 1$$

yielding the equation

> eq := "" = ";

$$(i+1)^3 = i^3 + 3i^2 + 3i + 1$$

We move the highest power terms to the left side and sum both sides of the equation over the chosen range of *i*.

> i^(k+1);

$$i^3$$

> eq - (" = ");

$$(i+1)^3 - i^3 = 3i^2 + 3i + 1$$

> eq2 := map(Sum,",i=1..n);

$$\sum_{i=1}^{n} \left((i+1)^3 - i^3\right) = \sum_{i=1}^{n} \left(3i^2 + 3i + 1\right)$$

As before, the summation on the left is easy to compute because of the cancellation of adjacent terms, so we obtain

> value(lhs(")) = expand(rhs("));

$$(n+1)^3 - 1 = 3\sum_{i=1}^{n} i^2 + 3\sum_{i=1}^{n} i + \sum_{i=1}^{n} 1$$

The desired sum *S* can again be isolated.

> isolate(",S);

$$\sum_{i=1}^{n} i^2 = \frac{1}{3}(n+1)^3 - \frac{1}{3} - \sum_{i=1}^{n} i - \frac{1}{3}\sum_{i=1}^{n} 1$$

This concludes our calculation of a formula for *S* because the right side is expressed entirely in terms of previously computed results.

> value(rhs("));

$$\frac{1}{3}(n+1)^3 + \frac{1}{6} - \frac{1}{2}(n+1)^2 + \frac{1}{6}n$$

This particular formula is often presented in the form

```
> factor(");
```

$$\frac{1}{6} n (n + 1) (2n + 1)$$

■

We can use this solution technique to reexpress a summation of any power i^k in terms of sums of lower powers, and because we have a formula for the case $k = 1$, this constitutes a proof by induction that there exists a formula for any sum of the form

```
> k := 'k':
> S;
```

$$\sum_{i=1}^{n} i^k$$

In this case, the proof that a formula exists also gives us a method of computing the formula. For example, we can now easily compute each of

```
> for k from 3 to 5 do S = factor( value(S) ) od;
```

$$\frac{1}{4} n^2 (n + 1)^2$$

$$\frac{1}{30} n (n + 1) (2n + 1) \left(3n^2 + 3n - 1\right)$$

$$\frac{1}{12} n^2 \left(2n^2 + 2n - 1\right) (n + 1)^2$$

5.3 Area

In this section we first consider areas beneath curves that lie entirely above the x-axis.[4] Specifically, the area to be computed is bounded by the horizontal axis, the curve f, and the vertical lines constructed at the endpoints of a given interval.

5.3.1 An Underestimate of the Area

Recall that for a function such as

```
> f := (x)-> x^2+3*x + 2 ;
```

$$x \mapsto x^2 + 3x + 2$$

[4]Later we find ways to relax this restriction.

an estimate for the area under f from 2 to 3 is obtained by constructing a sequence of rectangles that touch the curve.

```
> leftbox( f(x),x=2..3,10);
```

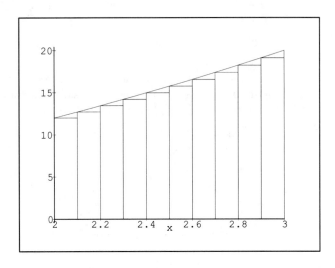

If we use 10 rectangles of equal width over this interval, the width of each is 1/10, and the ith rectangle is of height $f(i + \frac{i}{10})$. The area of the ith rectangle (summing from $i = 0$ to $i = 9$) is

```
> f(2+i/10)/10;
```

$$\frac{1}{10}\left(2 + \frac{1}{10}i\right)^2 + \frac{4}{5} + \frac{3}{100}i$$

The total estimate for the area is

```
> leftsum( f(x),x=0..1,10);
```

$$\frac{1}{10} \sum_{i=0}^{9} \left(\frac{1}{100}i^2 + \frac{3}{10}i + 2\right)$$

which can be rewritten as

```
> expand(");
```

$$\frac{1}{1000} \sum_{i=0}^{9} i^2 + \frac{3}{100} \sum_{i=0}^{9} i + \frac{1}{5} \sum_{i=0}^{9} 1$$

and has the value

```
> u10 := value(");
```

$$\frac{727}{200}$$

By using more rectangles, we can get a better estimate. For example, with 32 rectangles we have

> leftbox(f(x),x=2..3,32);

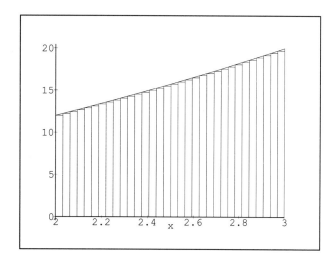

In this case, each rectangle is of width

> w := 1/32;

$$\frac{1}{32}$$

and the *i*th rectangle has an area of

> f(2 + i/32)*1/32;

$$\frac{1}{32}\left(2+\frac{1}{32}i\right)^2 + \frac{1}{4} + \frac{3}{1024}i$$

The total area of these rectangles is

> leftsum(f(x),x=2..3,32);

$$\frac{1}{32}\sum_{i=0}^{31}\left(\left(2+\frac{1}{32}i\right)^2 + 8 + \frac{3}{32}i\right)$$

which evaluates to

> u32 := value(");

$$\frac{32171}{2048}$$

All the rectangles constructed in this example fit completely under the curve.[5] Thus, the area estimates

---

[5]Whether or not these rectangles are over or under the curve very much depends on the shape of the chosen curve. For example, if the curve were monotonically decreasing, the rectangles used by **leftbox( )** would be overestimates.

```
> evalf( [u10,u32]);
```
$$[3.635000000, 15.70849609]$$

obtained in this way are underestimates of the total area. The second estimate is clearly larger than the first and is the better estimate for this particular area.

5.3.2 An Overestimate of the Area

If we choose the height of each rectangle in the previous example so that the right side of each rectangle touches the increasing curve f, the resulting rectangles completely cover the area under the curve and so overestimate the area. This is illustrated by the command

```
> rightbox( f(x),x=2..3,32);
```

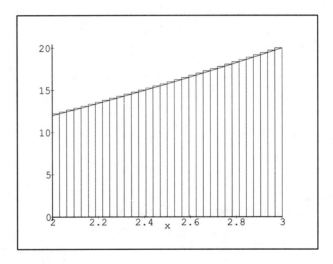

The ith rectangle (summing from $i = 1$ to $i = 32$ instead of from $i = 0$ to $i = 31$) is of height

```
> f(2 + i/32);
```

$$\left(2 + \frac{1}{32}i\right)^2 + 8 + \frac{3}{32}i$$

The total area of the rectangles is

```
> rightsum( f(x),x=2..3,32);
```

$$\frac{1}{32} \sum_{i=1}^{32} \left(\left(2 + \frac{1}{32}i\right)^2 + 8 + \frac{3}{32}i \right)$$

which evaluates to

```
> o32 := value(");
```

$$\frac{32683}{2048}$$

This estimate is an overestimate because the top of each rectangle is entirely above the curve. The exact area lies somewhere in the interval

```
> evalf( u32 < o32);
```

$$15.70849609 < 15.95849609$$

In fact, we already know the correct area to within

```
> evalf( o32 - u32 );
```

$$0.2500000000$$

5.3.3 Better Estimates

The more rectangles we use, the better are our estimates. This suggests using a limiting process. If we use n rectangles in our current example, then a lower bound for the total area is

```
> leftsum(f(x),x=2..3,n);
```

$$\sum_{i=0}^{n-1} \left(\left(2+\frac{i}{n}\right)^2 + 8 + \frac{3i}{n} \right) n^{-1}$$

As complicated as this looks, it really only involves the kinds of summations we evaluated earlier, multiplied by constants.

```
> expand(");
```

$$\frac{12 \sum_{i=0}^{n-1} 1}{n} + \frac{7 \sum_{i=0}^{n-1} i}{n^2} + \frac{\sum_{i=0}^{n-1} i^2}{n^3}$$

Replacing each of these summations by the formulas we developed earlier, we obtain a formula for the area of the rectangles.

```
> un := value(");
```

$$12 + 7\left(\frac{1}{2}n^2 - \frac{1}{2}n\right)n^{-2} + \left(\frac{1}{3}n^3 - \frac{1}{2}n^2 + \frac{1}{6}n\right)n^{-3}$$

This formula depends only on the number of rectangles used. Taking the limit as n tends to infinity, we obtain the lower bound $A1$ for the area under the curve f.

```
> A1 := limit(",n=infinity);
```

$$\frac{95}{6}$$

A similar computation can be made using overestimates. For n rectangles, an upper bound on the total area is

```
> rightsum( f(x),x=2..3,n);
```

$$\sum_{i=1}^{n} \left(\left(2+\frac{i}{n}\right)^2 + 8 + \frac{3i}{n} \right) n^{-1}$$

which evaluates to

> value(");

$$\left(12n + \frac{7}{2}\frac{(n+1)^2}{n} - \frac{7}{2}\frac{n+1}{n} + \frac{1}{3}\frac{(n+1)^3}{n^2} - \frac{1}{2}\frac{(n+1)^2}{n^2} + \frac{1}{6}\frac{n+1}{n^2}\right)n^{-1}$$

Letting n tend to infinity, we obtain the upper bound for the area under the curve f.

> A2 := limit(",n=infinity);

$$\frac{95}{6}$$

Because the upper and lower bound agree and are equal, this must be the true area.

The result of this process of constructing rectangles and computing the corresponding area as the number of rectangles goes to infinity is an example of a *definite integral*. Note that the bases of the rectangles partition the interval $[a, b]$ into a sequence of points $a = x_0 < x_1 < \cdots < x_n = b$, and the only real requirement is that maximum distance between adjacent pairs (x_i, x_{i+1}) tends to 0 as $n \to \infty$.

The basic estimation technique can be used to estimate the area numerically even in cases where we do not know how to compute the sum or the resulting limit.

Example 5.4 Estimate the area under the sin( ) curve from $x = 0$ to π.

Solution A partitioning of the interval into 20 rectangles is given by

> leftbox(sin(x), x=0..Pi, 20);

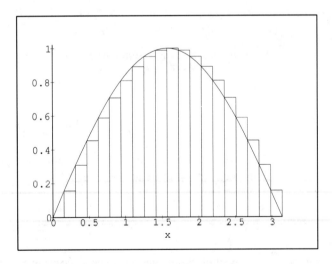

Note that the tops of some of the rectangles are under the curve and the tops of others are above the curve. This makes no difference to the validity of the technique though it does prevent us from drawing any quick conclusions about whether our estimate is over or under the true area.

The total area of these rectangles is

> leftsum(sin(x),x=0..Pi, 20);

$$\frac{1}{20}\pi \sum_{i=0}^{19} \sin(\frac{1}{20}i\pi)$$

which is approximately

```
> evalf(");
```

$$1.995885974$$

If we use 200 equal-width rectangles instead of 20, we obtain

```
> leftsum( sin(x),x=0..Pi,200);
```

$$\frac{1}{200}\pi \sum_{i=0}^{199} \sin(\frac{1}{200}i\pi)$$

```
> evalf(");
```

$$1.999958877$$

To obtain the true answer, we use n rectangles and let $n \to \infty$.

```
> leftsum( sin(x),x=0..Pi,n);
```

$$\pi \sum_{i=0}^{n-1} \sin(\frac{i\pi}{n}) n^{-1}$$

Although the evaluation of this sum goes beyond the techniques we have covered, the actual result is

```
> value(");
```

$$-\pi \sin(\frac{\pi}{n}) n^{-1} \left(\cos(\frac{\pi}{n}) - 1 \right)^{-1}$$

The true area under the curve is the limit

```
> limit(",n=infinity);
```

$$2$$

∎

5.4 The Definite Integral

A complete definition of a *definite integral* takes into account that the set of n rectangles used to approximate the area can be constructed with considerable freedom. All that is really required for a given set of rectangles is that the tops (bottoms if the curve is below the axis) of the rectangles intersect the curve f, the widths of the rectangles *partition* the horizontal distance involved, and, as $n \to \infty$, the maximum width of the rectangles tend to 0. Of course, with a poor choice of rectangles, the sums and the limits might be harder

DEFINITION 5.2 For each n let $\{H_i(n)|i = 1..n\}$ be a collection of n rectangles whose bases are on the x-axis and whose widths partition the interval $[a, b]$. If the maximum widths of the rectangles tend to 0 as $n \to \infty$, and if the following limit exists, then the *definite integral* of f over this interval is defined as

$$\int_a^b f(x)dx = \lim_{n\to\infty} \sum_{i=1}^n (A_i(n)) = \lim_{n\to\infty} \sum_{i=1}^n \text{width}((H_i(n))f(x_i(n)))$$

where $A_i(n) = \text{width}(H_i(n))f(x_i(n))$ and $x_i(n)$ is the x value corresponding to the intersection of the rectangle $H_i(n)$ with the curve f.

REMARK: When $f(x_i(n))$ is positive, the quantity $A_i(n)$ is the actual area of the ith rectangle. When $f(x_i(n))$ is negative, $A_i(n)$ is the negative of the area of the ith rectangle.

The notation $\int_a^b f(x)\,dx$ for the integral must specify

1. The variable x, which corresponds to the base of the rectangles (given by dx)
2. The expression used to define f as a function of x
3. The interval $[a, b]$.

The *Maple* notation for a definite integral is

```
> Int( f(x), x=a..b );
```

$$\int_a^b f(x)\,dx$$

The value of the integral will be the same no matter which set of rectangles are used. In practice, if we need to construct rectangles, we try to construct them in such a way that the sum of their areas and the corresponding limit are as easy to compute as possible.

5.4.1 Curves Above the Axis

For curves that do not cross below the x-axis on the specified region, the value of the definite integral can be used to compute the area between the curve and the x-axis.

Example 5.5 Find the area under the curve f defined by

```
> f := (x) -> x^3 - 6*x^2 + 11*x - 6:
```

between $x = 1$ and $x = 2$.

Solution The roots of $f(x) = 0$ are

```
> solve(f(x)=0);
```
$$1, 2, 3$$

This, and the graph

```
> plot(f,0..4,-2..2);
```

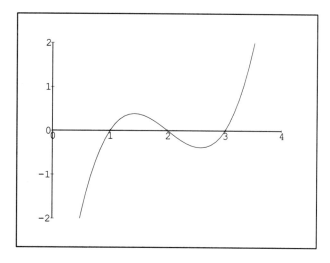

show that the curve f does not go below the x-axis for the entire interval [1, 2]. Thus, the area under the curve on this interval is

```
> Int(f(x),x=1..2);
```
$$\int_1^2 x^3 - 6x^2 + 11x - 6 \, dx$$

This definite integral evaluates[6] to

```
> value(");
```
$$\frac{1}{4}$$

∎

5.4.2 Curves Below the Axis

If the curve is entirely below the x-axis, then the definite integral is equal to the negative of the area between the curve and the x-axis.

Example 5.6 Use rectangles and limits to compute the area between f (defined in Example 5.5) and the x-axis on the interval [2, 3].

---
[6] You can also compute this value by using **leftbox( )** and **limit( )**, as in the next example.

Solution The curve f remains below the x-axis for the entire interval. The command

```
> leftbox(f(x),x=2..3,10);
```

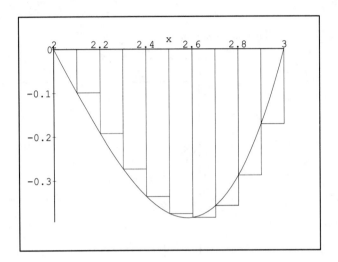

constructs rectangles corresponding to the general summation

```
> leftsum(f(x),x=2..3,n);
```

$$\sum_{i=0}^{n-1}\left(\left(2+\frac{i}{n}\right)^3 - 6\left(2+\frac{i}{n}\right)^2 + 16 + \frac{11i}{n}\right)n^{-1}$$

in the case $n = 10$. In the limit, as $n \to \infty$, this sum evaluates to

```
> value(");
```

$$\left(-\frac{1}{4}n + \frac{1}{4n}\right)n^{-1}$$

```
> limit(",n=infinity);
```

$$-\frac{1}{4}$$

which is the negative of the area between the curve f and x-axis on this interval. The actual area is

```
> -Int(f(x),x=2..3);
```

$$-\int_2^3 x^3 - 6x^2 + 11x - 6 \, dx$$

which evaluates to

```
> value(");
```

$$\frac{1}{4}$$

■

5.4.3 Curves That Cross the Axis

In the next example, we examine what happens if the curve is both above and below the x-axis in the interval [a, b].

Example 5.7 Find the area between the curve f (defined in Example 5.5) and the x-axis over the interval [3/2, 5/2].

Solution You might be tempted to find the solution as

> Int(f(x),x=3/2..5/2);

$$\int_{\frac{3}{2}}^{\frac{5}{2}} x^3 - 6x^2 + 11x - 6 \, dx$$

> value(");

$$0$$

We can see from the graph of the function in Example 5.5 and the results from the previous two examples that zero cannot possibly be the answer.

The following graph shows what happens.

> leftbox(f(x),x=3/2..5/2,10);

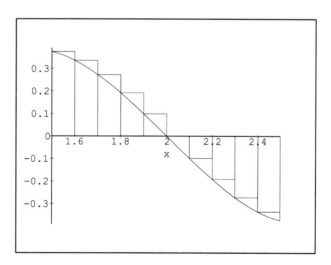

Some of the rectangles contribute a negative value to the estimate for the integral and others contribute a positive value. To find the total area, we break the integration problem into subproblems: one for each region that stays above the x-axis, and one for each region below the x-axis. We have already found the location of the axis crossings in this interval in Example 5.5. They are

> xintercept := solve(f(x) = 0 , x);

$$1, 2, 3$$

250   Chapter 5 Integration

Thus, the total area between the x-axis and the curve f is represented by

```
> Area := Int( f(x),x=3/2..2)
>            -Int( f(x),x=2..5/2);
```

$$\int_{\frac{3}{2}}^{2} x^3 - 6x^2 + 11x - 6\, dx - \int_{2}^{\frac{5}{2}} x^3 - 6x^2 + 11x - 6\, dx$$

which evaluates to

```
> value(Area);
```

$$\frac{7}{32}$$

∎

5.5 Shortcuts in Computation

In the previous section we introduced a method of approximation for finding the area under a curve and showed how to extend it to define the definite integral. When it comes to actual computation, the following should be observed:

1. The method of approximating an area by a sequence of rectangles is important in its own right. Many of the computational methods used to carry out definite integration by machine originate from this basic idea.
2. The computation of definite integrals can be streamlined by capitalizing on algebraic rules that correspond to the separation of an area into subareas.

The algebraic rules that allow us to rewrite integrals of complicated expressions in terms of integrals of simple expressions allow us to compute exact formulas for a wide variety of functions starting from just a few, very fundamental examples.

Most of the properties outlined in the following section are not surprising if you keep in mind that an area should be equal to the sum of its parts and that the underlying summations of rectangles approximating the areas can be regrouped in a large variety of ways.

5.5.1 The Basic Manipulations

Each of the following rules can be verified by defining the integral in terms of limits of summations and then rearranging the summations before taking the limit. In each case, the formula should make sense first of all for the area under the curve of a nonnegative function f. Many of these rules are defined only after loading the student package.

```
> with(student):
```

1. If we divide an area into two parts vertically[7] — the first part of a height defined by the function f and the second part of a height defined by the function g — then the fact that the two areas sum to the total area is represented by the fact that

---
[7] View one part as on top of the other.

5.5 Shortcuts in Computation

```
> Int(f(x) + g(x),x=a..b);
```

$$\int_a^b f(x) + g(x)\, dx$$

and

```
> expand(");
```

$$\int_a^b f(x)\, dx + \int_a^b g(x)\, dx$$

are equivalent.

2. We can also subtract one area from another. For integrals, this is reflected in the fact that

```
> Int(f(x) - g(x) , x=a..b);
```

$$\int_a^b f(x) - g(x)\, dx$$

is equivalent to

```
> expand(");
```

$$\int_a^b f(x)\, dx - \int_a^b g(x)\, dx$$

3. If we double the function value, this will double the area under the curve because

```
> Int(c*f(x),x=a..b);
```

$$\int_a^b cf(x)\, dx$$

expands to

```
> expand(");
```

$$c\int_a^b f(x)\, dx$$

4. We can partition the area below a curve by partitioning the set of x values $[a, b]$ into $[a, c]$ and $[c, b]$. This corresponds to placing a vertical line at $x = c$. In terms of integrals, this corresponds to the fact that

```
> Int(f(x),x=a..b) = Int(f(x),x=a..c) + Int(f(x),x=c..b);
```

$$\int_a^b f(x)\, dx = \int_a^c f(x)\, dx + \int_c^b f(x)\, dx$$

5. For constant functions, the approximating rectangles are exact so the integral is easy to evaluate. The definite integral

```
> Int(c,x=a..b);
```

$$\int_a^b c\, dx$$

can be rewritten as

> expand(");

$$c \int_a^b 1\, dx$$

and evaluates to

> value(");

$$c(b-a)$$

6. If the upper and lower bounds on the integral are reversed, this is interpreted as taking the negative of the width of each rectangle in the approximating sums. As a result, we have

> Int(f(x),x=b..a) = - Int(f(x),x=a..b);

$$\int_b^a f(x)\, dx = -\int_a^b f(x)\, dx$$

In each of the preceding rules where **expand( )** was used, the transformation can be undone by using **combine( )**.

These rules are true for all definite integrals whether or not they correspond directly to the computation of an area. Given exact formulas for a few very basic functions, we can use these rules to efficiently construct exact formulas for integrals of a much wider class of functions, and all without resorting to the computation of closed form formulas for summations or limits. Some examples of such basic formulas are listed here.[8]

> Int(x^k,x=a..b):
> " = value(");

$$\int_a^b x^k\, dx = \frac{b^{k+1}}{k+1} - \frac{a^{k+1}}{k+1}$$

> Int(sin(x),x=a..b):
> " = value(");

$$\int_a^b \sin(x)\, dx = -\cos(b) + \cos(a)$$

> Int(cos(x),x=a..b):
> " = value(");

$$\int_a^b \cos(x)\, dx = \sin(b) - \sin(a)$$

---

[8] In later sections we show how some of these formulas are derived.

```
> Int(sec(x),x=a..b):
> " = value(");
```

$$\int_a^b \sec(x)\,dx = \ln(\frac{1+\tan(b)\cos(b)}{\cos(b)}) - \ln(\frac{1+\tan(a)\cos(a)}{\cos(a)})$$

which simplifies to

```
> simplify(rhs("),trig);
```

$$\ln(\frac{1+\sin(b)}{\cos(b)}) - \ln(\frac{1+\sin(a)}{\cos(a)})$$

```
> Int(csc(x),x=a..b):
> " = value(");
```

$$\int_a^b \csc(x)\,dx = \ln(-\frac{-1+\cot(b)\sin(b)}{\sin(b)}) - \ln(-\frac{-1+\cot(a)\sin(a)}{\sin(a)})$$

```
> Int(tan(x),x=a..b):
> " = value(");
```

$$\int_a^b \tan(x)\,dx = -\ln(\cos(b)) + \ln(\cos(a))$$

```
> Int(cot(x),x=a..b):
> " = value(");
```

$$\int_a^b \cot(x)\,dx = \ln(\sin(b)) - \ln(\sin(a))$$

```
> Int(exp(x),x=a..b):
> " = value(");
```

$$\int_a^b e^x\,dx = e^b - e^a$$

```
> Int(ln(x),x=a..b):
> " = value(");
```

$$\int_a^b \ln(x)\,dx = -b + \ln(b)b + a - \ln(a)a$$

These few examples are already quite effective in conjunction with our algebraic rules for combining them into new integrals.

Example 5.8 Find a value for the integral

```
> Int( 3*x + 5* cos(x) , x = a..b);
```

$$\int_a^b 3x + 5\cos(x)\,dx$$

Solution This integral can be rewritten as

```
> expand(");
```

$$3\int_a^b x\,dx + 5\int_a^b \cos(x)\,dx$$

which, using the stated formulas, evaluates to

```
> value(");
```

$$\frac{3}{2}b^2 - \frac{3}{2}a^2 + 5\sin(b) - 5\sin(a)$$

■

5.5.2 Order Relationships

From the definition of the definite integral in terms of sums, it must follow that if $f(x) \geq 0$ on an interval $[a, b]$, then

```
> Int(f(x),x=a..b) >= 0;
```

$$0 \leq \int_a^b f(x)\,dx$$

and if $g(x) \geq f(x)$ on this interval, then

```
> Int(g(x),x=a..b) >= Int(f(x),x=a..b);
```

$$\int_a^b f(x)\,dx \leq \int_a^b g(x)\,dx$$

This allows us to estimate the value of integrals by comparing the given function with a function for which we know the value of the integral.

Example 5.9 Show that

```
> Int( 2*x^4 + 2 , x=1..3) < Int(4*x^4 + 2 , x=1..3);
```

$$\int_1^3 2x^4 + 2\,dx < \int_1^3 4x^4 + 2\,dx$$

without evaluating the integrals.

Solution We need only compare the given functions on the interval $[1, 3]$. They are

```
> f := (x) -> 4*x^4 + 2:
```

and

```
> g := (x) -> 2*x^4 + 2:
```

On the interval $[1, 3]$, we have $f(x) > g(x)$.

5.5 Shortcuts in Computation

```
> plot( {f,g} , 1..3 );
```

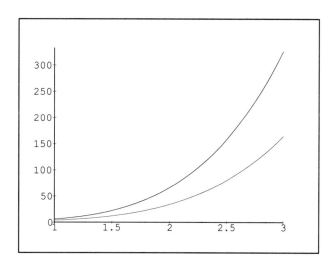

This technique can also be used to estimate the actual value of an integral by comparing a given integral with known ones.

Example 5.10 Estimate

```
> Int(sqrt(x),x=1..5);
```

$$\int_1^5 \sqrt{x}\, dx$$

Solution   The graph

```
> leftbox( sqrt(x),x=1..5,1);
```

Chapter 5 Integration

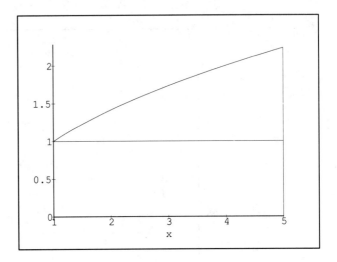

shows that

> `1 <= sqrt(x);`

$$1 \le \sqrt{x}$$

on the entire interval, so the area under the curve *sqrt* satisfies

> `map(Int,",x=1..5);`

$$\int_1^5 1\, dx \le \int_1^5 \sqrt{x}\, dx$$

The left side evaluates to the lower bound

> `value(lhs("));`

$$4$$

Similarly, the graph

> `rightbox( sqrt(x),x=1..5,1);`

5.5 Shortcuts in Computation

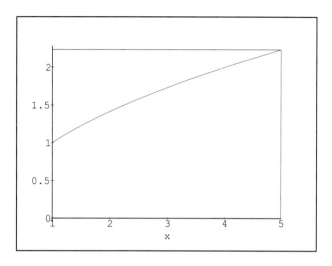

shows that

> `sqrt(x) <= sqrt(5);`

$$\sqrt{x} \leq \sqrt{5}$$

on the entire interval, so we have the relation

> `map(Int,",x=1..5);`

$$\int_1^5 \sqrt{x}\,dx \leq \int_1^5 \sqrt{5}\,dx$$

The right side evaluates to the upper bound

> `value(rhs("));`

$$4\sqrt{5}$$

A better lower estimate is given by the curve following the straight line from $[1, \sqrt{1}]$ to $[5, \sqrt{5}]$. The line is given by the equation

> `L1 := makeproc([1,1],[5,sqrt(5)]);`

$$x \mapsto \frac{1}{4}\sqrt{5}\,x - \frac{1}{4}\sqrt{5} - \frac{1}{4}x + \frac{5}{4}$$

> `plot( {sqrt,L1},1..5);`

Chapter 5 Integration

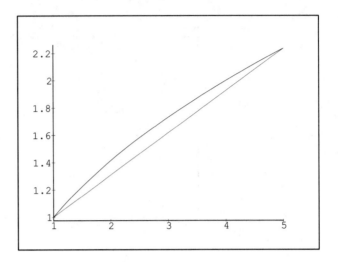

The integral

```
> Int( L1(x) , x=1..5);
```

$$\int_1^5 \frac{1}{4}\sqrt{5}\,x - \frac{1}{4}\sqrt{5} - \frac{1}{4}x + \frac{5}{4}\, dx$$

evaluates to

```
> value(");
```

$$2 + 2\sqrt{5}$$

and is a lower bound on the value of the integral

```
> Int( sqrt(x),x=1..5);
```

$$\int_1^5 \sqrt{x}\, dx$$

■

The summands in the sums used to estimate and define the definite integral are the exact values of integrals of constant functions on the appropriate subintervals. Thus, **leftsum( )** and **rightsum( )** compute the exact values of integrals of step functions and provide upper and lower bounds for integrals of monotonic functions. For example, with

```
> f := (x) -> sqrt(1 + x^2);
```

$$x \mapsto \sqrt{1 + x^2}$$

on the interval [1, 6], we have the lower bound

```
> leftsum(f(x),x=1..6,2);
```

$$\frac{5}{2}\sum_{i=0}^{1}\sqrt{1+\left(1+\frac{5}{2}i\right)^2}$$

which is the exact value of

> Int(f(1),x=1..7/2) + Int(f(7/2),x=7/2..6);

$$\int_{1}^{\frac{7}{2}} \sqrt{2}\,dx + \int_{\frac{7}{2}}^{6} \frac{1}{2}\sqrt{53}\,dx$$

Example 5.11 Show that

> Int(sqrt(1+x^2),x=1..6) > 15;

$$15 < \int_{1}^{6} \sqrt{1+x^2}\,dx$$

Solution From the graph

> plot(sqrt(1+x^2),x=1..6);

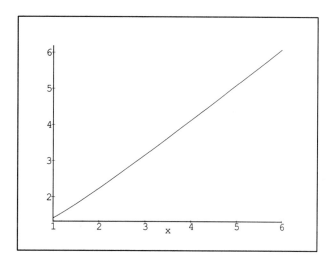

we see that the function is increasing so that

> leftsum(sqrt(1+x^2),x=1..6,n);

$$5\sum_{i=0}^{n-1}\sqrt{1+\left(1+\frac{5i}{n}\right)^2}\,n^{-1}$$

always produces a lower bound.

The value of **leftsum( )** is the exact value of the integral of a step function that is a lower bound for f. It remains to choose a step function that gives an accurate enough estimate. The case $n = 1$ is inappropriate for present purposes because

> value(leftsum(f(x) , x=1..6 , 1));

$$5\sqrt{2}$$

is smaller than 15.

A better lower bound is provided by using more intervals. With four intervals, the value of the integral of the corresponding step function is

```
> leftsum(sqrt(1+x^2),x=1..6,4);
```

$$\frac{5}{4} \sum_{i=0}^{3} \sqrt{1 + \left(1 + \frac{5}{4}i\right)^2}$$

which is approximately

```
> evalf(");
```

$$15.46325615$$

∎

Later we will discover that the exact result for

```
> Int( f(x),x=1..6);
```

$$\int_{1}^{6} \sqrt{1 + x^2}\, dx$$

is

```
> value(");
```

$$3\sqrt{37} + \frac{1}{2}\ln(6 + \sqrt{37}) - \frac{1}{2}\sqrt{2} - \frac{1}{2}\ln(\sqrt{2} + 1)$$

which is approximately

```
> evalf(");
```

$$18.34638395$$

5.6 The Fundamental Theorem of Calculus

Our ability to construct exact formulas for definite integrals depends to a large extent on the known formulas we begin with before combining them algebraically. An important extension to our set of known integrals is made possible by the fact that integration and differentiation are, in some sense, inverse operations.

To see this, we begin by using the operation of definite integration to define a function. We fix the lower endpoint of the definite integral and allow the other end to vary. Each choice of the upper endpoint yields a real number, so by varying x, we obtain the function

```
> A := (x) -> int(f(t),t=a..x):
```

If we reflect on the behavior of A for nonnegative functions and its interpretation in terms of areas, we discover that, at least in the case of $a < c < d$,

```
> A(d) - A(c);
```

$$\int_{a}^{d} f(t)\, dt - \int_{a}^{c} f(t)\, dt$$

is just the area under the curve f from $x = c$ to $x = d$. This is usually written as

> combine(");

$$\int_c^d f(t)\,dt$$

This interpretation is true regardless of the relative positions of a, c, and d, and even if f is allowed to take on negative values. An immediate and useful consequence of this interpretation is Theorem 5.1.

THEOREM 5.1 Let f be continuous on $[a,b]$ and suppose that A is defined by

```
> A := (x) -> int(f(t),t=a..x);
```

$$x \mapsto \int_a^x f(t)\,dt$$

Then for c and d in the interval $[a,b]$,

$$\int_c^d f(t)\,dt = A(d) - A(c)$$

The function A is often called an *antiderivative* of f. The terminology *antiderivative* comes from the following result.

THEOREM 5.2 Let f be continuous on $[a,b]$ and suppose that A is defined by

```
> A := (x) -> int(f(t),t=a..x);
```

$$x \mapsto \int_a^x f(t)\,dt$$

Then $D(A)(x) = f(x)$.

Proof We compute the derivative of A using the limit of the Newton quotient

```
> (A(x+h)-A(x))/h;
```

$$\frac{\int_a^{x+h} f(t)\,dt - \int_a^x f(t)\,dt}{h}$$

Based on our observations regarding addition and subtraction of areas, we can write the numerator of this expression as

```
> num := combine( A(x+h)-A(x) );
```

$$\int_x^{x+h} f(t)\,dt$$

Suppose $h > 0$. Since $f(x)$ is bounded on the region $[x, x+h]$, we have upper and lower rectangles of height m and M that bound the function f from below and above on this interval. (Think of **leftbox( )** and **rightbox( )** for an increasing function.)

Furthermore, m and M can be chosen[9] so that u and v are in $[x, x+h]$ and

---

[9]Take m and M to be the heights of rectangles that touch the curve f.

```
> {m=f(u),M=f(v)};
```
$$\{m = f(u), M = f(v)\}$$

We have
```
> r1 := Int(m,t=x..x+h) <= Int(f(t),t=x..x+h);
```
$$\int_x^{x+h} m\,dt \le \int_x^{x+h} f(t)\,dt$$

and
```
> r2 := Int(f(t),t=x..x+h) <= Int(M,t=x..x+h);
```
$$\int_x^{x+h} f(t)\,dt \le \int_x^{x+h} M\,dt$$

so that
```
> e1 := value(subs(m=f(u),r1/h));
```
$$\frac{f(u)(x+h) - f(u)x}{h} \le \frac{\int_x^{x+h} f(t)\,dt}{h}$$

and
```
> e2 := value(subs(M=f(v),r2/h));
```
$$\frac{\int_x^{x+h} f(t)\,dt}{h} \le \frac{f(v)(x+h) - f(v)x}{h}$$

must hold.

Now as $h \to 0$, $f(u)$ and $f(v)$ tend to $f(x)$, and the derivative of A at x is bounded by
```
> limit(lhs(e1),h=0) <= Limit( rhs(e1) , h= 0);
```
$$f(u) \le \lim_{h \to 0} \frac{\int_x^{x+h} f(t)\,dt}{h}$$

and
```
> Limit( lhs(e2),h=0) <= limit(rhs(e2), h=0 );
```
$$\lim_{h \to 0} \frac{\int_x^{x+h} f(t)\,dt}{h} \le f(v)$$

As $f(u)$ and $f(v)$ become equal in the limit, this proves the result. □

Example 5.12 Consider the increasing function f defined by
```
> f := (x) -> sqrt(x^2 + 2);
```
$$x \mapsto \sqrt{x^2 + 2}$$

5.6 The Fundamental Theorem of Calculus

Define and plot an antiderivative of f on the interval $[1, 3]$ and the Newton quotient at x for that antiderivative with $h = 0.2$. Note how close in value the Newton quotient at x and $f(x)$ are.

Solution An antiderivative is defined by

```
> A := (x) -> int( f(t),t=a..x);
```

$$x \mapsto \int_a^x f(t)\, dt$$

and the Newton quotient of A at x is computed by

```
> NQ := (x) -> 1/h*int( f(t), t= x..x+h );
```

$$x \mapsto \frac{\int_x^{x+h} f(t)\, dt}{h}$$

The function f, the antiderivative A, and the Newton quotient NQ, are all plotted here simultaneously as a function of x on the interval $[1, 3]$.

```
> a := 1; h := 0.2;
```

$$0.2$$

```
> plot( {A,NQ,f} , 1..3 );
```

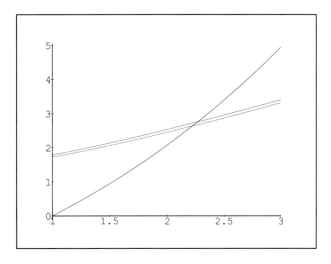

The curves for NQ and for f are almost parallel. To decide which is which, evaluate them at selected points.

```
> evalf( [NQ(2) , f(2)] );
```

$$[2.532003295, 2.449489743]$$

∎

The derivative of an antiderivative of f is f. Thus, in effect, we generate a formula for an antiderivative every time we compute a derivative.

Example 5.13 Show that

```
> Int(cos(x),x=a..b) = sin(b)-sin(a);
```

$$\int_a^b \cos(x)\, dx = \sin(b) - \sin(a)$$

Solution Use the fact that the derivative of sin is cos. ∎

Example 5.14 Show that

```
> Int( x^2,x=a..b) = b^3/3 - a^3/3;
```

$$\int_a^b x^2\, dx = \frac{1}{3}b^3 - \frac{1}{3}a^3$$

Solution Use the fact that the derivative of $x^3/3$ is x^2. ∎

Example 5.15 Given,

```
> f := (x) -> (x-3)/(x^2 + 4);
```

$$x \mapsto \frac{x-3}{x^2+4}$$

and

```
> g := D(f);
```

$$x \mapsto \left(x^2+4\right)^{-1} - \frac{2(x-3)x}{\left(x^2+4\right)^2}$$

show

```
> Int( g(x),x=1..3) = f(3)-f(1);
```

$$\int_1^3 \left(x^2+4\right)^{-1} - \frac{2(x-3)x}{\left(x^2+4\right)^2}\, dx = \frac{2}{5}$$

Solution This is just a restatement of

```
> f := 'f':  g :='g':
> Int( D(f)(x),x=1..3) = f(3)-f(1);
```

$$\int_1^3 D(f)(x)\, dx = f(3) - f(1)$$

with the given definitions for f and g. ∎

5.6.1 Indefinite Integrals

There is no unique antiderivative of $f(x)$. If A is an antiderivative of f, then each value of the constant C gives rise to a different function B, as defined by $(x) \to A(x) + C$. All these functions have derivative f and satisfy

$$\int_a^b f(x)\, dx = B(b) - B(a)$$

The expression

$$\int f(x)\, dx$$

(note the absence of specific bounds on the integral) is often used to denote *an antiderivative* of f. It is also referred to as the *indefinite integral* of $f(x)$ with respect to x. If its value is $A(x)$, then the set of all possible antiderivatives of $f(x)$ is denoted by $A(x) + C$ where C is an arbitrary, unspecified constant.

The equivalent notation in *Maple* is

```
> Int( f(x) , x );
```

$$\int f(x)\, dx$$

The set of all antiderivatives of $f(x)$ with respect to x is

```
> Int( f(x) ,x ) + C;
```

$$\int f(x)\, dx + C$$

These can often be evaluated by using **value( )**, or by using the command **int( )** directly. Later we investigate a variety of techniques for obtaining formulas for integrals directly from the integrand $f(x)$.

Example 5.16 Find a value for the indefinite integral $\int f(x)\, dx$ when f is defined by

```
> f := (x) -> (x-3)^3;
```

$$x \mapsto (x-3)^3$$

Solution For expanded polynomials, we can easily construct a formula whose derivative is the given polynomial. Thus, a suitable value for the indefinite integral

```
> Int(expand(f(x)),x);
```

$$\int x^3 - 9x^2 + 27x - 27\, dx$$

is the antiderivative

```
> a1 := value(");
```

$$\frac{1}{4}x^4 - 3x^3 + \frac{27}{2}x^2 - 27x$$

The set of all possible antiderivatives is

```
> " + C;
```

$$\frac{1}{4}x^4 - 3x^3 + \frac{27}{2}x^2 - 27x + C$$

■

Regardless of how such a formula is obtained, its correctness can be easily verified by differentiating it with respect to x and comparing the result with $f(x)$.

```
> diff(",x);
```

$$x^3 - 9x^2 + 27x - 27$$

```
> expand(" - f(x));
```

$$0$$

REMARK: The choice of the antiderivative constructed by the command **int( )** depends on the form in which $f(x)$ is given. For example, with f as defined in the previous example the solution

```
> Int(f(x),x);
```

$$\int (x-3)^3 \, dx$$

```
> a2 := value(");
```

$$\frac{1}{4}(x-3)^4$$

differs from $a1$ by the constant

```
> expand( a1 - a2);
```

$$-\frac{81}{4}$$

Both formulas are valid antiderivatives of $f(x)$ and, in particular, both

```
> A1 := makeproc( a1 , x );
```

$$x \mapsto \frac{1}{4}x^4 - 3x^3 + \frac{27}{2}x^2 - 27x$$

and

```
> A2 := makeproc( a2 , x );
```

$$x \mapsto \frac{1}{4}(x-3)^4$$

can be used to correctly compute

```
> Int( f(x),x=a..b);
```

$$\int_a^b (x-3)^3 \, dx$$

since

```
> A1(b) - A1(a) = expand( A2(b) - A2(a));
```

$$\frac{1}{4}b^4 - 3b^3 + \frac{27}{2}b^2 - 27b - \frac{1}{4}a^4 + 3a^3 - \frac{27}{2}a^2 + 27a =$$
$$\frac{1}{4}b^4 - 3b^3 + \frac{27}{2}b^2 - 27b - \frac{1}{4}a^4 + 3a^3 - \frac{27}{2}a^2 + 27a$$

Beware of Discontinuities

The proper domains of antiderivatives must be closely observed when using them to evaluate definite integrals. For example, consider the problem of finding the area under the curve defined by

```
> f := (x) -> 1/x^2;
```

$$x \mapsto x^{-2}$$

on the interval $[-1, 1]$, as shown by

```
> plot(f,-1..1,0..10);
```

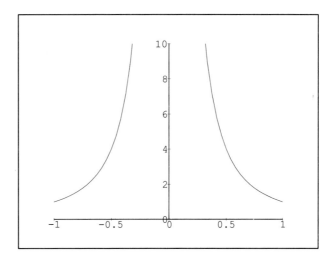

The computation

```
> int(1/x^2,x);
```

$$-x^{-1}$$

yields the valid antiderivative

```
> A := makeproc(",x);
```

$$x \mapsto -x^{-1}$$

but

```
> A(1) - A(-1);
```
$$-2$$

is not the value of

```
> Int(f(x),x=-1..1);
```
$$\int_{-1}^{1} x^{-2}\, dx$$

The vertical asymptote occurring at $x = 0$ hints at the problem. The correct value [10] is

```
> value(");
```
$$\infty$$

REMARK: Before using the fundamental theorem to evaluate a definite integral, make sure that the function f is continuous.[11]

5.7 Applications of the Fundamental Theorem

Before studying in any detail techniques that can be used to construct integrals, we first examine a few of the applications that arise.

Example 5.17 A record is kept of the velocity of a given particle as it moves along a line. It is discovered that this velocity is given by the function

```
> v := (t) -> t^2 - 8*t + 5;
```
$$t \mapsto t^2 - 8t + 5$$

where the time t is given in seconds and velocity is given in meters per second. Find a formula for the position of the particle as a function of time.

Solution Observe the velocity during the first 10 seconds.

```
> plot( v , 0..10 );
```

---

[10] To compute this, consider the positive and negative domains separately, and compute, for example, $\int_a^1 \frac{1}{x^2}\, dx$. Then let $a \to 0^+$.

[11] *Maple* uses the function **iscont( )** to check for continuity.

5.7 Applications of the Fundamental Theorem

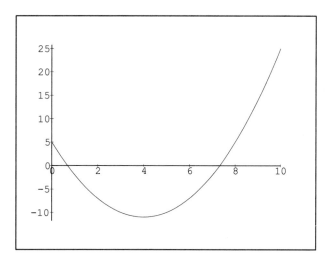

Note that the initial velocity is positive, at time $t = 4$ it is negative, and then it becomes positive again. The particle moves forward, backward, and then forward again.

We know that velocity is the derivative of the position vector as a function of time, so the desired position function is one of the antiderivatives of v.

```
> Int(v(t),t) + C;
```

$$\int t^2 - 8t + 5 \, dt + C$$

```
> value(");
```

$$\frac{1}{3}t^3 - 4t^2 + 5t + C$$

```
> d := makeproc(",t);
```

$$t \mapsto \frac{1}{3}t^3 - 4t^2 + 5t + C$$

■

In the preceding solution, the constant C is still arbitrary. By choosing the constant, we can force the particle to be in a particular position at a given time. For example, to force $d(4) = 0$, we solve the equation

```
> d(4) = 0;
```

$$-\frac{68}{3} + C = 0$$

for C.

```
> isolate(",C);
```

$$C = \frac{68}{3}$$

Example 5.18 Find the total displacement of the particle in Example 5.17 from the time $t = 1$ to time $t = 5$.

Solution When we are asking for displacement, we are comparing the two positions of the particle. At the end of the time interval, we are at position

```
> d(5);
```

$$-\frac{100}{3} + C$$

At the beginning, we are at

```
> d(1);
```

$$\frac{4}{3} + C$$

So the answer is

```
> d(5) - d(1);
```

$$-\frac{104}{3}$$

Note that here it does not matter which antiderivative (i.e., which constant C) is used because the same constant appears in both $d(5)$ and $d(1)$. ∎

REMARK: The total displacement in the preceding example is the definite integral

```
> Int( v(t),t=1..5);
```

$$\int_1^5 t^2 - 8t + 5 \, dt$$

This automatically takes into account the fact that the total displacement decreases when the velocity is negative.

Example 5.19 Find the total displacement of a particle from $t = 0$ to $t = 2$ if its velocity at time t (in seconds) is given by

```
> v := (t) -> exp(t^3)/(t+1);
```

$$t \mapsto \frac{e^{t^3}}{t+1}$$

5.7 Applications of the Fundamental Theorem

Solution A graph of the velocity of the particle is shown here.

> plot(v,0..2);

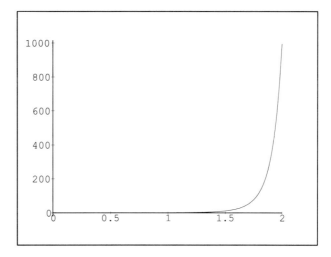

The total displacement is the area under this curve from $t = 0$ to $t = 2$. It is the definite integral

> Int(v(t),t=0..2);

$$\int_0^2 \frac{e^{t^3}}{t+1} \, dt$$

but *Maple* is unable to obtain an exact value for this.

> value(");

$$\int_0^2 \frac{e^{t^3}}{t+1} \, dt$$

Although a proof that no antiderivative exists goes well beyond the scope of this course, there really is no antiderivative[12] of $v(t)$ so that this integral cannot be easily represented as

> A(2) - A(0);

$$A(2) - A(0)$$

We can still approximate the definite integral by rectangles. The approximation

---

[12] Here we are restricting the allowable formulas to those involving only quotients of polynomials, logarithms, and exponentials.

```
> leftbox( v(t),t=0..2,20);
```

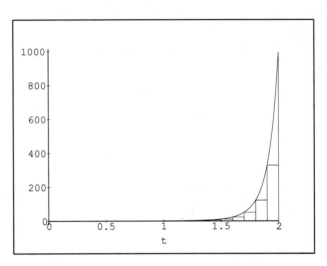

gives a lower bound of

```
> leftsum( v(t),t=0..2,20);
```

$$\frac{1}{10} \sum_{i=0}^{19} e^{\frac{1}{1000}i^3} \left(\frac{1}{10}i+1\right)^{-1}$$

```
> evalf(");
```

$$56.02024444$$

By using 200 rectangles we obtain the better lower bound

```
> evalf( leftsum( v(t), t=0..2,200) );
```

$$91.38966162$$

Using *Maple*'s sophisticated built-in approximation routines, we find that the true answer is closer to

```
> evalf( Int( v(t) , t=0..2 ));
```

$$96.25633931$$

∎

REMARK: In Example 5.19, the function

```
> A := (t) -> int( v(u),u=0..t);
```

$$t \mapsto \int_0^t \frac{e^{u^3}}{u+1}\, du$$

is still an antiderivative of v. In particular, it is still the case that its derivative is

```
> diff(A(t),t);
```

$$\frac{e^{t^3}}{t+1}$$

This all works even though we have no simple formula for $A(x)$ in terms of familiar functions.

EXERCISE SET 5

Some of the important commands that have been used in this chapter are shown in Table 5.1. Keep them in mind as you explore the examples from the chapter and the exercises that are listed here.

1. Find the area under the curve

   ```
   > f := (x) -> 3*x^2 + x + 3:
   ```

   from $x = 2$ to $x = 6$ by carrying out the following steps.
   a. Use the command **leftbox( )** to generate graphs showing rectangular approximations to the total area using 4, 8, and 64 intervals.
   b. For each of the previous approximations, calculate the sum of the areas of the rectangles by using the command **leftsum( )**, and find the value of the resulting summations. Find the area of the 3rd, the 7th, and the 43rd rectangles of the respective sums.
   c. Construct the summation representing the area of the rectangles in the case where n rectangles are used. You can do this by using the command **leftsum( )**. What is the area of the of the kth rectangle?
   d. Obtain a formula in terms of n for the sum of the area corresponding to n rectangles. Compute the limit of this value at $n = \infty$.

   How accurate are the estimates based on 4, 8, and 64 rectangles? Are they overestimates or underestimates?

2. Repeat the steps outlined in Exercise 1, but use the commands **rightbox( )** and **rightsum( )**.

3. A better estimate for the area under the curve f can be obtained by averaging the result of **rightsum( )** and **leftsum( )**. Compute this estimate for the area specified in Exercise 1 in each of the cases of 4, 8, and 64 rectangles.

4. Use rectangles to estimate the area under the cosine curve from $x = -\pi/2$ to $\pi/2$. How many rectangles are required before the estimate is correct to within 0.001? (You can obtain the correct result by using the **int( )**.)

5. Reexpress the summation

   ```
   > Sum(2*i^2 + 3*i + 1,i=1..n);
   ```

   $$\sum_{i=1}^{n} (2i^2 + 3i + 1)$$

Chapter 5 Integration

Command	Description
(t)->exp(t^3)/(t+1);	A function definbfion
combine(s1+s2);	Regroup terms of a sum or product
diff(A(t),t);	differentiate an expression
evalf(Int(v(t),t=0..2));	Approximate an integral
evalf(leftsum(v(t),t=0..2,8));	Approximate a sum
expand(s3);	Expand an algebraic expression
factor(e1);	Factor an algebraic expression
int(1/x,x);	Evaluate an antiderivative
Int(c*f(x),x=a..b);	Represent a definite integral
isolate(",C);	Solve for C
leftbox(f(x),x=1..5,50);	Graph an approximation to an integral
leftsum(f(x),x=1..6,n);	Construct the sum corresponding to an n step approximation
limit(f(n),n=infinity);	Compute a limit
Limit(f(n),n=1);	Represent a limit
makeproc(f(t),t);	Construct a procedure from an expression
makeproc([1,1],[5,7]);	Construct a procedure for the line defined by two points
map(Int,",x=1..5);	Integrate term by term
plot({A,NQ,f},1..3);	Plot a set of functions
plot({f(x),[5,0,5,28]},x=0..6);	Plot an expression and a line segment
restart:with(student):	Reset Maple to its starting state
rhs(a=b);	The right-hand side of an equation
rightbox(f(x),x=2..3,32);	Graph an approximation to an integral
rightsum(f(x),x=2..3,n);	Construct the sum corresponding to an n step approximation
value(e1);	Evaluate an inert form
simplify(e1,trig);	Simplify using trig identities
solve(f(x)=0);	Solve an equation
solve(f(x)=0,x);	Solve an equation for x
Sum(f(x[i]),i=1..k);	Represent a k term sum
sum(i,i=1..10000);	Compute a sum
value(leftsum(f(x),x=1..6));	Find the exact value of an integral approximation

Table 5.1 Some of the Commands Used in Chapter 5

in terms of the summations

```
> Sum( i^2,i=1..n), Sum(i,i=1..n);
```

$$\sum_{i=1}^{n} i^2, \sum_{i=1}^{n} i$$

Find its value.

6. Mathematical induction can be used to prove that the value of

```
> S1 := Sum(i^4 + i^5,i=1..n);
```

$$\sum_{i=1}^{n} \left(i^4 + i^5\right)$$

is correctly computed by the function

```
> s := (n) -> 1/60*n*(n+2)*(n+1)*(10*n^3+12*n^2-n-1);
```

$$n \mapsto \frac{1}{60} n(n+2)(n+1)\left(10n^3 + 12n^2 - n - 1\right)$$

Proceed as follows:
a. Verify that the formula is correct at least once. For example, check the results for $n = 1$ or $n = 2$.
b. Verify that $s(n+1) - s(n)$ increases by the value of the $(n+1)$st term of the sum.

7. Use the techniques discussed in this chapter to express the sum

```
> Sum( i^4,i=1..n);
```

$$\sum_{i=1}^{n} i^4$$

in terms of

```
> Sum( i^3,i=1..n), Sum(i^2,i=1..n), Sum(i,i=1..n);
```

$$\sum_{i=1}^{n} i^3, \sum_{i=1}^{n} i^2, \sum_{i=1}^{n} i$$

Find its value in terms of n.

8. Express the area under the curve f defined by

```
> f := (x) -> x^3-9*x^2+26*x-24:
```

from $x = 2$ to $x = 3$ in terms of integrals, and find its value.

9. Express the area between the curve f defined by

```
> f := (x) -> x^3-9*x^2+26*x-24:
```

and the x-axis, from $x = 1$ to $x = 5/2$, in terms of integrals and find its value.

10. Find the exact area between the curves f and g defined by

```
> f := (x) -> x^3-9*x^2+26*x-24:
> g := (x) -> -x^3+9*x^2-26*x+22:
```

from $x = 2$ to $x = 4$, and from $x = 1$ to $x = 4$. Verify your answer by approximating this value using either **leftbox( )** or **evalf( )**.

11. Approximate the value of the integral

```
> Int(1/x,x=1..12);
```

$$\int_1^{12} x^{-1}\,dx$$

by using rectangles and also **evalf( )**. Compare your answer with ln(12).

12. Rewrite

```
> Int( 7*x + 2*sin(x) , x = a..b);
```

$$\int_a^b 7x + 2\sin(x)\,dx$$

as a sum of integrals.

13. Show that

```
> Int( x^5 + x , x=2..4) < Int(2*x^5 + x , x=2..4);
```

$$\int_2^4 x^5 + x\,dx < \int_2^4 2x^5 + x\,dx$$

without evaluating the integrals.

14. Show that the value of the integral

```
> Int((x+2)^(3/2),x=1..5);
```

$$\int_1^5 (x+2)^{\frac{3}{2}}\,dx$$

is at least 20, by comparing this integrand with $x^{3/2}$.

15. The value of the integral

```
> i1 := Int( cos(cos(x^2)),x=0..4);
```

$$\int_0^4 \cos(\cos(x^2))\,dx$$

is approximately

```
> evalf(");
```

2.952914758

How many rectangles are required before the command **leftsum( )** obtains this answer accurately to three decimal places? Examine a plot of this integrand to see why so many are needed.

16. Consider the increasing function f defined by

```
> f := (x) -> sqrt(3*x^2 + 5);
```

$$x \mapsto \sqrt{3x^2 + 5}$$

Define and plot an antiderivative of f on the interval $[2,4]$ and the Newton quotient for that antiderivative with $h = 0.2$. Experiment with different values of h to discover how the size of h affects the accuracy of the approximation.

17. By considering the value of

    > e1 := Diff(cos(cos(x^2)),x);

    $$\frac{d}{dx}\cos(\cos(x^2))$$

    and its relationship to the integrand of

    > i1 := Int(a*sin(cos(x^2))*sin(x^2)*x, x=c..d);

    $$\int_c^d a\sin(\cos(x^2))\sin(x^2)x\,dx$$

    express the value of $i1$ in terms of $e1$.

18. Use

    > diff(u^6,u);

    $$6u^5$$

    to construct an antiderivative for

    > f := (x) -> (2*x-5)^5;

    $$x \mapsto (2x-5)^5$$

19. A record is kept of the velocity of a given particle as it moves along a line. It is discovered that this velocity is given by the function

    > v := (t) -> t^2 - 10*t + 14;

    $$t \mapsto t^2 - 10t + 14$$

    where the time t is given in seconds (s) and velocity is given in m/s. Find a formula for the position of the particle as a function of time.

20. If

    > v := (t) -> t^2 - 10*t + 34;

    $$t \mapsto t^2 - 10t + 34$$

    find the total displacement of the particle from the time $t = 1$ to time $t = 5$.

21. Find the total displacement of a particle from $t = 0$ to $t = 2$ if its velocity at time t (in seconds) is given by

    > v := (t) -> exp(-t^3)/(t+1);

    $$t \mapsto \frac{e^{-t^3}}{t+1}$$

6 Applications of Integration

6.1 Areas Between Curves

The rectangles that were used to approximate areas beneath a curve also approximate the area between two curves. The height of each rectangle is the positive difference of the two functions.

Example 6.1 Let f and g be defined by

```
> f := (x) -> sin(x/2);
```

$$x \mapsto \sin(\frac{1}{2}x)$$

```
> g := (x) -> 1/x^2;
```

$$x \mapsto x^{-2}$$

and find the area between these two curves from $x = \frac{3}{2}$ to $x = 2$.

Solution The region under consideration is shown by the command

```
> plot( {f,g},3/2..3);
```

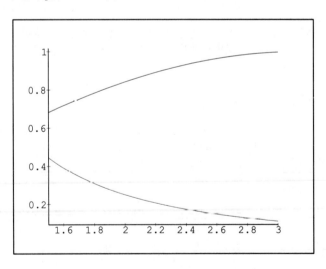

We have $f(x) > g(x)$ in this region, so the area is given by the definite integral

```
> Int( f(x) - g(x),x=3/2..3);
```

$$\int_{\frac{3}{2}}^{3} \sin(\frac{1}{2}x) - x^{-2} \, dx$$

This evaluates to

```
> value(");
```

$$-2\cos(\frac{3}{2}) - \frac{1}{3} + 2\cos(\frac{3}{4})$$

which is approximately

```
> evalf(");
```

$$0.9885700014$$

∎

REMARK: *Maple* uses the antiderivative

```
> Int( f(x)-g(x),x);
```

$$\int \sin(\frac{1}{2}x) - x^{-2}\, dx$$

to compute the preceding exact value. The techniques required to construct such antiderivatives will be discussed later.

Sometimes the two curves cross in the region under consideration. This happens for f and g in the interval [1, 3], at the x value given by

```
> a := fsolve( f(x) - g(x) = 0 , x , 1..2);
```

$$1.289794951$$

When this happens, we use the absolute value function to force each rectangle in the approximating sum to contribute a positive value to the total sum.

Example 6.2 Find the area between f and g from $x = 1$ to $x = 3$.

Solution The region is shown by the graph

```
> plot( {f,g},1..3);
```

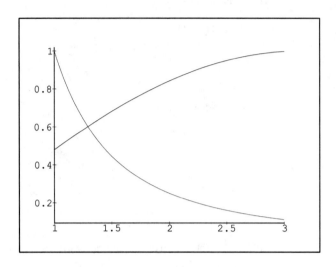

Because the curves cross, we use absolute values. The area is

```
> Int( abs( f(x) - g(x) ) , x = 1 .. 3 );
```

$$\int_1^3 \left| \sin(\tfrac{1}{2}x) - x^{-2} \right| dx$$

which has an approximate value[1] of

```
> evalf(");
```

$$1.082705333$$

∎

REMARK: We can also compute this area as

```
> - Int( f(x) - g(x) , x=1..a )
> + Int( f(x) - g(x) , x=a..3);
```

$$-\int_1^{1.289794951} \sin(\tfrac{1}{2}x) - x^{-2} \, dx + \int_{1.289794951}^3 \sin(\tfrac{1}{2}x) - x^{-2} \, dx$$

This evaluates (using the fundamental theorem of calculus) to

```
> value(");
```

$$2.979344860 - 2\cos(\tfrac{1}{2}) - 2\cos(\tfrac{3}{2})$$

If we knew the value of the root a of $f(x) - g(x) = 0$ exactly, then this answer would be exact. Nevertheless, it is a good approximation, as shown by the command

[1] For now, think of **evalf(Int(...))** as equivalent to approximating the area by rectangles.

```
> evalf(");
```
$$1.082705333$$

Example 6.3 Find the exact area between the sine and cosine curves on on the domain $[0, \frac{\pi}{2}]$.

Solution A graph of these two curves is given by

```
> plot({sin,cos},0..Pi/2);
```

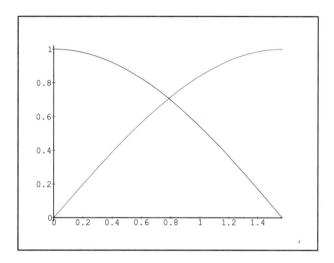

The area between them is

```
> A1 := Int( abs( cos(x) - sin(x)),x=0..Pi/2);
```

$$\int_0^{\frac{1}{2}\pi} |\cos(x) - \sin(x)| \, dx$$

which is approximately

```
> evalf(");
```

$$0.8284271247$$

To obtain an exact solution, we first find the exact location where the two curves cross, as given by

```
> solve( cos(x) - sin(x) = 0 ,{x} );
```

$$\left\{ x = \frac{1}{4}\pi \right\}$$

We use it to rewrite the integral for the total area as

```
> Area := Int( cos(x) - sin(x) ,x = 0..Pi/4 )
> + Int( sin(x) - cos(x) , x= Pi/4..Pi/2 );
```

$$\int_0^{\frac{1}{4}\pi} \cos(x) - \sin(x)\, dx + \int_{\frac{1}{4}\pi}^{\frac{1}{2}\pi} -\cos(x) + \sin(x)\, dx$$

Both of these integrals can be computed exactly by using the antiderivative

```
> int( sin(x)-cos(x) ,x  );
```
$$-\sin(x) - \cos(x)$$

If F is defined by

```
> F := makeproc(",x);
```
$$x \mapsto -\sin(x) - \cos(x)$$

then the exact area is

```
> F(Pi/4) - F(0) - ( F(Pi/2) - F(Pi/4));
```
$$-2\sqrt{2} + 2$$

∎

Example 6.4 Find the area between the curves defined by the equations

```
> eq1 := y = x-1; eq2 := y^2 = 2*x + 6;
```
$$y = x - 1$$
$$y^2 = 2x + 6$$

Solution Until now, we have worked through all the details required to plot curves defined by equations such as $y^2 = 2x + 6$. Now, we are interested primarily in using these computational tools to study the problem at hand. To this end, we introduce a new command **implicitplot( )**, which can be used to generate graphs directly from the defining equation in one step. These two curves are shown directly by

```
> with(plots):

> implicitplot( {eq1,eq2} , x=-4..6,y=-4..4);
```

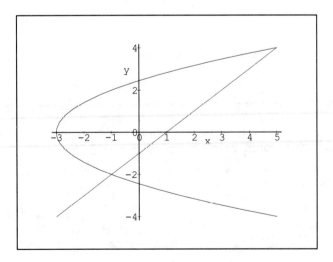

They intersect at the x values given by

6.1 Areas Between Curves

```
> sols := [solve( {eq1,eq2})];
```

$$[\{y = -2, x = -1\}, \{x = 5, y = 4\}]$$

We name these x values $x1$ and $x2$.

```
> x1 := -1: x2 := 5:
```

The rightmost point of the region occurs at $x2 = 5$.

To find the leftmost point of the region, we first regard x as a function of y in $eq2$ by writing equation $eq2$ in the form

```
> isolate(eq2,x);
```

$$x = \frac{1}{2}y^2 - 3$$

We can find the minimum x value by minimizing the right-hand side of this equation. Once again, to focus on the problem at hand, we introduce a new *Maple* command **extrema( )**, which finds the local extreme values in a single step. The first argument to **extrema** is the expression to be optimized. The second argument indicates a set of additional constraints — in this case, an empty set. The last argument indicates the variable that has been used to define the expression. The command

```
> extrema(rhs("),{},y);
```

$$\{-3\}$$

reports the values of the local extreme values. In this case there is only one local extreme value. From the other things that we already know about this problem, we conclude that this must be the minimum possible value for x.

We are now ready to express the solution to this problem as two integrals. On the domain $[-3, -1]$, the two solutions to

```
> solve( eq2 , y );
```

$$\sqrt{2}\sqrt{x+3},\ -\sqrt{2}\sqrt{x+3}$$

define the upper and lower boundaries of the region. The positive distance between these two boundaries is

```
> f1 := (x) -> 2*sqrt(2*(x+3));
```

$$x \mapsto 2\sqrt{2x+6}$$

and the area of the region between these two curves on this domain is

```
> I1 := Int( f1(x),x=3..1);
```

$$\int_3^1 2\sqrt{2x+6}\,dx$$

On the domain $[-1, 5]$, the upper curve is

```
> u2 := (x) -> sqrt(2*(x+3));
```

$$x \mapsto \sqrt{2x+6}$$

284 Chapter 6 Applications of Integration

and the lower curve is the straight line

```
> l2 := makeproc( [-1,-2], [5,4] );
```
$$x \mapsto x - 1$$

The area between these two curves on this domain is

```
> I2 := Int( u2(x) - l2(x) , x=-1..5);
```
$$\int_{-1}^{5} \sqrt{2x+6} - x + 1 \, dx$$

The total area is

```
> value( I1 + I2 );
```
$$-16\sqrt{3} + \frac{32}{3}\sqrt{2} + \frac{38}{3}$$

■

There is a better approach to this problem. By interchanging the roles of x and y in the cartesian plane so that x is regarded as function of y, we can greatly simplify the above solution.

Example 6.5 Find the area between the curves defined by the equations

```
> eq1 := y = x-1:
> eq2 := y^2 = 2*x + 6;
```
$$y^2 = 2x + 6$$

by regarding x as a function of y.

Solution With the role of x and y interchanged, the graph becomes

```
> implicitplot( {eq1,eq2}, y=-4..4, x=-4..6);
```

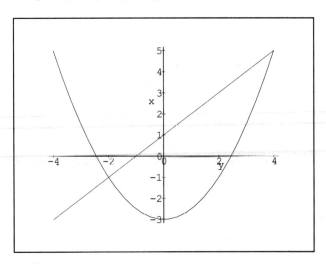

For these equations, it is easy to write x as a function of y. We obtain the equations

```
> map( isolate , {eq1,eq2} , x );
```

$$\left\{ x = y + 1, x = \frac{1}{2}y^2 - 3 \right\}$$

The upper curve is defined by

```
> u3 := (y) -> y + 1 ;
```

$$y \mapsto y + 1$$

and the lower curve is defined by

```
> l3 := (y) -> y^2/2 - 3;
```

$$y \mapsto \frac{1}{2}y^2 - 3$$

and the integral for the area between these curves is

```
> I3 := Int( u3(y) - l3(y) ,y = -2 .. 4 );
```

$$\int_{-2}^{4} y + 4 - \frac{1}{2}y^2 \, dy$$

```
> value(");
```

$$18$$

∎

REMARK: Always choose the roles of x and y so that the upper and lower curves are as easy as possible to describe.

6.2 Volume

A three-dimensional object can be described in Cartesian coordinates as a collection of points. For example, a sphere of radius r centered at the origin is the set of points

$$\{(x, y, z) | x^2 + y^2 + z^2 \leq r^2\}$$

The surface of this object corresponds to the points $\{(x, y, z) | x^2 + y^2 + z^2 = r^2\}$. For a specific radius, such as $r = 1$, this sphere can be displayed in three dimensions by the commands

```
> with(plots):
> eq1 := x^2 + y^2 + z^2 = 1:
> opts := x=-1..1, y=-1..1, z=-1..1, axes=boxed:
```

```
> implicitplot3d( eq1 , opts );
```

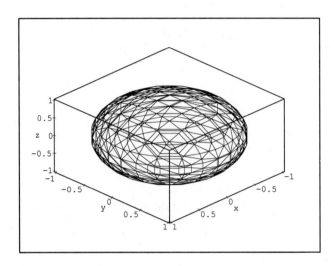

When computing the area between two curves, we approximated the area by constructing a sequence of rectangles orthogonal to the horizontal axis whose heights were determined by the two curves. Similarly, the volume of a three-dimensional object can be approximated by a sequence of thin layers, or cross sections, orthogonal to one of the axis. If the layers are orthogonal to the x-axis, then each layer corresponds to a specific value of x. For example, a layer corresponding to $x = \frac{1}{2}$ is suggested by the graph

```
> implicitplot3d({eq1, x=1/2},opts );
```

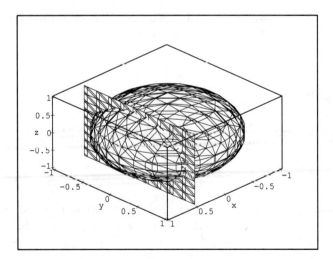

The volume of such a layer[2] is its thickness multiplied by its surface area.

---

[2] We force the edges of the layer to be orthogonal to the layer.

Assume that the points of the given object all have x values in the interval $[a, b]$. Slice the object into layers orthogonal to the x-axis so that the boundary of the ith layer is the intersection of the plane at, say, $x = x_i^*$ with the surface of the object, and the thickness of the layer is Δx_i.

The chosen layers partition $[a, b]$, and the more layers there are, the thinner they tend to be. If P is that partitioning, then denote the maximum thickness of the layers of P by $||P||$.

DEFINITION 6.1 Let S be a solid that lies between the planes P_a and P_b and let the cross-sectional area of S in the plane P_x be $A(x)$. If the limit

$$V = \lim_{||P|| \to 0} \sum_{i=1}^{n} A(x_i^*) \Delta x_i = \int_a^b A(x)\, dx$$

exists, then it is the volume of S.

REMARK: If you graph the cross-sectional area of the object, as it changes with x, then the *volume* of the object is the *area* under the resulting curve.

We have complete freedom as to how to orient the object in 3-space. A good choice can make it much easier to find the cross-sectional area as a function of x and may result in a much easier definite integral.

Example 6.6 Find the volume of a sphere of radius r.

Solution The surface of a sphere of radius r centered at the origin is given by

```
> eq1 := x^2 + y^2 + z^2 = r^2;
```
$$x^2 + y^2 + z^2 = r^2$$

The cross section corresponding to any specific value of x satisfies

```
> isolate(eq1,y^2);
```
$$y^2 = r^2 - x^2 - z^2$$

```
> map(sqrt,");
```
$$y = \sqrt{r^2 - x^2 - z^2}$$

The radius of the cross section at x is just the value of y when $z = 0$. This is

```
> radius := subs(z=0,rhs("));
```
$$\sqrt{r^2 - x^2}$$

The domain of x is $[-r, r]$, so the volume is

```
> V := Int( Pi*radius^2, x=-r..r);
```

$$\int_{-r}^{r} \pi \left(r^2 - x^2\right) dx$$

which evaluates to

```
> value(");
```

$$\frac{4}{3}\pi r^3$$

∎

6.3 Solids of Revolution

Whenever there is an axis of symmetry, say the x-axis, the volume can be described simply by indicating the radius r as a function of x and the chosen domain $[a, b]$ for x. The cross-sectional area at x will be $\pi r^2(x)$, and the volume will be

$$V = \int_a^b \pi r^2(x)\, dx$$

We view this object as having been constructed by "rotating the region beneath the curve r about the x-axis."

Example 6.7 Find the volume of the solid obtained by rotating about the x-axis the region under the curve

```
> f := (x) -> sqrt(5*x);
```

$$x \mapsto \sqrt{5}\sqrt{x}$$

from $x = 0$ to $x = 1$.

Solution A convenient way to visualize this volume is to use the command **tubeplot( )**. The axis of symmetry is expressed in parametric form by a list of three coordinates, and the radius is expressed as a function of the chosen parameter. The region to be rotated is bounded by the line

```
> ll := [[1,0],[1,f(1)]];
```

$$[[1,0],[1,\sqrt{5}]]$$

and the curve f.

```
> plot( {ll,f(x)},x=0..1);
```

6.3 Solids of Revolution

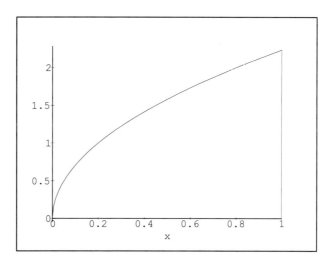

Rotating this about the axis denoted by [x, 0, 0] results in the volume

```
> with(plots):
> PltStyle := axes=NORMAL,style=PATCHNOGRID:
> tubeplot( [x,0,0],x=0..1,radius=f(x),PltStyle);
```

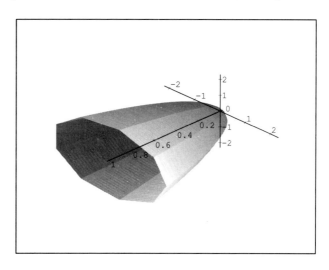

The cross-sectional area of each layer as a function of x is

```
> Area := (x) -> Pi*f(x)^2;
```

$$x \mapsto 5\pi x$$

The volume is

```
> Int( Area(x) ,x =0..1);
```

$$\int_0^1 5\pi x\, dx$$

which evaluates to

```
> value(");
```

$$\frac{5}{2}\pi$$

∎

The axis of symmetry can also be one of the coordinate axes.

Example 6.8 Find the volume of the solid obtained by rotating about the *y*-axis the region bounded by

```
> eq1 := y = x^3/2; eq2 := y = 4; eq3 := x = 0;
```

$$y = \frac{1}{2}x^3$$

$$y = 4$$

$$x = 0$$

Solution The region of interest is shown in the following graph.

```
> with(plots):
```

```
> implicitplot( {eq.(1..3)} , x=0..3,y=0..6);
```

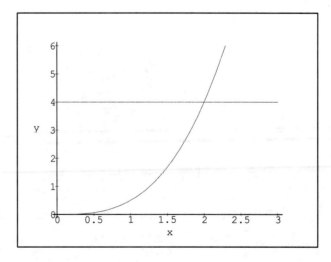

This region is rotated about the *y*-axis, so the radius of the cross sections should be expressed as a function of *y*. From *eq*1 we have

```
> isolate( eq1 , x );
```

$$x = \sqrt[3]{2}\sqrt[3]{y}$$

so the surface area of each layer is

```
> A := makeproc( Pi*rhs(")^2 , y );
```

$$y \mapsto \pi 2^{\frac{2}{3}} y^{\frac{2}{3}}$$

The volume is

```
> V := Int( A(y),y=0..4);
```

$$\int_0^4 \pi 2^{\frac{2}{3}} y^{\frac{2}{3}} \, dy$$

which evaluates to

```
> value(") = evalf(");
```

$$\frac{12}{5} 4^{\frac{2}{3}} \pi 2^{\frac{2}{3}} = 30.15928947$$

∎

The success of this approach depends only on our being able to find a nice formula for the area of each of the cross sections. The cross sections can be more general than circles.

Example 6.9 Find the volume of the solid obtained by rotating about the x-axis the region bounded by

```
> eq1 := y = x^2/3;
```

$$y = \frac{1}{3}x^2$$

and

```
> eq2 := y = 2*x;
```

$$y = 2x$$

Solution Because only two curves are given, they must intersect twice if there is to be a closed region. These intersection points are given by

```
> solve( {eq1,eq2} );
```

$$\{x = 0, y = 0\}, \{x = 6, y = 12\}$$

The region to be rotated is given by

292   Chapter 6 Applications of Integration

```
> implicitplot( { eq1, eq2 },x=0..6,y=0..12);
```

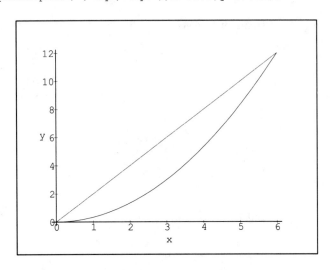

and results in a solid that can be thought of as the difference of the two solids

```
> p1 := tubeplot( [x,0,0],x=0..6,radius=x^2/3 ):
> p2 := tubeplot( [x,0,0],x=0..6,radius=2*x ):
```

In the following plot we display only the sample points generated by these commands.

```
> display( [p1,p2],axes=NORMAL,style=POINT );
```

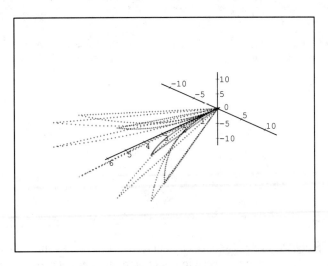

The cross-sectional area for a particular x value is an annulus, between an inner circle of radius

```
> r1 := rhs(eq1);
```

$$\frac{1}{3}x^2$$

6.3 Solids of Revolution

and an outer circle of radius

```
> r2 := rhs(eq2);
```

$$2x$$

The area of each annulus as a function of x is computed by the function

```
> A := makeproc( Pi*r2^2 - Pi*r1^2, x);
```

$$x \mapsto 4\pi x^2 - \frac{1}{9}\pi x^4$$

The total volume is

```
> V := Int( A(x), x=0..6);
```

$$\int_0^6 4\pi x^2 - \frac{1}{9}\pi x^4 \, dx$$

which evaluates to

```
> value(") = evalf(");
```

$$\frac{576}{5}\pi = 361.9114737$$

∎

A solid of revolution may also be formed by revolving a region about a line parallel to one of the axes.

Example 6.10 Find the volume obtained by revolving the region in Example 6.9 about the line $y = 15$ in the x-y plane.

Solution The region is the same as in the previous example, but the new inner and outer radii are

```
> r3 := 15 - r2;
```

$$15 - 2x$$

and

```
> r4 := 15 - r1;
```

$$15 - \frac{1}{3}x^2$$

and the solid is

```
> p3 := tubeplot( [x,15,0], x=0..6, radius=r3 ):
> p4 := tubeplot( [x,15,0], x=0..6, radius=r4 ):
```

```
> display( [p3,p4],axes=FRAMED );
```

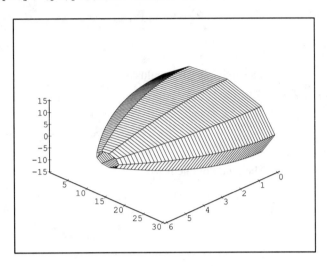

The cross-sectional area at x is computed by the function

```
> A := makeproc( Pi*r4^2 - Pi*r3^2 , x );
```

$$x \mapsto \pi \left(15 - \frac{1}{3}x^2\right)^2 - \pi (15 - 2x)^2$$

and the total volume is

```
> V := Int( A(x),x=0..6);
```

$$\int_0^6 \pi \left(15 - \frac{1}{3}x^2\right)^2 - \pi (15 - 2x)^2 \, dx$$

which evaluates as

```
> value(") = evalf(");
```

$$\frac{1224}{5}\pi = 769.0618816$$

∎

6.4 Generalized Cross Sections

The techniques we have been using to compute volumes are not restricted to solids of revolution. In this section, we investigate their application to objects of a more general shape.

An important aspect of any investigation of volumes is the ability to visualize the actual object. The *Maple* routine **surfdata( )** can be used to help with such visualizations.

An arbitrary surface can be approximated as a list of lists of points. Each list of points represents a curve that lies in the surface. For example,

```
> l1 := [[0,0,0],[0,1,0],[0,2,0],[0,3,0]];
```

$$[[0,0,0],[0,1,0],[0,2,0],[0,3,0]]$$

and

> l2 := [[1,0,1],[1,1,1],[1,2,1],[1,3,1]];

$$[[1,0,1],[1,1,1],[1,2,1],[1,3,1]]$$

and

> l3 := [[2,0,0],[2,1,0],[2,2,0],[2,3,0]];

$$[[2,0,0],[2,1,0],[2,2,0],[2,3,0]]$$

represent three distinct lines.

These curves can be tied together in sequence to yield a picture of the surface. The **surfdata( )** routine accepts a list of such lines and produces an appropriate image.

> with(plots):

> surfdata([l1,l2,l3],axes=framed);

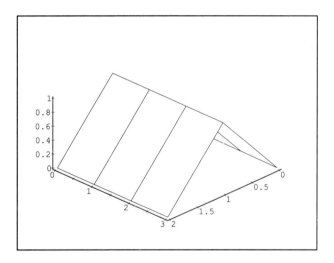

Example 6.11 Construct a graph of the solid whose base is a circle of radius 1 centered at the origin and whose vertical cross sections orthogonal to the x-axis are rectangles whose height is one-half their base.

Solution The cross section of this solid at $x = \frac{1}{2}$ is displayed by connecting the two lines

> l1 := [[1/2,-h(1/2),0],[1/2,h(1/2),0]];

$$[[\frac{1}{2}, -h(\frac{1}{2}), 0], [\frac{1}{2}, h(\frac{1}{2}), 0]]$$

and

> l2 := [[1/2,-h(1/2),h(1/2)],[1/2,h(1/2),h(1/2)]];

$$[[\tfrac{1}{2}, -h(\tfrac{1}{2}), h(\tfrac{1}{2})], [\tfrac{1}{2}, h(\tfrac{1}{2}), h(\tfrac{1}{2})]]$$

where the y values on the edge of the circular base and the y and z values on the upper surface are given in terms of

```
> h := (x) -> sqrt( 1 - x^2);
```

$$x \mapsto \sqrt{1-x^2}$$

The cross section is

```
> surfdata( [l1,l2],axes=framed);
```

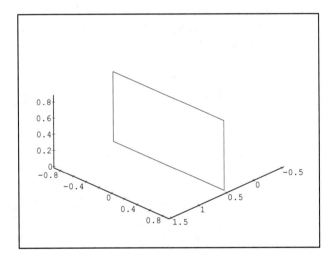

and it has an area of

```
> 2*h(1/2)^2;
```

$$\frac{3}{2}$$

The entire surface of the object can be constructed from two separate surfaces: the circular base and a top surface matched to the circular base at a height of $h(x)$ above the circle.

To construct the top and bottom surfaces we construct a sequence of points across the surfaces, one for each x value.

```
> xvals := [seq(-1+i*2/10,i=0..10)];
```

$$[-1, -\tfrac{4}{5}, -\tfrac{3}{5}, -\tfrac{2}{5}, -\tfrac{1}{5}, 0, \tfrac{1}{5}, \tfrac{2}{5}, \tfrac{3}{5}, \tfrac{4}{5}, 1]$$

This results in

```
> n := nops(xvals);
```

11

6.4 Generalized Cross Sections

lines across the base and the top.

Each point is represented by an [x, y, z] triple. The following *Maple* procedure constructs a sequence of 11 points, forming a line across the base at a given value of x.

```
> baseline := proc(x) local j;
> [seq( [ x , -h(x) + j*2*h(x)/10 , 0],j=0..10 )];
> end:
```

The corresponding line across the top of this object includes an additional point of height $z = 0$ at the beginning and at the end, which are used to construct the sides of the object.

```
> topline := proc(x) local j;
> [[x,-h(x),0],
> seq( [ x , -h(x) + j*2*h(x)/10 , h(x)],j=0..10 ),
> [x,h(x),0]];
> end:
```

Each x value gives rise to such a line. The circular base corresponds to

```
> circle := [ seq( baseline(xvals[i]), i=1..n ) ]:
```

and the top corresponds to

```
> top := [ seq( topline(xvals[i]) , i=1..n )]:
```

A picture of the resulting object is generated by the commands

```
> with(plots):
```

```
> surfdata( {circle,top} );
```

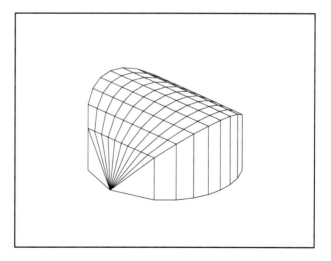

■

To compute the volume of such an object, we simply integrate the cross-sectional area as a function of x over the appropriate interval for x.

Example 6.12 Find the volume of the solid whose base is a circle of radius 1 centered at the origin and whose vertical cross sections orthogonal to the x-axis are rectangles whose heights are one-half their base.

Solution The cross-sectional area as a function of x is

```
> A := (x) -> 2*h(x)*h(x);
```

$$x \mapsto 2 - 2x^2$$

The total volume is

```
> V := Int( A(x),x=-1..1);
```

$$\int_{-1}^{1} 2 - 2x^2 \, dx$$

which evaluates to

```
> value(") = evalf(");
```

$$\frac{8}{3} = 2.666666667$$

■

6.5 Cylindrical Shells

There are other decompositions of objects that do not rely on cross-sectional areas orthogonal to an axis. One that is especially effective for a solid of revolution is in terms of thin cylindrical shells of increasing radius. The following graph shows two concentric cylinders.

```
> with(plots):
> opts := tubepoints=20,numpoints=5,z=0..2,axes=framed:

> tubeplot({[0,0,z,radius=1/2], [0,0,z,radius=5/8]}, opts);
```

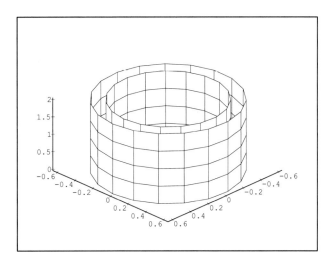

The volume of a shell of thickness Δr at radius r and of height h is approximately

$$V = 2\pi r h \Delta r$$

which is

$$V = [\text{circumfrence}][\text{height}][\text{thickness}]$$

If the radius of the object ranges from r_1 to r_2, then the total volume is approximately

$$V = \sum_{r_1}^{r_2} 2\pi r h \Delta r$$

As the maximum thickness of the layers tends to 0, this becomes

$$V = \int_{r_1}^{r_2} 2\pi r h \, dr$$

Note that in the limit the Δx becomes dx.

Example 6.13 Find the volume of the solid obtained by rotating about the y-axis the region bounded by

```
> eq1 := y = x*(x-2)^2;   eq2 := y = 2;
```
$$y = x(x-2)^2$$
$$y = 2$$

Solution The region to be rotated is shown in the next graph.

```
> with(student):
> with(plots):
```

```
> implicitplot({eq1,eq2},x=0..3,y=-3..3);
```

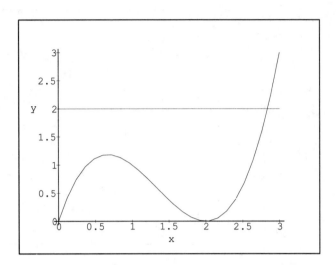

The axis of symmetry is the y-axis, and a cross-sectional decomposition would result in horizontal ring-shaped cross sections and circles. For a given value of y, the inner and outer radii of the rings or of the circles would be x values obtained by solving the cubic equation $eq1$. Just one of these roots is

```
> solve( eq1 , x ):
> "[1];
```

$$\sqrt[3]{-\frac{8}{27}+\frac{1}{2}y+\frac{1}{18}\sqrt{y}\sqrt{-32+27y}\sqrt{3}}+\frac{4}{9}\frac{1}{\sqrt[3]{-\frac{8}{27}+\frac{1}{2}y+\frac{1}{18}\sqrt{y}\sqrt{-32+27y}\sqrt{3}}}+\frac{4}{3}$$

so this formulation is quite messy.

A cylindrical decomposition leads to a much simpler formulation of the problem. The radius of each cylinder is x, the height of each cylinder is given by

```
> h := makeproc( rhs(eq2) - rhs(eq1) , x );
```

$$x \mapsto 2 - x(x-2)^2$$

and the radius x ranges from 0 to the the real solution[3] of the equation

```
> subs(eq2,eq1);
```

$$2 = x(x-2)^2$$

If we want only a number rather than a general formula, we can use the approximate solution

---

[3] Also try **solve({eq1,eq2})**.

```
> upperbnd := fsolve(",x);
```
$$2.839286755$$

The volume is approximately
```
> V := Int( 2*Pi*x*h(x) , x=0..upperbnd );
```
$$\int_0^{2.839286755} 2\pi x \left(2 - x(x-2)^2\right) dx$$

which evaluates to
```
> value(") = evalf(");
```
$$35.35643588 = 35.35643586$$

∎

6.5.1 Visualizing Cylinders

A picture of a cylindrical object can be drawn by connecting points that lie on the cylinders used in the cylindrical decomposition. We shall build the picture from $m = 10$ concentric cylinders whose radii occur at a separation of

```
> m := 10: d := upperbnd/m;
```
$$0.2839286755$$

and are
```
> for i from 0 to m do r.i := i*d; od:
```
$$2.839286755$$

```
> radii := [ r.(0..m) ]:
> evalf(",3);
```
$$[0, 0.284, 0.568, 0.852, 1.14, 1.42, 1.70, 1.99, 2.27, 2.56, 2.84]$$

Each cylinder is concentric about the z-axis. We can construct a picture of the cylinder by first building a circle at the top and bottom of the cylinder from points equally spaced along the perimeter of the cylinder. The points of these two circles are then joined by lines.

For m equally spaced points about the z-axis at the origin, the angles between the x-axis and the radii to the points are

```
> for i from 0 to m do a.i := i*(2*Pi)/m od:
> angles := [a.(0..m)];
```
$$[0, \frac{1}{5}\pi, \frac{2}{5}\pi, \frac{3}{5}\pi, \frac{4}{5}\pi, \pi, \frac{6}{5}\pi, \frac{7}{5}\pi, \frac{8}{5}\pi, \frac{9}{5}\pi, 2\pi]$$

A point P at the top of the cylinder of radius r has a z coordinate of 2. If a is the angle that the radius to the point P would make with the x-axis if projected down onto the plane $z = 0$, then the x and y coordinates of P are $r\cos(a)$ and $r\sin(a)$ respectively. The procedure

```
> Pt1 := (a,r) -> [ r*cos(a) , r*sin(a) , 2 ];
```

$$(a, r) \mapsto [r\cos(a), r\sin(a), 2]$$

computes the coordinates of that point. For example, the point

```
> P := Pt1(0,r.m);
```

$$[2.839286755, 0, 2]$$

lies at the top of the outer cylinder of the object.

A point Q at the bottom of the cylinder has a z value of $2 - h(r)$. Its coordinates are computed by the procedure

```
> Pt2 := (a,r) -> [r*cos(a) , r*sin(a) , 2-h(r)];
```

$$(a, r) \mapsto [r\cos(a), r\sin(a), r(r-2)^2]$$

The point

```
> Q := Pt2(a.0,r.m);
```

$$[2.839286755, 0, 1.999999999]$$

lies directly below P.

The circle at the top of a cylinder of radius, say, $r3$ is represented by the list of points (to 4 decimal digits accuracy)

```
> map(Pt1,angles,r3): evalf( " , 4 );
```

$$[0.8518, 0, 2.0], [0.6889, 0.5005, 2.0], [0.2631, 0.8099, 2.0],$$
$$[-0.2631, 0.8099, 2.0], [-0.6889, 0.5005, 2.0], [-0.8518, 0, 2.0],$$
$$[-0.6889, -0.5005, 2.0], [-0.2631, -0.8099, 2.0], [0.2631, -0.8099, 2.0],$$
$$[0.6889, -0.5005, 2.0], [0.8518, 0, 2.0]$$

so the command

```
> for j from 0 to m do t.j := map(Pt1,angles,r.j) od:
```

assigns to $t.i$ the points representing the top of the cylinder of radius $r.i$. Similarly,

```
> for j from 0 to m do b.j := map(Pt2,angles,r.j) od:
```

assigns to $b.j$ the list of points representing the bottom of the cylinder of radius $r.j$. The top of the entire object is represented by the list of curves

```
> top := [t.(0..m)]:
```

and the bottom is represented by the list of curves

```
> bottom := [b.(0..m)]:
```

We obtain a picture of the object by displaying these two surfaces together.

```
> surfdata( [top,bottom] );
```

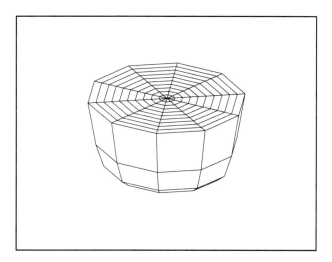

The resulting object can be rotated interactively, to study its surface in greater detail. The view from the bottom of the object reveals the general shape predicted by the cross section.

6.5.2 Cylindrical Decompositions Off the Axis

There is no need to restrict this cylindrical technique to solids of revolution about the z-axis. The generality of this approach is illustrated by the following example.

Example 6.14 Find the volume of the solid obtained by rotating about the line $x = 2$ the region bounded by the curves

```
> eq1 := y = 4*x - 2*x^2; eq2 := y = 0;
```

$$y = 4x - 2x^2$$

$$y = 0$$

Solution The region to be rotated about the line $x = 2$ is shown below.

```
> implicitplot( eq1 , x=-1..3,y=-1..3);
```

There is a cylinder corresponding to each value of x from $x = 0$ to $x = 2$. For a given value of x, the height of the cylinder will be $h(x)$ where

> h := makeproc(rhs(eq1) , x);

$$x \mapsto 4x - 2x^2$$

The thickness of each cylinder will still be Δx or in the limiting case dx. However, instead of using x as the radius, we now use $2 - x$. For a cylindrical decomposition, the volume is always

> Int(2*Pi*radius*height ,x =0..maxrad);

$$\int_0^{maxrad} 2\pi\, radius\, height\, dx$$

which in this case becomes

> V := subs({radius=(2-x), height=h(x),maxrad = 2}, ");

$$\int_0^2 2\pi\,(2-x)\left(4x - 2x^2\right)\,dx$$

This evaluates to

> value(") = evalf(");

$$\frac{16}{3}\pi = 16.75516082$$

The object itself is displayed as

```
> tubeplot( [x,2,0],x=0..3,radius=h(x),axes=framed);
```

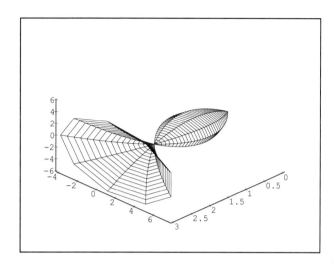

REMARK: For a cylindrical decomposition, integration formalizes the notion of summing up the volumes of the surface of the cylinders. It makes no difference if this is done from the smallest to the largest radius or vice versa, so long as all the relevant radii are included.

6.6 Work

According to Newton's second law of motion, the force F acting on an object satisfies the relationship

$$F = ma$$

where the mass of the object is m its acceleration is $a = \frac{d^2}{dt^2}s$. A force of 1 Newton (N) acting on a mass of 1 kg produces an acceleration of $1\frac{\text{meters}}{\text{seconds}^2}$.

Under constant acceleration, Work (W) satisfies the relationship

$$W = Fd$$

where d denotes distance. One joule (J) of work corresponds to exerting a force of 1 Newton through a distance of 1 meter.

Example 6.15 Find the work required to lift a 2-kg book off the floor onto a shelf that is 1.5 meters high.

Solution The acceleration caused by gravity is 9.8 meters/sec² so

```
> F = m*a;
```
$$F = ma$$

becomes

```
> subs( m=2*kg,a=9.8*meters/sec^2,");
```
$$F = \frac{19.6\,\text{kg meters}}{\text{sec}^2}$$

This is the number of Newtons of force required to oppose the force of gravity acting on the book. This force is applied through a distance of 1.5 meters, so the work in joules is given by

```
> "*1.5*meters;
```
$$1.5\,\text{meters}\,F = \frac{29.40\,\text{meters}^2\text{kg}}{\text{sec}^2}$$

■

To handle the case of nonconstant acceleration, we first approximate the work done under nonconstant acceleration by partitioning the distance d into small distances and pretend that the acceleration is constant on each subinterval. But this is precisely how we computed the area under the curve defined by treating force as a function of distance.

DEFINITION 6.2 The *work* done in moving an object from a to b along a line is

$$W = \int_a^b f(x)dx$$

Example 6.16 According to Hooke's law, the force required to maintain a spring at position x relative to its natural resting position is

$$f(x) = kx$$

Find the work required to stretch a spring from 10 cm to 15 cm if a force of 30 N is required to hold it stretched at a length of 10 cm and it is at rest at 5 cm.

Solution Convert all units to meters and consider the 5 cm position to be 0. The function defining force is

```
> f := (x) -> k*x;
```
$$x \mapsto kx$$

Since

```
> f(.10) = 30;
```

$$0.10\,k = 30$$

we can determine the value of k.

```
> isolate(",k);
```

$$k = 300.0$$

```
> assign(");
```

$$300.0 = 300.0$$

The work involved in going from 10 cm to 15 cm is

```
> Int( f(x),x= 0.10 ... 0.15 );
```

$$\int_{0.10}^{0.15} 300.0\,x\,dx$$

which evaluates to

```
> value(");
```

$$1.8750$$

■

Work is done to pump fluids into or out of tanks. To put this in the context of motion through a distance, we consider each "layer" of water separately. The entire thin layer of water at depth x and of thickness dx moves the same distance. The force involved is just the mass of this layer of water times acceleration caused by gravity.

Example 6.17 Compute the amount of work required to drain a tank of water that is an inverted hemisphere of radius 3 meters if the the tank is filled to a depth of 2 meters and the water is to be pumped out over the top. Assume that the water has a density of 1000 kg/m^3.

Solution The tank is displayed in Figure 6.1. Let $x = 0$ correspond to the top of the tank (i.e., the center of the sphere) so that $x = 1$ corresponds to the surface of the water. Each "layer" of water must travel through a different distance. The volume of water at depth x is of thickness Δx and forms a disk of radius r where

```
> eq1 := x/r = tan(theta);
```

$$\frac{x}{r} = \tan(\theta)$$

and θ is the angle that a line from the center of the sphere to the surface of the water makes with the top of hemisphere. Furthermore,

```
> eq2 := x = 3*sin(theta);
```

$$x = 3\sin(\theta)$$

To compute the volume of the ith layer as a function of x, we must know r as a function of x. This is accomplished as

```
> isolate(eq1 , r );
```

$$r = \frac{x}{\tan(\theta)}$$

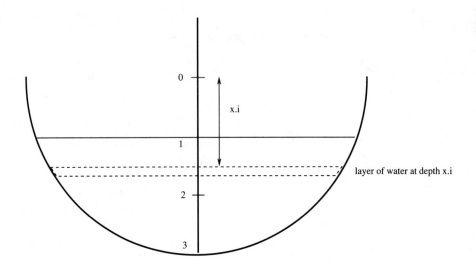

Figure 6.1 Draining a tank

```
> isolate( eq2 , theta );
```
$$\theta = \arcsin(\frac{1}{3}x)$$

```
> value(subs(",""));
```
$$r = \sqrt{9 - x^2}$$

Thus, the surface area of the layer at x is

```
> A := makeproc( subs( ", Pi*r^2 ) , x );
```
$$x \mapsto \pi \left(9 - x^2\right)$$

The mass associated with a layer of water at level $x.i$ and of thickness $Dx.i$ is its volume multiplied by its density, which is

```
> 1000*A(xi)*Dx.i;
```
$$1000\pi \left(9 - \xi^2\right) Dxi$$

To obtain the force involved, multiply this by the acceleration $a = 9.8$, as in

```
>    " * 9.8;
```
$$9800.0 \pi \left(9 - \xi^2\right) Dxi$$

The work done to move this one layer[4] is

$$x_i 9800.0 \pi \left(9 - x_i^2\right) Dxi$$

[4]Note that if we needed to lift the layer to a level other than to the top of the tank, only this distance would change.

The total work is approximated by summing this work over all layers.

$$\sum_{x_i} x_i 9800.0\pi\left(9-x_i^2\right) Dx_i$$

The actual work is obtained by letting the thickness of the layers tend to 0. The total work done is

```
> Wrk := Int( 9.8 * 1000 * A(x) * x , x = 1 .. 3 );
```

$$\int_1^3 9800.0\pi\left(9-x^2\right) x\, dx$$

which is approximately

```
> evalf(");
```

$$492601.7281$$

∎

EXERCISE SET 6

Some of the important commands that have been used in this chapter are shown in Table 6.1. Keep them in mind as you explore the examples from the chapter and the exercises that are listed here.

1. Let f and g be defined by

    ```
    > f := (x) -> cos(x/2);
    ```

    $$x \mapsto \cos(\tfrac{1}{2}x)$$

    ```
    > g := (x) -> 1/x^3;
    ```

    $$x \mapsto x^{-3}$$

    Show that these two curves intersect at approximately $a = 1.049$ and $b = 3.07$. Approximate the area between these two curves on the interval $[a, b]$.
2. Find the area between the curves f and g from $x = \tfrac{3}{2}$ to $x = 4$.
3. Both the functions

    ```
    > f := (x) -> sin(2*x);
    ```

    $$x \mapsto \sin(2x)$$

    and

    ```
    > g := (x) -> cos(3*x);
    ```

    $$x \mapsto \cos(3x)$$

Command	Description
(x)→ sin(x/2);	A procedure definition
assign(");	Convert equations to assignments
axes=boxed	Plot option
axes=FRAMED	Plot option
axes=NORMAL;	Plot option
display([p1,p2]);	Superimpose two plots
display([p3,p4],insequence=true);	Animate two plots
evalf(e1);	Approximate
extrema(eq1,{},y);	Find extreme values
fsolve(f(x)-g(x)=0,x,1..2);	Find an approximate solution
implicitplot({eq1,eq2},x=-4..6,y=-4..4);	An implicit plot
implicitplot3d({eq1,x=1/2},opts);	A 3D implicit plot
Int(f(x)-g(x),x=3/2..3);	Represent an integral
isolate(eq2,x);	Solve for x
makeproc(",x);	Use an expression to build a procedure
map(isolate,{eq1,eq2},x);	Map a procedure with extra parameters
numpoints=5	Plot option
plot({f,g},1..3);	Plot a set of functions
solve(eq2,y);	Solve an equation
style=PATCHNOGRID;	A plot style
style=POINT	A plot style
surfdata({circle,top});	Draw a set of surfaces from lists of curves
tubeplot([x,0,0],x=0..1,radius=f(x));	Generate a graph of a solid of revolution
tubepoints=20	A plot option
value(I1+I2);	Evaluate an inert expression
with(plots);	Load the plots package
[seq(-1+i*2/10,i=0..10)];	Generate a list of numbers
[solve({eq1,eq2})];	List solutions to a set of equations

Table 6.1 Some of the Commands Used in Chapter 6

are periodic. Find the exact area between these two curves over an entire period including $x = 0$. (You can assume that the function $f - g$ is also periodic.) Repeat the computation of the area between the curves for at least two more periods of your choice.

4. Find the area between the curves defined by the equations

```
> eq1 := y^2+2*y+1 = 2*x+8;
```
$$y^2 + 2y + 1 = 2x + 8$$

```
> eq2 := y+1 = x;
```
$$y + 1 = x$$

by defining and evaluating appropriate integrals with respect to x. (You may wish to use commands such as **implicitplot( )** and **solve( )** and **extrema( )** to find appropriate intervals for the integrals.)

5. Repeat the computation of the area found in Exercise 4 by defining and evaluating appropriate integrals with respect to y.

6. Use the approach outlined in Example 6.6 to find the volume of the three-dimensional object defined by the equation

```
> eq3 := x^2 + 2*y^2 + 2*z^2 = r^2;
```
$$x^2 + 2y^2 + 2z^2 = r^2$$

7. Generate a three-dimensional graph of the object obtained by rotating about the x-axis the curve defined by

```
> f := (x) -> sqrt(7*x) + 3;
```
$$x \mapsto \sqrt{7}\sqrt{x} + 3$$

from $x = 0$ to $x = 1$. Find the volume of that solid.

8. Repeat Exercise 7 for the case where the curve is rotated about the line $y = 1$.

9. Generate a picture of the region bounded by the curves defined by the equations

```
> {y+2 = 4, x = 0, y+2 = 1/2*(x-1)^3};
```
$$\left\{ y + 2 = \frac{1}{2}(x-1)^3, x = 0, y + 2 = 4 \right\}$$

Find the volume of the solid obtained by rotating this region about the y-axis.

10. Generate a graph of the region bounded by the curves defined by

```
> {y+2 = 2*x-2, y+2 = 1/3*(x-1)^2};
```
$$\left\{ y + 2 = \frac{1}{3}(x-1)^2, y + 2 = 2x - 2 \right\}$$

Find the volume of the solid obtained by rotating that region about the line $y = -5$. Generate a picture of the solid.

11. Find the volume generated by rotating the region in Exercise 10 about the y-axis.

12. Construct a graph of the solid whose base is a circle of radius 1 centered at the origin and whose vertical cross sections orthogonal to the x-axis are triangles whose height is one-half their base. Find its volume.

13. Find the volume of the solid whose base is a circle of radius 1 centered at the origin and whose vertical cross sections orthogonal to the x-axis are rectangles whose heights are one-half their base.
14. Find the volume of the solid obtained by rotating about the y-axis the region bounded by

    ```
    > eq1 := y = x*(x-2)^2;   eq2 := y = 2;
    ```

15. Find the volume of the solid obtained by rotating about the line $x = 2$ the region bounded by the curves

    ```
    > eq1 := y = 4*x - 2*x^2; eq2 := y = 0;
    ```

16. Find the work required to lift a 3 kg book off a shelf that is 2 meters high down onto the floor.
17. (Adapted from the *Maple Calculus Work Book, Problems and Solutions*, by Geddes, Marshman, McGee, Ponzo, and Char) A circular swimming pool of radius 30 ft is constructed by building a wall that is 5 feet high. The inside surface of the wall is vertical, If the pool wall is regarded as being the solid of revolution described by rotating about the y-axis the region bounded by the x-axis, the line $x = 30$, and the curve g defined by

    ```
    > g := (x) -> -10/27*(-x+33)^3+5/3*(-x+33)^2;
    ```

    $$x \mapsto -\frac{10}{27}(-x+33)^3 + \frac{5}{3}(-x+33)^2$$

    find the volume of cement required to build the wall. Generate a picture of the pool. Is the the pool wall flat on top?
18. Assuming that the wall in the previous problem begins and ends on the ground as the radius ranges from 30 feet to 33 feet, and that the cross sectional shape is governed by a cubic equation, find the volume of cement required if the maximum height of 5 feet occurs exactly 1/2 way through the wall.
19. Let a denote the position of the maximum height of the wall from Exercise 18. Determine a formula for the volume of cement required to build the wall, as a function of the position a. (Hint: Place the cross section along the x-axis with the center of the pool at the origin. Then roughly speaking, a can range in value from $x = 30$ to $x = 33$.)
20. Find the work required to stretch a spring from 5 cm to 7 cm if a force of 20 N is required to hold it stretched at a length of 5 cm and it is at rest at 3 cm.
21. Compute the amount of work required to drain water from a tank in the shape of an inverted pyramid with a square base of 4 meters to a side and a height of 4 meters, if the tank is filled to a depth of 3 meters and the water is to be pumped out through an opening placed 2 meters above the tank.

7 Integration Techniques

7.1 Introduction

From the very limited perspective of simply finding a mechanism for the computation of an area or a volume, there would appear to be little reason to go beyond just formulating a solution to a mathematical problem in terms of a definite integral and computing an approximation to the value of that integral (perhaps using rectangles). Given the power of today's computers, such approximations can be constructed quickly and accurately using the equivalent of thousands of rectangles. As well, in any real situation, we are given only approximate measurements for the boundaries anyway.

In spite of the enormous advances made in hardware design, exact formulas for integrals and techniques for their derivation continue to play an important role. Some of the many reasons are:

1. A short exact formula can be evaluated more quickly than an approximation involving a large sum. Such differences are drastic for hand computation, and they remain important for computation on very fast machines. This is because even in moderate-sized applications we need to compute such quantities thousands or even millions of times. Recent history suggests that it requires about 10 years of development to increase machine speed by a factor of 100. That same improvement can be achieved at almost no expense simply by using a better mathematical approach.

2. Error is an intrinsic part of arithmetic involving approximations. For example, if 5.1 and 3.4 are both smaller than intended by .1, then their product will be too small by .85. The cumulative effect of such errors can be disastrous. Depending on the precise details of how the floating-point arithmetic is carried out by machine, with only 2 digits of precision at each step, the answer to $\sum_{i=1}^{1000}$ given by

```
> Digits := 2; total := 0;
```

$$0$$

```
> for i from 1 to 1000 do total := total + 1.0*1/i: od:
```

is

```
> total;
```

$$3.9$$

This answer is much too small from a mathematical perspective. A close examination of what happens shows that after 20 steps the sum has reached a value of 3.9. From the 21st step onward, we are adding a number to the sum that is at most 0.048. However, the result must continue to be 3.9 because we are allowed only two digits in the answer, and 3.9 + 0.048 = 3.948 is rounded off to 3.9. This phenomenon is known as *roundoff error*. In this case, the mathematically correct answer to this sum is

```
>   i := 'i': Sum( 1/i, i=1..1000):
```

is

```
> value(");
```

$$\Psi(1001) + \gamma$$

which is closer to

```
> Digits := 10;
```

$$10$$

```
> evalf("",10);
```

$$7.485470861$$

3. Exact integration formulas allow us to confine any errors in the computation of an integral to the original measurements.
4. Approximation techniques are often based on performing exact or accurate computations on well-behaved functions such as polynomials, which are constructed close to the original curve.
5. Mathematical transformations of the original problem provide useful ways of avoiding certain types of numerical errors.
6. A numerical algorithm solves a specific instance of a problem, yielding a number. An exact algebraic solution with undetermined parameters can represent a whole class of solutions and can be used to build solutions to more complex problems.
7. Mathematical insight comes from formulating the same problem or computation in two or more different ways. Techniques for integration play a crucial role in generating these alternative viewpoints.

Finally, exact integration is an important and accessible example of a certain approach to mathematical problem solving. In this problem domain, problems for which there are no obvious solutions can systematically be transformed into familiar problems whose solutions can then be transformed back into solutions to the original problem.

7.2 Changing Variables

We have discovered how to rewrite an integral of a sum as a sum of integrals. Our objective in this section is to discover similar rules for rewriting integrals of compositions or products. It is helpful to examine carefully known differentiation techniques involving products. For example, compositions and products are involved every time we use the chain rule.

Example 7.1 Let f and g and h be defined by

```
> f := (u) -> 2/3*u^(3/2): g := (x) -> 1 + x^2:
> h := f@g;
```

$$f \circ g$$

By the chain rule, the derivative of h at x is

```
> D(h)(x);
```

$$2\sqrt{1+x^2}\,x$$

In the reverse direction, we have

```
> Int( D(h)(x) ,x ) = (f@g)(x);
```

$$\int 2\sqrt{1+x^2}\,x\,dx = \frac{2}{3}\left(1+x^2\right)^{\frac{3}{2}}$$

∎

In practice, we know only $D(f)$, $D(g)$, and g, but not h or f. Still, the relationships remain valid even when f and g are unspecified, and they lead to a general technique of integration.

For the undefined case,

```
> f := 'f': g := 'g': h := f @ g:
```

the fact that $D(h)(x)$ evaluates as

```
> D(h)(x);
```

$$D(f)(g(x))\,D(g)(x)$$

corresponds to the fact that

```
> Int( " , x ) = h(x);
```

$$\int D(f)(g(x))\,D(g)(x)\,dx = f(g(x))$$

If we let $g(x) = u$ so that $D(g)(x)\,dx = du$, we can rewrite the preceding as

```
> changevar(g(x)=u, " , u );
```

$$\int D(f)(u)\,du = f(u)$$

The *Maple* command **changevar( )** has been used here to specify the change of variables defined by $g(x) = u$. The third argument indicates that the final result should be written in terms of u. This command can be read as "express the original integral in terms of the variable u where $u = g(x)$." Its net effect is to reduce the original problem to that of finding an antiderivative for $D(f)$.

REMARK: The transformation is more than the replacement of $g(x)$ by u. The expression dx must also be rewritten in terms of du, and, if necessary, the equation $x = g^{(-1)}(u)$ must be used to remove any further references to x.

7.2.1 Using a Change of Variables to Integrate

The *change of variables* technique is used as follows. Consider

```
> I1 := Int( 2*x*sqrt(1+x^2),x);
```

$$\int 2\sqrt{1+x^2}\, x\, dx$$

Given the relation $1 + x^2 = u$, we have $du = 2x\, dx$. This change of variables reduces the integration problem, to that of solving

```
> I2 := changevar(1+x^2=u,I1,u);
```

$$\int \sqrt{u}\, du$$

This new integration problem is essentially equivalent to the original one. However, we have simplified our task to that of finding an antiderivative for $D(f)$, which in this case is

```
> value(");
```

$$\frac{2}{3}u^{\frac{3}{2}}$$

To recover the value of the original integral, we do a direct substitution using the relation $u = 1 + x^2$.

```
> subs(u=1+x^2,");
```

$$\frac{2}{3}\left(1+x^2\right)^{\frac{3}{2}}$$

REMARK: An important aspect of this technique is that there is an independent way of verifying that this is a correct solution. All we need do is differentiate the final solution with respect to x, as in

```
> diff(",x);
```

$$2\sqrt{1+x^2}\, x$$

We must get back something that can be simplified to the original integrand.

7.2.2 Summary

The change of variables technique involves four steps:

1. Select a suitable transformation $u = g(x)$.
2. Rewrite the integral in terms of u, using both the relations $u = g(x)$ and $du = D(g)(x)dx$.
3. Solve the new integral.
4. Transform the solution of the new integral back into a solution of the original problem by using $u = g(x)$.

REMARK: The first three steps may be repeated several times before a final answer is obtained.

A wide variety of equations $u = g(x)$ can be used to define the transformation, although only some choices will lead to a simplification of the problem. Indeed, it is the first step that requires practice and insight because the second step is completely mechanical for any given transformation.

Each change of variables allows you to look at a problem from a different point of view. Some points of view are more helpful than others so, as with any tool, you must learn how to use it effectively.

7.2.3 Additional Examples

One method of generating the kind of sample problems found in most traditional textbooks is to do a change of variables in reverse, starting with a familiar integration problem such as

> I2;

$$\int \sqrt{u}\, du$$

For example, each change of variables

> I3 := changevar(3+2*x=u , I2 , x);

$$\int 2\sqrt{3 + 2x}\, dx$$

> I4 := changevar(arccos(v)=u , I2 , v);

$$\int -\sqrt{\arccos(v)} \frac{1}{\sqrt{1 - v^2}}\, dv$$

or

> I5 := changevar(2 +t^2/3=u, I2 , t);

$$\int \frac{2}{3}\sqrt{2 + \frac{1}{3}t^2}\, t\, dt$$

results in a new equivalent integration problem.[1] To transform these new problems back to the original integration problem, we use the indicated transformations in reverse. We have the relations

> r2 := isolate(u=3+2*x,x);

$$x = \frac{1}{2}u - \frac{3}{2}$$

---

[1] In practice, we need to restrict our attention to domains where transformations such as $u(x) = 1 + x^2$ are invertible.

```
> r3 := isolate( u=arccos(v),v);
```
$$v = \cos(u)$$

```
> r4 := isolate( u=2 + x^2/3,x);
```
$$x = \sqrt{|3u - 6|}$$

which can be used as

```
> changevar( r2 , I3 , u );
```
$$\int \sqrt{u}\, du$$

REMARK: The command **changevar( )** does not require that the equation $u = 3 + 2*x$ be in any particular form. The third argument indicates clearly in which direction the transformation is intended to proceed. Thus, the two commands

```
> changevar( u = 3 + 2*x , I3 , u );
> changevar( 3 + 2*x = x , I3 , u );
```

have essentially the same effect on the integral. In either case, the necessary equations for u and du are constructed by *Maple*.

7.2.4 The Effective Use of Change of Variables

Our primary goal is to restructure an integral into a form for which we can easily find an antiderivative. Our success at this depends on our ability to recognize appropriate patterns.

A good choice for $u = g(x)$ will usually correspond to $D(g)(x)$ occurring as a factor of the integrand and $g(x)$ occurring as an argument to a function that is present.

Example 7.2 Find $\int x^3 \cos(x^4 + 2)\, dx$.

Solution Consider the integral

```
> I1 := Int(4*x^3*cos(x^4+2),x);
```
$$\int 4x^3 \cos(x^4 + 2)\, dx$$

We require a transformation of the form $g(x) = u$ that will transform this integration problem into one for which we already know an antiderivative. Because x^3 occurs as a factor of the integrand and x^4 occurs in the argument to cosine, a good candidate for defining the transformation is

```
> g1 :=   (x) -> x^4;
```
$$x \mapsto x^4$$

The relations

```
> [u = g1(x), D(u) = D(g1(x))];
```

$$[u = x^4, D(u) = 4D(x)x^3]$$

define the change of variables

> I2 := changevar(g1(x)=u , I1 , u);

$$\int \cos(u + 2)\, du$$

This new integral appears simpler in structure but is not yet one that we have already solved. To complete the solution, we can use a second change of variables to further simplify the argument to cosine.

> changevar(v=u+2,I2,v);

$$\int \cos(v)\, dv$$

This new result can be easily integrated to yield

> value(");

$$\sin(v)$$

The solution to the original problem is then recovered by the substitutions[2]

> sol := subs(v=u+2,u=g1(x) , ");

$$\sin(x^4 + 2)$$

∎

REMARK: Although there may not be a unique transformation that will work here, there is a better choice. Consider the transformation defined by

> g2 := (x) -> x^4 + 2:

The relations

> [u = g2(x),D(u)=D(g2(x))];

$$[u = x^4 + 2, D(u) = 4D(x)x^3]$$

define the transformation

> changevar(u=g2(x),I1,u);

$$\int \cos(u)\, du$$

which can be evaluated immediately.

REMARK: The answer *sol* is correct because differentiating with respect to x results in the original integrand.

> diff(sol,x);

$$4x^3 \cos(x^4 + 2)$$

---

[2] Had we used a set of equations, $\{v = u + 2, u = g1(x)\}$, in **subs( )**, this would indicate that u and v should be replaced simultaneously. The format used has the same effect as first doing **subs(u=v+2,...)** and then **subs(u=g1(x),")**.

Example 7.3 Find

> I3 := Int(sqrt(3*x+4),x);

$$\int \sqrt{3x+4}\, dx$$

Solution A change of variables $u = g(x)$ can be effective even when $D(g)(x) = 1$. The transformation is defined by the relations

> [u = 3*x + 4, D(u)=D(3*x+4)];

$$[u = 3x + 4, D(u) = 3 D(x)]$$

Because $D(u)$ does not appear explicitly, we use

> isolate("[2],D(x));

$$D(x) = \frac{1}{3} D(u)$$

Together these relations lead to the change of variables

> changevar(u=3*x + 4, I3 , u);

$$\int \frac{1}{3}\sqrt{u}\, du$$

∎

Example 7.4 Find

> I4 := Int(x/sqrt(1-4*x^2),x);

$$\int x \frac{1}{\sqrt{1-4x^2}}\, dx$$

Solution Consider the transformation defined by

> [u = sqrt(1-4*x^2), D(u)=D(sqrt(1-4*x^2))];

$$[u = \sqrt{1 - 4x^2}, D(u) = -4 D(x) x \frac{1}{\sqrt{1-4x^2}}]$$

From the second equation we have

> isolate("[2],D(x));

$$D(x) = -\frac{1}{4} D(u) \sqrt{1 - 4x^2}\, x^{-1}$$

so the change of variables is

> changevar(u=sqrt(1-4*x^2),I4,u);

$$\int -\frac{1}{4}\, du$$

The solutions to this and the original integral are

```
> value(");
```

$$-\frac{1}{4}u$$

and

```
> subs(u=sqrt(1-4*x^2),");
```

$$-\frac{1}{4}\sqrt{1-4x^2}$$

∎

7.2.5 Definite Integrals

An important aspect of the change of variables $u = g(x)$ is that certain properties of the function defined by the integrand

```
> h(x)*D(g)(x);
```

$$f(g(x))D(g)(x)$$

are preserved by the transformation.

One such property is area under the curve. Definite integrals have a precise interpretation in terms of area under the graph of the given function. The total area under the original curve (defined by the integrand as a function of x) from $x = a$ to $x = b$ is

```
> Int( h(x)*D(g)(x),x=a..b);
```

$$\int_a^b f(g(x))D(g)(x)\,dx$$

This is exactly equal to the area under the curve f from $u = g(a)$ to $u = g(b)$ and as given by

```
> changevar(u=g(x),",u);
```

$$\int_{g(a)}^{g(b)} f(u)\,du$$

This is in spite of the fact that the shape of the curve may have been changed by the transformation.

Example 7.5 Show that the area under the curve

```
> f1 := (x) -> (x^2 + 1)*2*x;
```

$$x \mapsto 2\left(1+x^2\right)x$$

from $x = 2$ to $x = 3$ is preserved by the change of variables defined by $u = x^2 + 1$.

322 Chapter 7 Integration Techniques

Solution The area under the curve corresponding to

> I1 := Int(f1(x),x=2..3);

$$\int_2^3 2\left(1+x^2\right)x\,dx$$

is bounded by the x-axis, the two lines

> l1 := [[2,0],[2,f1(2)]]:
> l2 := [[3,0],[3,f1(3)]]:

and the curve $f1$.

> plot({l1,l2,f1(x)},x=2..3,y=0..60);

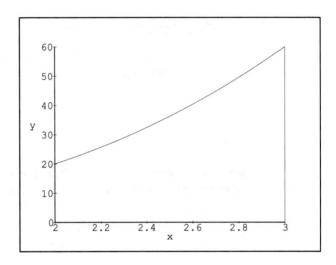

Now consider the change of variables

> g := (x)->x^2+1:
> I2 := changevar(u=g(x) , I1 , u);

$$\int_5^{10} u\,du$$

The area represented by this integral is bounded by the u-axis, the curve

> f2 := makeproc(op(1,I2) , u);

$$u \mapsto u$$

and the lines

> l3 := [[5,0],[5,f2(5)]]:
> l4 := [[10,0],[10,f2(10)]]:

as in

```
> plot( {f2(u),l3,l4},u=5..10,'h'=0..10);
```

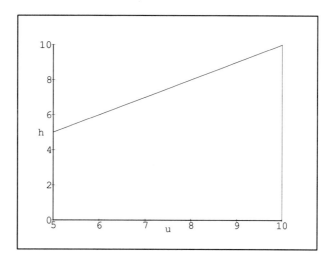

Note that 2 → g(2) and 3 → g(3). In both cases the area is exactly

```
> value( I2 );
```

$$\frac{75}{2}$$

∎

REMARK: For definite integrals, always remember to transform the starting and endpoints of the interval when doing a change of variables.

Example 7.6 Find the area between the curve $x^3 \cos(x^4 + 2)$ and the x-axis from $x = 0$ to $x = 1$.

Solution The integrand defines a function

```
> f := (x) -> x^3*cos(x^4+2);
```

$$x \mapsto x^3 \cos(x^4 + 2)$$

This time the area of interest is bounded by the curve *h* and the vertical line

```
> [[1,0],[1,f(1)]];
```

$$[[1, 0], [1, \cos(3)]]$$

as in

```
> plot({'',f(x)},x=0..1);
```

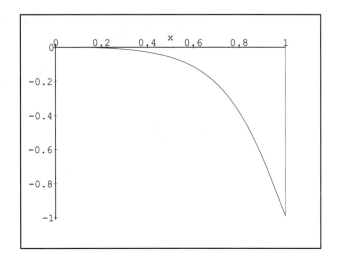

The total area is

> I1 := -Int(f(x), x=0..1);

$$-\int_0^1 x^3 \cos(x^4 + 2)\, dx$$

which has an approximate value of

> evalf(");

$$0.1920443547$$

Using the transformation $u = g(x)$, where

> g := (x) -> x^4 + 2;

$$x \mapsto x^4 + 2$$

the new definite integral is

> I2 := changevar(u=g(x) , I1 , u);

$$-\int_2^3 \frac{1}{4} \cos(u)\, du$$

which has an exact value

> value(I2);

$$-\frac{1}{4}\sin(3) + \frac{1}{4}\sin(2)$$

Compare this with the approximate value of $I1$.

> evalf(I1) = evalf(I2);

$$0.1920443547 = 0.1920443547$$

∎

Functional Notation

In the preceding sections we relied heavily on the D operator, which transforms functions into functions, to compute the appropriate relation between derivatives. This interpretation was valid for the following reason.

Consider the equation

```
> eq1 := u=x^2 + 1;
```

$$u = x^2 + 1$$

defining a typical change of variables transformation. This equation is more than a statement about two fixed parameters x and u. It must be true for *all possible pairs* $[x, u]$. If x and u vary with, say, time, then for all t, the equation

```
> eq1(t);
```

$$u(t) = x(t)^2 + 1$$

must hold.

The operator D must be applied to functions, and it assumes that eq is a relationship between functions. If two functions are equal, then their derivatives are equal, as in

```
> D(eq1);
```

$$D(u) = 2 D(x) x$$

As well as equations like $u = x^2 + 1$, we need to be able to interpret equations such as

```
> eq2 := u = sin(x);
```

$$u = \sin(x)$$

as a relationship between the functions x and u. The correct interpretation is that

```
> u(t) = sin(x(t));
```

$$u(t) = \sin(x(t))$$

must hold for all values of t. The right-hand side of this equation is the result of applying the functional composition $\sin \circ x$ to t.

To avoid any possible confusion over the meaning of $\sin(x)(t)$ in

```
> (eq2)(t);
```

$$u(t) = \sin(x)(t)$$

we rewrite $\sin(x)$ as the $\sin \circ x$ before interpreting $eq2$ as a relation between functions. This is accomplished in *Maple* by the command

```
> eq3 := convert( eq2 , '@');
```

$$u = \sin \circ x$$

and it leads to the correct functional interpretations

```
> eq3(t);
```

$$u(t) = \sin(x(t))$$

and

```
> D(eq3);
```

$$D(u) = \cos \circ x \, D(x)$$

At any time, we can rewrite sin ∘x as sin(x) by using the command

```
> convert( sin @ x , nested );
```

$$\sin(x)$$

In particular, *eq3* can be rewritten as

```
> eq4 := convert( D(eq3) , nested );
```

$$D(u) = \cos(x) \, D(x)$$

If we interpret *du* as $D(u)$ and *dx* as $D(x)$, then the integral

```
> I1 := Int( sin(x)^2*cos(x) , x );
```

$$\int \sin(x)^2 \cos(x) \, dx$$

can be transformed, by using the relationships

```
> [eq2,eq4];
```

$$[u = \sin(x), D(u) = \cos(x) \, D(x)]$$

to the integral

```
> changevar(eq2,I1,u);
```

$$\int u^2 \, du$$

7.3 Integration by Parts

We continue to seek rules that will help us when we are confronted with an integral of a product. The next transformation we consider is known as *integration by parts*. It arises from a careful examination of the standard rule for computing derivatives of products.

Let the differentiable functions F, G, and H satisfy

```
> eq1 := F*G = H;
```

$$FG = H$$

Their derivatives $D(F)$, $D(G)$, and $D(H)$ are related by the equation

```
> map(D,");
```

$$D(F)G + F \, D(G) = D(H)$$

obtained by differentiating both sides of this equation. When these functions are applied to an argument x, the resulting expressions in x are related by the equation

```
> eq2 := "(x);
```

$$D(F)(x)G(x) + F(x)D(G)(x) = D(H)(x)$$

The result of integrating both sides of *eq2* with respect to x is a new equation that must be interpreted as "equal to within a constant."[3]

```
> map(Int, " , x );
```

$$\int D(F)(x)G(x) + F(x)D(G)(x)\, dx = \int D(H)(x)\, dx$$

This in turn can be rewritten as

```
> expand(") :
> isolate(", Int( D(G)(x)*F(x),x));
```

$$\int F(x)D(G)(x)\, dx = \int D(H)(x)\, dx - \int D(F)(x)G(x)\, dx$$

By the fundamental theorem, this evaluates to

```
> value(");
```

$$\int F(x)D(G)(x)\, dx = H(x) - \int D(F)(x)G(x)\, dx$$

or equivalently

```
> value( subs( H=F*G,"));
```

$$\int F(x)D(G)(x)\, dx = F(x)G(x) - \int D(F)(x)G(x)\, dx$$

Note that one of the terms in the product, in our case $D(F)(x)$, is thought of as "a derivative" and the other is thought of as an "antiderivative." This transformation rule is usually described as follows.

THEOREM 7.1 (Integration by Parts) Let f and g be integrable functions, then

$$\int f(x)g'(x)\,dx = f(x)g(x) - \int f'(x)g(x)\,dx$$

Because, for $u = f(x)$ and $v = g(x)$, we have $dv = g'(x)dx$, this can also be written as

$$\int u\, dv = uv - \int v\, du$$

---

[3] Recall that there is an infinite number of antiderivatives for any given function, all differing by a constant. The representative found by **Int( )** often depends on the form (expanded, factored, etc.) of the original integrand.

REMARK: Such a transformation may not actually bring us any closer to a solution in the sense of evaluating the integral. Integration by parts is a tool that, like change of variables, can be used to examine an integration problem from different points of view. We must still learn how to use the tool effectively, however.

In *Maple*, integration by parts is available as part of the *student* package. It accepts two arguments. The first is an expression containing an unevaluated integral and the second indicates the part that is to be regarded as already being an antiderivative.

```
> I1 := Int( F(x)* D(G)(x) , x);
```

$$\int F(x) D(G)(x)\, dx$$

```
> intparts(I1,F(x));
```

$$F(x)G(x) - \int \frac{d}{dx} F(x) G(x)\, dx$$

Recall that $\frac{d}{dx} F(x) = D(F)(x)$.

It remains our choice whether to regard a particular term of a product in x as the derivative or the antiderivative. When the integrand is given as $f(x)g(x)$, the rule becomes

```
> I2 := Int( f(x)*g(x) ,x ):
> I2 = intparts(",f(x));
```

$$\int f(x)g(x)\, dx = f(x) \int g(x)\, dx - \int \frac{d}{dx} f(x) \int g(x)\, dx\, dx$$

The only real change here is notational. We have used

```
> Int( g(x),x): " = value(");
```

$$\int g(x)\, dx = \int g(x)\, dx$$

instead of

```
> Int( D(G)(x),x): " = value(");
```

$$\int D(G)(x)\, dx = G(x)$$

7.3.1 Applications of Integration by Parts

Integration by parts is used most often where we have been unable to find a helpful change of variables.

7.3 Integration by Parts

Example 7.7 Evaluate

```
> I1 := Int(x*sin(x),x);
```

$$\int x \sin(x)\, dx$$

Solution The derivative of the sine function involves the cosine function, which does not appear in the integrand, so a straightforward change of variables is unlikely to be effective in this example. Because we know how to integrate and differentiate both factors x and $\sin(x)$ of the integrand, we have a number of options for integration by parts. Our objective is to simplify the integrand and our obvious choices are

```
> intparts(I1,sin(x));
```

$$\frac{1}{2}\sin(x)x^2 - \int \frac{1}{2}\cos(x)x^2\, dx$$

and

```
> intparts(I1,x);
```

$$-x\cos(x) - \int -\cos(x)\, dx$$

The second of these is clearly the most helpful, yielding a value of

```
> value(");
```

$$-x\cos(x) + \sin(x)$$

∎

REMARK: For integrals of the form $p(x)\sin(x)$ where $p(x)$ is a polynomial in x, we can reduce the degree of the polynomial by regarding $p(x)$ as the antiderivative.

It is sometimes helpful to regard a single term as a product multiplied by 1.

Example 7.8 Evaluate

```
> I2 := Int( ln(x) , x );
```

$$\int \ln(x)\, dx$$

Solution If we regard the integrand as the product $1\,\ln(x)$ and take $\ln(x)$ as the antiderivative, then we obtain

```
> intparts( I2,ln(x))= value(I2);
```

$$\ln(x)x - \int 1\, dx = \ln(x)x - x$$

The new integrand produced by the integration by parts was really

```
> Int( 1 ,x )*Diff(ln(x),x);
```

$$\int 1\,dx \frac{d}{dx}\ln(x)$$

which simplifies to

> value(");

$$1$$

∎

Example 7.9 Find

> I3 := Int(x^2*exp(x),x);

$$\int x^2 e^x \, dx$$

Solution To reduce the degree of a polynomial in a product involving a polynomial and almost any other type of elementary function, we can use integration by parts. This process is repeated until the polynomial has been reduced to a constant.

> intparts(",x^2);

$$x^2 e^x - \int 2xe^x \, dx$$

> intparts(",x);

$$x^2 e^x - 2xe^x + \int 2e^x \, dx$$

> value(");

$$x^2 e^x - 2xe^x + 2e^x$$

∎

Example 7.10 Evaluate

> I4 := Int(exp(x)*sin(x),x);

$$\int e^x \sin(x) \, dx$$

Solution If you repeatedly take the derivative of sin(x), you eventually get back to sin(x). Furthermore, exp(x) is unchanged by integration or differentiation with respect to x. This means that after repeated application of integration by parts, the original integral will appear as part of the answer.

> intparts(",exp(x));

$$-e^x \cos(x) - \int -e^x \cos(x) \, dx$$

```
> intparts(",exp(x));
```

$$-e^x \cos(x) + e^x \sin(x) + \int -e^x \sin(x)\, dx$$

The original integral is part of this expression, so a solution is

```
> isolate(I4 = expand("),I4);
```

$$\int e^x \sin(x)\, dx = -\frac{1}{2} e^x \cos(x) + \frac{1}{2} e^x \sin(x)$$

∎

It is difficult to avoid finding the solution for this example because we can either integrate or differentiate. Consider

```
> I4;
```

$$\int e^x \sin(x)\, dx$$

```
> intparts(",sin(x));
```

$$e^x \sin(x) - \int e^x \cos(x)\, dx$$

```
> intparts(",cos(x));
```

$$-e^x \cos(x) + e^x \sin(x) + \int -e^x \sin(x)\, dx$$

Again we have constructed the original integral as part of this expression, and the complete solution is

```
> isolate(I4 = expand("),I4);
```

$$\int e^x \sin(x)\, dx = -\frac{1}{2} e^x \cos(x) + \frac{1}{2} e^x \sin(x)$$

7.3.2 Definite Integrals

An indefinite integral such as

```
> Int(x*sin(x),x);
```

$$\int x \sin(x)\, dx$$

is transformed via integration by parts to

```
> intparts(",x);
```

$$-x \cos(x) - \int -\cos(x)\, dx$$

The first term in this expression is part of the antiderivative. When the fundamental theorem is used to evaluate a corresponding definite integral, both parts of the antiderivative must be evaluated at the appropriate x values and subtracted. Thus, the corresponding definite integral

```
> Int( x*sin(x),x=a..b);
```

$$\int_a^b x \sin(x)\, dx$$

becomes

```
> intparts(",x);
```

$$-b\cos(b) + a\cos(a) - \int_a^b -\cos(x)\, dx$$

Example 7.11 Calculate

```
> Int(arctan(x),x=0..1);
```

$$\int_0^1 \arctan(x)\, dx$$

Solution There are no obvious factors of the integrand, but, by regarding arctan(x) as the antiderivative, we can make progress.

```
> s := intparts(",arctan(x));
```

$$\frac{1}{4}\pi - \int_0^1 \frac{x}{1+x^2}\, dx$$

Contrast this with the change of variables on the corresponding indefinite integral.

```
> intparts( Int( arctan(x),x) , arctan(x));
```

$$\arctan(x)x - \int \frac{x}{1+x^2}\, dx$$

The constant in s is just

```
> subs(x=1,op(1,")) - subs(x=0,op(1,"));
```

$$\frac{1}{4}\pi$$

To complete the solution we do the change of variables

```
> changevar(u = x^2+1,s,u);
```

$$\frac{1}{4}\pi - \int_1^2 \frac{1}{2u}\, du$$

which evaluates to

```
> value(");
```

$$\frac{1}{4}\pi - \frac{1}{2}\ln(2)$$

This simplifies to

```
> simplify(",arctrig);
```

$$\frac{1}{4}\pi - \frac{1}{2}\ln(2)$$

∎

7.3.3 Reduction Formulas

Example 7.12   Prove the reduction formula

$$\int \sin^n x\, dx = \frac{1}{n}\cos x \sin^{n-1} x + \frac{n-1}{n}\int \sin^{n-2} x\, dx$$

where $n \geq 2$ is an integer.

Solution   We illustrate this for a specific value of n.

```
> I1 := Int(sin(x)^8,x);
```

$$\int \sin(x)^8\, dx$$

```
> intparts(",sin(x)^(7));
```

$$-\sin(x)^7 \cos(x) - \int -7\sin(x)^6 \cos(x)^2\, dx$$

If we rewrite this entirely in terms of $\sin(x)$

```
> powsubs(cos(x)^2=1-sin(x)^2,");
```

$$-\sin(x)^7 \cos(x) - \int -7\sin(x)^6 \left(1 - \sin(x)^2\right) dx$$

then the original integral appears on the right again so that we have

```
> isolate(I1 = expand("),I1);
```

$$\int \sin(x)^8\, dx = -\frac{1}{8}\sin(x)^7 \cos(x) + \frac{7}{8}\int \sin(x)^6\, dx$$

Thus, if we can integrate $(\sin(x))^{18}$, we can integrate $(\sin(x))^{20}$.   ∎

There is no unique way to approach an integration problem such as found in Example 7.12. The following approach to the case $n = 8$ is just as valid.

```
> intparts(I1,sin(x)^4);
```

$$\sin(x)^4 \left(-\frac{1}{4}\sin(x)^3 \cos(x) - \frac{3}{8}\cos(x)\sin(x) + \frac{3}{8}x\right) -$$
$$\int 4\sin(x)^3 \cos(x) \left(-\frac{1}{4}\sin(x)^3 \cos(x) - \frac{3}{8}\cos(x)\sin(x) + \frac{3}{8}x\right) dx$$

```
> subs(cos(x)^2=1-sin(x)^2,"):
> expand(");
```

$$-\frac{1}{4}\sin(x)^7\cos(x) - \frac{3}{8}\sin(x)^5\cos(x) + \frac{3}{8}\sin(x)^4 x + \int \sin(x)^6 \cos(x)^2\, dx +$$
$$\frac{3}{2}\int \sin(x)^4 \cos(x)^2\, dx - \frac{3}{2}\int \sin(x)^3 \cos(x)x\, dx$$

We can still get a final answer, provided we can compute all the integrals involving lesser powers of sin(x).

An interesting computational question is to discover which is more efficient: reducing the powers two at a time via the preceding reduction formula, or carrying out some sort of binary decomposition (i.e., dividing the total powers of sin(x) in half at each stage).

7.4 Trigonometric Substitutions

Trigonometric identities enable us to reformulate a great many problems. For example, the identity

```
> sin(x)^2 + cos(x)^2 = 1:
```

can be used to effect the substitutions

```
> s1 := isolate(",sin(x)^2);
```

$$\sin(x)^2 = 1 - \cos(x)^2$$

```
> s2 := isolate("",cos(x)^2);
```

$$\cos(x)^2 = 1 - \sin(x)^2$$

and the identity

```
> cos(2*x) = expand( cos(2*x) ):
```

can be used to effect the substitution

```
> s3 := isolate(",cos(x)^2);
```

$$\cos(x)^2 = \frac{1}{2}\cos(2x) + \frac{1}{2}$$

These substitutions allow us to reduce expressions such as $\sin^k(x)$ and $\cos^k(x)$ to sums involving $\sin(u)$ and $\cos(u)$ for various values of u, which are linear in x. For example, the expression

```
> e1 := (sin^3)(x);
```

$$\sin(x)^3$$

becomes

```
> powsubs( (sin^2)(x)=1-(cos^2)(x) , " ):
> expand(",cos);
```

7.4 Trigonometric Substitutions

$$\sin(x) - \sin(x)\cos(x)^2$$

which in turn can be written as

```
> e2 := expand( subs(s3,"), cos );
```

$$\frac{1}{2}\sin(x) - \frac{1}{2}\sin(x)\cos(2x)$$

(We have specified **cos** as a second argument to **expand( )** to force **expand( )** not to expand $\cos(2x)$.) This can be even further simplified by using the identity:

```
> a0 := sin(a+b) = expand(sin(a+b));
```

$$\sin(a+b) = \sin(a)\cos(b) + \cos(a)\sin(b)$$

```
> a1 := subs( a=2*x,b=x,a0);
```

$$\sin(3x) = \sin(2x)\cos(x) + \sin(x)\cos(2x)$$

and

```
> b1 := subs( a=2*x,b=-x,a0);
```

$$\sin(x) = \sin(2x)\cos(x) - \sin(x)\cos(2x)$$

Together they imply

```
> (a1 - b1)/2;
```

$$\frac{1}{2}\sin(3x) - \frac{1}{2}\sin(x) = \sin(x)\cos(2x)$$

so that *e2* simplifies to

```
> e3 := powsubs(rhs(")=lhs("),e2);
```

$$\frac{3}{4}\sin(x) - \frac{1}{4}\sin(3x)$$

This type of simplification can be done all at once by the command **combine(...,trig)**.

```
> combine((sin^3)(x),trig);
```

$$\frac{3}{4}\sin(x) - \frac{1}{4}\sin(3x)$$

Similarly, we can reformulate

```
> (cos^5)(x) = combine( (cos^5)(x) , trig );
```

$$\cos(x)^5 = \frac{1}{16}\cos(5x) + \frac{5}{16}\cos(3x) + \frac{5}{8}\cos(x)$$

or

```
> (sin^8)(x) = combine( (sin^8)(x) , trig );
```

$$\sin(x)^8 = \frac{35}{128} + \frac{1}{128}\cos(8x) - \frac{1}{16}\cos(6x) + \frac{7}{32}\cos(4x) - \frac{7}{16}\cos(2x)$$

Each of the trig functions in the result has an argument that is linear in x, so any integral of the original expression can be completed by performing some simple changes of variables.

Example 7.13 Find

```
> I1 := Int(sin(x)^2,x);
```

$$\int \sin(x)^2 \, dx$$

Solution This integration problem is equivalent to

```
> combine(",trig);
```

$$\int \frac{1}{2} - \frac{1}{2}\cos(2x) \, dx$$

which can be solved by the change of variables

```
> changevar(2*x=v,",v);
```

$$\int \frac{1}{4} - \frac{1}{4}\cos(v) \, dv$$

```
> value(");
```

$$\frac{1}{4}v - \frac{1}{4}\sin(v)$$

In terms of the original variable x, this is

```
> subs(v=2*x,");
```

$$\frac{1}{2}x - \frac{1}{4}\sin(2x)$$

∎

Example 7.14 Find

```
> I2 := Int(sin(x)^4,x);
```

$$\int \sin(x)^4 \, dx$$

Solution This integral is equivalent to

```
> combine(I1,trig);
```

$$\int \frac{1}{2} - \frac{1}{2}\cos(2x) \, dx$$

and can be written as the sum of simpler integrals, as in

```
> I2a := expand(",cos);
```

$$\frac{1}{2}\int 1\,dx - \frac{1}{2}\int \cos(2x)\,dx$$

The change of variables

```
> changevar(2*x=v,",v);
```

$$\frac{1}{2}\int \frac{1}{2}\,dv - \frac{1}{2}\int \frac{1}{2}\cos(v)\,dv$$

yields the antiderivative

```
> value(");
```

$$\frac{1}{4}v - \frac{1}{4}\sin(v)$$

which can be reexpressed in terms of x as

```
> subs( v=2*x,");
```

$$\frac{1}{2}x - \frac{1}{4}\sin(2x)$$

∎

An alternative approach for this example is to use integration by parts.

Example 7.15 Evaluate

```
> I3 := Int(cos(x)^3,x=0..Pi/2);
```

$$\int_0^{\frac{1}{2}\pi} \cos(x)^3\,dx$$

Solution This is equivalent to the integral

```
> combine(I3,trig);
```

$$\int_0^{\frac{1}{2}\pi} \frac{1}{4}\cos(3x) + \frac{3}{4}\cos(x)\,dx$$

which simplifies to

```
> I3a := expand(",cos);
```

$$\frac{1}{4}\int_0^{\frac{1}{2}\pi} \cos(3x)\,dx + \frac{3}{4}\int_0^{\frac{1}{2}\pi} \cos(x)\,dx$$

The change of variables $u = 3x$ can be used to obtain

```
> value(I3a);
```

$$\frac{2}{3}$$

∎

REMARK: A alternative approach to Example 7.15 is to use identity $s2$ directly, as in

```
> s2;
```

$$\cos(x)^2 = 1 - \sin(x)^2$$

```
> powsubs(s2,I3);
```

$$\int_0^{\frac{1}{2}\pi} \left(1 - \sin(x)^2\right) \cos(x)\, dx$$

The solution can then be computed by the change of variables

```
> changevar(sin(x)=v,",v);
```

$$\int_0^1 1 - v^2\, dv$$

We could also use integration by parts, as in

```
> intparts( I3 , cos(x)^2);
```

$$-\int_0^{\frac{1}{2}\pi} -2\cos(x)\sin(x)^2\, dx$$

```
> changevar(sin(x)=u,",u);
```

$$-\int_0^1 -2u^2\, du$$

7.4.1 Mixtures of Sines and Cosines

The trigonometric identities mentioned in the preceding section can be used to systematically reformulate any product of powers of sines and cosines into a sum of sines or cosines. The computations are routine, if somewhat tedious, but they give us a systematic approach to a wide class of integrals.

Example 7.16 Find

```
> I4 := Int(sin(x)^5*cos(x)^2,x);
```

$$\int \sin(x)^5 \cos(x)^2\, dx$$

Solution This simplifies to

```
> combine(",trig);
```

$$\int \frac{1}{64}\sin(7x) + \frac{1}{64}\sin(3x) - \frac{3}{64}\sin(5x) + \frac{5}{64}\sin(x)\, dx$$

which evaluates to

```
> value(");
```

$$-\frac{1}{448}\cos(7x) - \frac{1}{192}\cos(3x) + \frac{3}{320}\cos(5x) - \frac{5}{64}\cos(x)$$

∎

The more traditional approach to these problems is to use the minimal number of transformations possible to get the original expression into a form that can be integrated. For example, the substitutions *s1* and *s2* also allow us to eliminate even powers of **sin( )** or **cos( )**. Thus, for an odd power of **sin( )** (or **cos( )**) all but one of the powers can be eliminated using *s1* or *s2*, and this extra term is used by a change of variables. For *I4* we have

```
> powsubs(s1,I4);
```
$$\int \left(1 - \cos(x)^2\right)^2 \sin(x) \cos(x)^2 \, dx$$

```
> changevar(v=cos(x),",v);
```
$$\int -\left(1 - v^2\right)^2 v^2 \, dv$$

```
> value(");
```
$$-\frac{1}{7}v^7 + \frac{2}{5}v^5 - \frac{1}{3}v^3$$

which can be reexpressed in terms of *x* as

```
> subs(v=cos(x),");
```
$$-\frac{1}{7}\cos(x)^7 + \frac{2}{5}\cos(x)^5 - \frac{1}{3}\cos(x)^3$$

Can you show that these two answers differ by at most a constant?

7.4.2 Identities for Secant and Tangent

For these examples, we follow a more traditional approach based directly on identities corresponding to *s1* and *s2* on page 334 but for **sec( )** and **tan( )**.

```
> s4 := tan(x)^2=sec(x)^2-1;
```
$$\tan(x)^2 = \sec(x)^2 - 1$$

```
> s5 := isolate(",sec(x)^2);
```
$$\sec(x)^2 = \tan(x)^2 + 1$$

These identities are used to eliminate one of **sec( )** or **tan( )** and to force the appropriate expression

```
> diff(tan(x),x);
```
$$\tan(x)^2 + 1$$

or

```
> diff(sec(x),x);
```
$$\sec(x)\tan(x)$$

to be present as a term in the integrand. This sets things up properly for a change of variables such as $v = \tan(x)$ or $v = \sec(x)$.

Example 7.17 Find

```
> I5 := Int(tan(x)^3,x);
```

$$\int \tan(x)^3 \, dx$$

Solution We can rewrite this expression as

```
> powsubs(s4,I5):
> expand(");
```

$$\int \tan(x) \sec(x)^2 \, dx - \int \tan(x) \, dx$$

It remains to evaluate the two integrals

```
> I5a := op(1,"):
> I5b := op(2,""):
```

Because

```
> diff(sec(x),x);
```

$$\sec(x) \tan(x)$$

appears as part of the integrand of *I5a*, we can use the change of variables

```
> I5a := changevar(v=sec(x),I5a,v);
```

$$\int v \, dv$$

The complete answer is

```
> I5 = subs(v=sec(x),value(")) + value(I5b);
```

$$\int \tan(x)^3 \, dx = \frac{1}{2} \sec(x)^2 + \ln(\cos(x))$$

■

Example 7.18 Evaluate

```
> I6 := Int( tan(x)^6*sec(x)^4,x);
```

$$\int \tan(x)^6 \sec(x)^4 \, dx$$

Solution This time we can eliminate all occurrences of tan by using *s4*, as in

```
> powsubs( s4,");
```

$$\int \left(\sec(x)^2 - 1\right)^3 \sec(x)^4 \, dx$$

7.4 Trigonometric Substitutions

Two of the powers of sec(x) are used by the change of variables $v = \tan(x)$ because $\frac{d}{dx}\tan(x) = \sec^2(x)$, which must appear as a factor of the integrand. The result is

```
> changevar(v=tan(x),",v);
```

$$\int v^6 \left(v^2 + 1\right) dv$$

which evaluates to

```
> value(");
```

$$\frac{1}{9}v^9 + \frac{1}{7}v^7$$

∎

Example 7.19 Find

```
> I7 := Int(tan(x)^5*sec(x)^7,x);
```

$$\int \tan(x)^5 \sec(x)^7 \, dx$$

Solution Again we use s4 to rewrite as much of the integrand as possible in terms of sec(x).

```
> powsubs(s4,I7);
```

$$\int \left(\sec(x)^2 - 1\right)^2 \tan(x) \sec(x)^7 \, dx$$

(We map **expand**( ) onto the expression, forcing the command to affect only the integrand.)

```
> map(expand,");
```

$$\int \tan(x)\sec(x)^{11} - 2\tan(x)\sec(x)^9 + \tan(x)\sec(x)^7 \, dx$$

The factor $\frac{d}{dx}\sec(x) = \sec(x)\tan(x)$ is now present, so the change of variables

```
> changevar(v=sec(x),"",v);
```

$$\int \left(v^2 - 1\right)^2 v^6 \, dv$$

evaluates to

```
> value(");
```

$$\frac{1}{11}v^{11} - \frac{2}{9}v^9 + \frac{1}{7}v^7$$

∎

7.4.3 Integrating Secant

Not all changes of variables are so easy to recognize. The following example is one that, short of a lot of experimentation with derivatives, would be hard to guess. It is included here primarily as an example of an obscure transformation.

Example 7.20 Show that

```
> I8 := Int(sec(x),x):
> " = value(");
```

$$\int \sec(x)\,dx = \ln(\sec(x) + \tan(x))$$

Solution Note that this is the kind of solution that would result if somehow the integral could be reduced to

```
> Int(1/u,u);
```

$$\int u^{-1}\,du$$

with $u = \sec(x) + \tan(x)$. This suggests trying the change of variables defined by

```
> eq := u = sec(x) + tan(x);
```

$$u = \sec(x) + \tan(x)$$

The change of variables done naively produces the very messy

```
> changevar(u=sec(x)+tan(x),I8,u);
```

$$\int \sec(2\arctan(\frac{1}{2}\frac{-2+2u}{u+1}))$$
$$\left(\sec(2\arctan(\frac{1}{2}\frac{-2+2u}{u+1}))\tan(2\arctan(\frac{1}{2}\frac{-2+2u}{u+1})) + 1 + \tan(2\arctan(\frac{1}{2}\frac{-2+2u}{u+1}))^2\right)^{-1}\,du$$

which, nevertheless, does simplify to

```
> normal(expand("));
```

$$\int u^{-1}\,du$$

as required. ∎

REMARK: By hand, these computations benefit from some clever observations. For example, given the transformation

```
> u = sec(x) + tan(x);
```

$$u = \sec(x) + \tan(x)$$

the functional relation between $D(u)$ and $D(x)$ is given by

```
> convert(",`@`):
> map(D,");
```

$$D(u) = \sec\tan\circ x\,D(x) + D(x)$$

```
> convert(",nested);
```
$$D(u) = \sec(x)\tan(x) D(x) + \left(\tan(x)^2 + 1\right) D(x)$$

The integrand in this problem can then be written as

```
> expand(rhs(eq)*sec(x))/rhs(eq);
```
$$\frac{\sec(x)^2 + \sec(x)\tan(x)}{\sec(x) + \tan(x)}$$

```
> subs( s5 , " );
```
$$\frac{\sec(x)\tan(x) + 1 + \tan(x)^2}{\sec(x) + \tan(x)}$$

As a result, many of the terms cancel, thereby avoiding the complicated simplification we did by machine.

Example 7.21   Find

```
> I9 := Int(sec(x)^3,x);
```
$$\int \sec(x)^3\, dx$$

Solution   For even powers of sec(x) we can rewrite all but two of the powers in terms of tan(x) and then use a change of variables of the form $u = \tan(x)$. Here, we introduce an approach that is applicable to all odd powers of sec(x). The current integral is

```
> n := 3:
> I9 := Int( sec(x)^n,x);
```
$$\int \sec(x)^3\, dx$$

We first use integration by parts, treating $\sec^2(x)$ as the derivative.

```
> intparts(",sec(x)^(n-2));
```
$$\frac{\sec(x)\sin(x)}{\cos(x)} - \int \frac{\sec(x)\tan(x)\sin(x)}{\cos(x)}\, dx$$

The identities $\dfrac{\sin}{\cos} = \tan$ and s4 lead to the expression

```
> powsubs(sin(x)/cos(x)=tan(x),s4,");
```
$$\sec(x)\tan(x) - \int \sec(x)\left(\sec(x)^2 - 1\right)\, dx$$

A recursive decomposition in terms of lower powers is now available since *I9* is part of the preceding expression. We have

```
> isolate( I9=expand("), I9);
```

$$\int \sec(x)^3 \, dx = \frac{1}{2} \sec(x)\tan(x) + \frac{1}{2} \int \sec(x) \, dx$$

which evaluates to

```
> value(rhs("));
```

$$\frac{1}{2} \sec(x)\tan(x) + \frac{1}{2} \ln(\sec(x) + \tan(x))$$

∎

REMARK: This recursive decomposition can be used to produce a solution for any odd n with $n > 1$. For example, try it for $n = 5$ and $n = 7$.

7.4.4 Sums and Differences of Angles

These examples are all handled by the group of transformations carried out by **combine(...,trig)**.

Example 7.22 Evaluate

```
> I10 := Int(sin(4*x)*cos(5*x),x);
```

$$\int \sin(4x)\cos(5x) \, dx$$

Solution To solve this integral, we do

```
> combine(I10,trig);
```

$$\int \frac{1}{2}\sin(9x) - \frac{1}{2}\sin(x) \, dx$$

Each of these remaining integrals can be handled by a simple change of variables. They evaluate to

```
> value(");
```

$$-\frac{1}{18}\cos(9x) + \frac{1}{2}\cos(x)$$

∎

7.5 Square Roots of Quadratics

The earlier trigonometric identities

```
> s1, s2;
```

$$\sin(x)^2 = 1 - \cos(x)^2, \quad \cos(x)^2 = 1 - \sin(x)^2$$

and

> s4, s5;

$$\tan(x)^2 = \sec(x)^2 - 1, \ \sec(x)^2 = \tan(x)^2 + 1$$

can also be used to simplify integrands involving $\sqrt{1-x^2}$, $\sqrt{x^2-1}$, and $\sqrt{1+x^2}$. A general strategy is

1. Complete the square under the square root.
2. Perform a change of variables, followed by a removal of common factors, in order to rewrite the completed square as $1 \pm u^2$.
3. Use one of these identities to turn the expression into a perfect square.

Example 7.23 Evaluate

> I1 := Int(sqrt(9-x^2)/x^2,x);

$$\int \sqrt{9-x^2} \, x^{-2} \, dx$$

Solution In this case, the expression under the square root is already in the form produced by completing the square. Still, to use one of the four identities just mentioned, we must have a 1 instead of 9 under the square root sign. The change of variables

> changevar(x=3*u,I1,u);

$$\int \frac{1}{3} \sqrt{9 - 9u^2} \, u^{-2} \, du$$

among other things, replaces x^2 by $9u^2$ so that 9 becomes a common factor. The integrand can be restructured in many ways, by using the commands **factor( )**, **simplify( )**, **expand( )**, **combine( )**, and **normal( )**. The command **normal( )** has as its primary role the elimination of common factors and does the required task.

> normal(");

$$\int \sqrt{1-u^2} \, u^{-2} \, du$$

Next, the change of variables

> changevar(u=sin(v),",v);

$$\int \sqrt{1-\sin(v)^2} \, \cos(v) \sin(v)^{-2} \, dv$$

sets the stage for use of the identity

> powsubs(1-sin(v)^2=cos(v)^2,");

$$\int \frac{\cos(v)^2}{\sin(v)^2} \, dv$$

This in turn can be expressed in terms of the tangent function.

> powsubs(cos(v)/sin(v)=tan(v),");

$$\int \tan(v)^2 \, dv$$

The resulting expression evaluates to

```
> value(");
```

$$\tan(v) - \arctan(\tan(v))$$

which can be simplified to

```
> simplify(",arctrig);
```

$$\tan(v) - v$$

∎

Integrals involving square roots of a quadratic frequently occur when computing areas of conic sections.

Example 7.24 Find the area enclosed by the ellipse

```
> eq := x^2/a^2 + y^2/b^2 = 1;
```

$$\frac{x^2}{a^2} + \frac{y^2}{b^2} = 1$$

Solution This is the general equation for an ellipse. For example, with $a = 3$ and $b = 5$, we have

```
> with(plots):
```

```
> implicitplot(subs(a=3,b=5,eq),x=-5..5,y=-5..5);
```

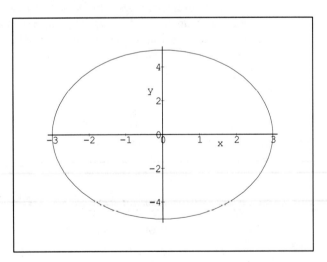

In general we know

```
> assume(a>0): assume(b>0):
> constants := constants,a,b;
```

$$\textit{false}, \gamma, \infty, \textit{true}, \textit{Catalan}, e, \pi, a, b$$

so that an explicit representation of y in terms of x is

```
> isolate(eq,y);
```
$$y = \sqrt{|-a^2 + x^2|}\, ba^{-1}$$

```
> e1 := rhs(");
```
$$\sqrt{|-a^2 + x^2|}\, ba^{-1}$$

There are other solutions, one for each root of the quadratic in y. This solution corresponds to one part of the curve. The ellipse is symmetric about both the x-axis and the y-axis so that the integral

```
> I2 := Int(e1,x=0..a);
```
$$\int_0^a \sqrt{|-a^2 + x^2|}\, ba^{-1}\, dx$$

is one-fourth of the total area. Because $0 \le x \le a$ on this interval, the argument to **abs( )** is always negative Thus, we can simplify this to

```
> subs(abs = ((x)-> -x),I2):
> I2 := ";
```
$$\int_0^a \sqrt{a^2 - x^2}\, ba^{-1}\, dx$$

which can be evaluated by a change of variables. We have

```
> changevar(x=u*a,",u);
```
$$\int_0^1 \sqrt{a^2 - u^2 a^2}\, b\, du$$

the value of which is

```
> value(");
```
$$\frac{1}{4} a\pi b$$

■

Example 7.25 Find

```
> I3 := Int( 1/x^2/sqrt( x^2 + 4 ) , x );
```
$$\int \frac{1}{x^2 \sqrt{x^2 + 4}}\, dx$$

Solution We proceed by converting the integral to one involving $\sqrt{u^2 + 1}$.

```
> changevar(x=2*u,",u);
```
$$\int \frac{1}{2u^2 \sqrt{4u^2 + 4}}\, du$$

348 Chapter 7 Integration Techniques

```
> simplify(");
```

$$\int \frac{1}{4u^2\sqrt{u^2+1}}\,du$$

The presence of $u^2 + 1$ dictates that we should consider the tangent substitution.

```
> changevar(u=tan(v),",v);
```

$$\int 1/4 \frac{\sqrt{1+\tan(v)^2}}{\tan(v)^2}\,dv$$

```
> powsubs(1+tan(v)^2=sec(v)^2,");
```

$$\int 1/4 \frac{\sec(v)}{\tan(v)^2}\,dv$$

This simplifies when we rewrite it in terms of sine and cosine. The substitution

```
> subs(sec=1/cos,tan=sin/cos,"):
```

evaluates to

```
> ";
```

$$\int 1/4 \frac{\cos(v)}{\sin(v)^2}\,dv$$

and the change of variables

```
> changevar(z=sin(v),",z);
```

$$\int \frac{1}{4z^2}\,dz$$

leads to the solution

```
> value(");
```

$$-\frac{1}{4z}$$

■

Example 7.26 Find

```
> I4 := Int(x/sqrt(x^2 + 4),x);
```

$$\int \frac{x}{\sqrt{x^2+4}}\,dx$$

Solution Always be on the lookout for easy changes of variables. In this case, the integral reduces to

```
> changevar(u=x^2+4,I4,u);
```

$$\int \frac{1}{2\sqrt{u}}\,du$$

A solution based on completing the square is still viable and makes a good exercise. ∎

Example 7.27 Evaluate

```
> I5 := Int(x/sqrt(x^2-4),x);
```

$$\int \frac{x}{\sqrt{x^2 - 4}}\, dx$$

Solution As with all these examples, our strategy is to put the square root part of the integrand into the form $\sqrt{u^2 - 1}$ or $\sqrt{u^2 + 1}$. This is accomplished by first changing x^2 into $4u^2$ and then factoring out the 4.

```
> changevar(x=2*u,I5,u);
```

$$\int \frac{4u}{\sqrt{4u^2 - 4}}\, du$$

```
> simplify(");
```

$$\int \frac{2u}{\sqrt{u^2 - 1}}\, du$$

The change of variables

```
> changevar(u=sec(v),",v);
```

$$\int \frac{2\sec(v)^2 \tan(v)}{\sqrt{\sec(v)^2 - 1}}\, dv$$

and the substitution

```
> subs(sec(v)^2-1=tan(v)^2,");
```

$$\int 2\sec(v)^2\, dv$$

lead to the solution

```
> value(");
```

$$\frac{2\sin(v)}{\cos(v)}$$

∎

Many quadratic expressions with half integer rational exponents are amenable to this technique.

Example 7.28 Find

```
> I6 := Int(x^3*(4*x^2 + 9)^(3/2),x);
```

$$\int x^3 \left(4x^2 + 9\right)^{3/2} dx$$

Solution First we use a change of variables to make it possible to factor out the unwanted constant.

```
> changevar(x=3*v/2 , " , v);
```

$$\int \frac{81}{16} v^3 \left(9v^2 + 9\right)^{3/2} dv$$

```
> simplify(");
```

$$\int \frac{2187}{16} v^3 \left(v^2 + 1\right)^{3/2} dv$$

The change of variables

```
> changevar(v=tan(t),",t);
```

$$\int \frac{2187}{16} \tan(t)^3 \left(1 + \tan(t)^2\right)^{5/2} dt$$

and the identity

```
> 1 + tan(t)^2 = sec(t)^2;
```

$$1 + \tan(t)^2 = \sec(t)^2$$

simplify this to

```
> subs(","");
```

$$\int \frac{2187}{16} \tan(t)^3 \sec(t)^5 dt$$

which evaluates as

```
> value(");
```

$$\frac{2187}{112} \frac{\sin(t)^4}{\cos(t)^7} + \frac{6561}{560} \frac{\sin(t)^4}{\cos(t)^5} + \frac{2187}{560} \frac{\sin(t)^4}{\cos(t)^3} - \frac{2187}{560} \frac{\sin(t)^4}{\cos(t)} - \frac{2187}{560} \sin(t)^2 \cos(t) - \frac{2187}{280} \cos(t)$$

∎

Because we can complete the square, this kind of approach works generally for polynomials of degree 2 under the square root sign.

Example 7.29 Evaluate

```
> I7 := Int( x/sqrt( 3 - 2*x - x^2 ), x );
```

$$\int \frac{x}{\sqrt{3 - 2x - x^2}} dx$$

7.5 Square Roots of Quadratics

Solution   We must first complete the square under the square root sign.

> completesquare(",x);

$$\int \frac{x}{\sqrt{-(x+1)^2+4}}\, dx$$

Then the changes of variables

> changevar(u=x+1,",u);

$$\int \frac{-1+u}{\sqrt{-u^2+4}}\, du$$

> changevar(u=2*v,",v);

$$\int \frac{-2+4v}{\sqrt{-4v^2+4}}\, dv$$

lead to the simplification

> normal(");

$$\int \frac{-1+2v}{\sqrt{-v^2+1}}\, dv$$

To eliminate the square root, we use the change of variables

> changevar(v=sin(z),",z);

$$\int \frac{(-1+2\sin(z))\cos(z)}{\sqrt{-\sin(z)^2+1}}\, dz$$

followed by the substitution

> subs(1-sin(z)^2=cos(z)^2,");

$$\int -1+2\sin(z)\, dz$$

The value of the resulting integral is

> value(");

$$-z - 2\cos(z)$$

∎

7.6 Partial Fraction Decompositions

An extremely important algebraic technique for the reformulation of integrals and many other mathematical problems is that of a *partial fraction decomposition*. This decomposition should be considered whenever a quotient of polynomials is involved.

Recall the integral

```
> Int(1/(a*x + b),x):
> " = value(");
```

$$\int (ax+b)^{-1} \, dx = \frac{\ln(ax+b)}{a}$$

A partial fraction decomposition often allows us to rewrite a general quotient of polynomials as a sum of quotients of the form $\frac{1}{ax+b}$. A typical example is

```
> f := (x^3 - 3*x^2 + 2*x - 1)/(x-1)/(x-2)/(x-3)/(x-4);
```

$$\frac{x^3 - 3x^2 + 2x - 1}{(x-1)(x-2)(x-3)(x-4)}$$

```
> f = convert(f,parfrac,x);
```

$$\frac{x^3 - 3x^2 + 2x - 1}{(x-1)(x-2)(x-3)(x-4)} = \frac{1}{6(x-1)} - \frac{1}{2(x-2)} - \frac{5}{2(x-3)} + \frac{23}{6(x-4)}$$

The corresponding integral is

```
> map(Int,",x);
```

$$\int \frac{x^3 - 3x^2 + 2x - 1}{(x-1)(x-2)(x-3)(x-4)} \, dx = \int \frac{1}{6(x-1)} - \frac{1}{2(x-2)} - \frac{5}{2(x-3)} + \frac{23}{6(x-4)} \, dx$$

7.6.1 Computing a Partial Fraction Decomposition

If we know the general form of the decomposition, then finding it becomes very routine. Consider

```
> f;
```

$$\frac{x^3 - 3x^2 + 2x - 1}{(x-1)(x-2)(x-3)(x-4)}$$

Is there a decomposition of the form

```
> eq := ( f = a/(x-1) + b/(x-2) + c/(x-3) + d/(x-4) );
```

$$\frac{x^3 - 3x^2 + 2x - 1}{(x-1)(x-2)(x-3)(x-4)} = \frac{a}{x-1} + \frac{b}{x-2} + \frac{c}{x-3} + \frac{d}{x-4}$$

(Note the role of the factors in the denominator.)

To find out, we search for values of a, b, c, and d. Multiplying through by a common denominator

```
> eq*denom(lhs(")): eq1 := simplify(");
```

$$x^3 - 3x^2 + 2x - 1 = ax^3 - 9ax^2 + 26ax - 24a + bx^3 - 8bx^2 + 19bx - 12b + cx^3 - 7cx^2 + 14cx - 8c + dx^3 - 6dx^2 + 11dx - 6d$$

The result is a polynomial equation in x that must be true for all values of x. Such an equation is known as a *polynomial identity in x*, and it can be used to find the values of a, b, c, and d.

We have already encountered such equations. Recall how we completed the square by "satisfying a polynomial identity." Both polynomials must be equal for all values of x, and this can be true only if the polynomial coefficients are all identical. In this case there are four nontrivial coefficients yielding four linear equations in four unknowns.

We construct equations from the coefficients as follows. By subtracting the right side from both sides of the equation, the right-hand polynomial becomes zero. Thus, all the coefficients on the left-hand side of the new equation must be zero.

```
> (lhs-rhs)(eq1);
```
$$x^3 - 3x^2 + 2x - 1 - ax^3 + 9ax^2 - 26ax + 24a - bx^3 + 8bx^2 - 19bx + 12b - cx^3 + 7cx^2 - 14cx + 8c - dx^3 + 6dx^2 - 11dx + 6d$$

After collecting like powers of x by the command

```
> collect(",x);
```
$$(-b - c + 1 - d - a)x^3 + (7c + 6d + 8b - 3 + 9a)x^2 - \\ -(-14c - 11d - 19b + 2 - 26a)x + 8c - 1 + 6d + 24a + 12b$$

the set of coefficients of the powers of x on the left is

```
> cf := { coeffs(",x) };
```
$$\{8c - 1 + 6d + 24a + 12b, -14c - 11d - 19b + 2 - 26a, 7c + 6d + 8b - 3 + 9a, \\ -b - c + 1 - d - a\}$$

All these must be zero. The command

```
> solve( cf );
```
$$\left\{ a = \frac{1}{6}, b = -\frac{1}{2}, c = -\frac{5}{2}, d = \frac{23}{6} \right\}$$

finds values of a, b, c, and d for which this is true.

(See if you can solve this system by solving the equations one at a time and substituting the results back into the original system.)

Polynomial Identities in *Maple*

For each partial fraction decomposition we compute, we have a polynomial identity to solve. Rather than carry out the full procedure as just outlined in every case, we can proceed as follows.

Example 7.30 Find a partial fraction decomposition for f.

```
> f := (x^3 - 3*x^2 + 2*x - 5)/(x-1)/(x-2)/(x-3)/(x-4);
```
$$\frac{x^3 - 3x^2 + 2x - 5}{(x-1)(x-2)(x-3)(x-4)}$$

Solution We must use the equation

```
> f = rhs(eq);
```

$$\frac{x^3 - 3x^2 + 2x - 5}{(x-1)(x-2)(x-3)(x-4)} = \frac{a}{x-1} + \frac{b}{x-2} + \frac{c}{x-3} + \frac{d}{x-4}$$

The polynomial identity in x is

```
> simplify( eq*denom(f) );
```

$$x^3 - 3x^2 + 2x - 1 = ax^3 - 9ax^2 + 26ax - 24a + bx^3 - 8bx^2 + 19bx - 12b + cx^3 - 7cx^2 + 14cx - 8c + dx^3 - 6dx^2 + 11dx - 6d$$

We can tell the **solve( )** command that this is an "identity for x" as follows.

```
> identity(",x);
```

$$identity(x^3 - 3x^2 + 2x - 1 = ax^3 - 9ax^2 + 26ax - 24a + bx^3 - 8bx^2 + 19bx - 12b + cx^3 - 7cx^2 + 14cx - 8c + dx^3 - 6dx^2 + 11dx - 6d, x)$$

The solution to the identity is obtained as

```
> solve(");
```

$$\left\{ x = x, a = \frac{1}{6}, b = -\frac{1}{2}, c = -\frac{5}{2}, d = \frac{23}{6} \right\}$$

and the partial fraction decomposition is

```
> subs(",rhs(eq));
```

$$\frac{1}{6(x-1)} - \frac{1}{2(x-2)} - \frac{5}{2(x-3)} + \frac{23}{6(x-4)}$$

∎

7.6.2 Patterns for Partial Fractions

The key to finding a partial fraction decomposition is identifying the patterns to look for. Typically, they are a sum of one or more terms of the form

```
> e1 := A/(a + x);
```

$$\frac{A}{a+x}$$

```
> e2 :=A/(a + x)^2;
```

$$\frac{A}{(a+x)^2}$$

```
> e3 :=A/(a + x)^3;
```

$$\frac{A}{(a+x)^3}$$

7.6 Partial Fraction Decompositions

```
> e4 := (B + C*x)/(a + b*x + c*x^2);
```

$$\frac{B + Cx}{a + bx + cx^2}$$

```
> e5 := (B + C*x)/(a + b*x + c*x^2)^2;
```

$$\frac{B + Cx}{\left(a + bx + cx^2\right)^2}$$

The clue as to which of these to choose lies in the factors of the denominator of f. Linear factors yield terms such as $e1$, $e2$, and $e3$, depending on how often they are repeated as linear factors. Quadratic factors yield terms such as $e4$ and $e5$.

For

```
> g := (x^3 - 3*x^2 + 2*x - 5)/(x^2-x-1)/(x-3)/(x-4);
```

$$\frac{x^3 - 3x^2 + 2x - 5}{\left(x^2 - x - 1\right)(x - 3)(x - 4)}$$

the pattern is

```
> convert(g,parfrac,x);
```

$$-\frac{1}{5(x-3)} + \frac{19}{11(x-4)} - \frac{1}{55}\frac{43 + 29x}{x^2 - x - 1}$$

As another example, note that the expression g^2 has repeated factors. The partial fraction decomposition of g^2 is

```
> convert(g^2,parfrac,x);
```

$$\frac{1}{25(x-3)^2} + \frac{22}{25(x-3)} + \frac{361}{121(x-4)^2} - \frac{2128}{1331(x-4)} + \frac{1}{33275}\frac{33705 + 23918x}{x^2 - x - 1} + \frac{1}{605}\frac{538 + 667x}{\left(x^2 - x - 1\right)^2}$$

The numerator needs to accommodate degrees up to one less than the degree of the single square-free factor used to define the denominator.

REMARK: Every example we have given so far had a numerator with a degree that was lower than the denominator. We can use long division to break other types of quotients into a polynomial plus a quotient of this form. We do this in some of the later examples.

7.6.3 Integrals with Quadratics in the Denominator

In the examples that follow, we need to evaluate integrals of the general form represented by

```
> I1 := Int(1/(x^2 - 4*x - 1),x):
> " = value(");
```

$$\int (x^2 - 4x - 1)^{-1} \, dx = -\frac{1}{5}\sqrt{5}\, arctanh(\frac{1}{10}(2x-4)\sqrt{5})$$

and

```
> I2 := Int( 1/(x^2-4*x + 9),x);
```

$$\int (x^2 - 4x + 9)^{-1} \, dx$$

```
> " = value(");
```

$$\int (x^2 - 4x + 9)^{-1} \, dx = \frac{1}{5}\sqrt{5}\, \arctan(\frac{1}{10}(2x-4)\sqrt{5})$$

Because the derivatives have the structure

```
> diff(arctan(v),v);
```

$$(1 + v^2)^{-1}$$

and

```
> diff(arctanh(v),v);
```

$$(1 - v^2)^{-1}$$

the functions arctan and arctanh arise naturally from the quadratic factors resulting from partial fraction decompositions.

For example, consider

```
> I2 = completesquare(I2,x);
```

$$\int (x^2 - 4x + 9)^{-1} \, dx = \int ((x - 2)^2 + 5)^{-1} \, dx$$

After the change of variables

```
> changevar(u=x-2,rhs("),u);
```

$$\int (u^2 + 5)^{-1} \, du$$

the unwanted constants are removed as common factors by the substitution

```
> changevar(u=sqrt(5)*v,",v);
```

$$\int \frac{\sqrt{5}}{5v^2 + 5} \, dv$$

and simplification

```
> normal(");
```

$$\int \frac{1}{5} \frac{\sqrt{5}}{1 + v^2} \, dv$$

7.6.4 Partial Fractions in Action

In the examples that follow, we systematically reduce any rational function whose denominator has at most quadratic factors to integrals involving ln, arctan, and arctanh.

Example 7.31 Find $\int \frac{x^3+x}{x-1} dx$.

Solution The integrand is

```
> f1 := (x^3+x)/(x-1);
```

$$\frac{x^3+x}{x-1}$$

Long division can be used to rewrite this as

```
> quo(numer(f1),denom(f1),x);
```

$$x^2+x+2$$

plus

```
> rem( numer(f1),denom(f1),x)/denom(f1);
```

$$\frac{2}{x-1}$$

The polynomial part is easy to integrate and in this case the fractional part is already a partial fraction decomposition as produced by

```
> convert(f1,parfrac,x);
```

$$x^2+x+2+\frac{2}{x-1}$$

The integral is then

```
> Int(",x);
```

$$\int x^2+x+2+\frac{2}{x-1} dx$$

```
> expand(");
```

$$\int x^2 \, dx + \int x \, dx + 2\int 1 \, dx + 2\int (x-1)^{-1} \, dx$$

which evaluates[4] as

```
> changevar(u=x-1,",u);
```

$$\int (1+u)^2 \, du + \int 1+u \, du + 2\int 1 \, du + 2\int u^{-1} \, du$$

---

[4]When computing by hand, we would apply the change of variables only to $\int \frac{2}{x-1} dx$. However, the change of variables has no adverse effect on the other integrals.

> value(");

$$\frac{1}{3}(1+u)^3 + 3u + \frac{1}{2}u^2 + 2\ln(u)$$

In terms of the original variable, the antiderivative is

> subs(u=x-1,");

$$\frac{1}{3}x^3 + 3x - 3 + \frac{1}{2}(x-1)^2 + 2\ln(x-1)$$

■

Example 7.32 Evaluate

$$\int \frac{4x^2 - 3x + 2}{4x^2 - 4x + 3} dx$$

Solution The integrand is

> f6 := (4*x^2 - 3*x + 2)/(4*x^2 - 4*x + 3);

$$\frac{4x^2 - 3x + 2}{4x^2 - 4x + 3}$$

which does not factor any further.

> factor(");

$$\frac{4x^2 - 3x + 2}{4x^2 - 4x + 3}$$

Thus, the partial fraction decomposition is

> convert(",parfrac,x);

$$1 + \frac{x-1}{4x^2 - 4x + 3}$$

The integral is

> Int(",x);

$$\int 1 + \frac{x-1}{4x^2 - 4x + 3} dx$$

which has the value

> expand(");

$$\int 1\, dx + \int \frac{x}{4x^2 - 4x + 3} dx - \int (4x^2 - 4x + 3)^{-1} dx$$

7.6 Partial Fraction Decompositions

```
> value(");
```

$$x + \frac{1}{8}\ln(4x^2 - 4x + 3) - \frac{1}{8}\sqrt{2}\arctan(\frac{1}{8}(8x-4)\sqrt{2})$$

■

Example 7.33 Write the form of the partial fraction decomposition of the function

$$\frac{x^3 + x^2}{x(x-1)(x^2+x+1)(x^2+1)^3}$$

Solution The integrand is

```
> f7 := (x^3 + x^2)/(x*(x-1)*(x^2 + x + 1)*(x^2 +1)^3);
```

$$\frac{x^3 + x^2}{x(x-1)\left(x^2+x+1\right)\left(x^2+1\right)^3}$$

```
> factor(");
```

$$\frac{x(x+1)}{(x-1)\left(x^2+x+1\right)\left(x^2+1\right)^3}$$

Thus, the partial fraction decomposition is

```
> convert(",parfrac,x);
```

$$\frac{1}{12(x-1)} - \frac{1}{3}\frac{x+2}{x^2+x+1} + \frac{1}{4}\frac{x+1}{x^2+1} - \frac{1}{2}\frac{x-1}{\left(x^2+1\right)^2} - \frac{x}{\left(x^2+1\right)^3}$$

The integral is

```
> Int(",x);
```

$$\int \frac{1}{12(x-1)} - \frac{1}{3}\frac{x+2}{x^2+x+1} + \frac{1}{4}\frac{x+1}{x^2+1} - \frac{1}{2}\frac{x-1}{\left(x^2+1\right)^2} - \frac{x}{\left(x^2+1\right)^3}\,dx$$

which has the value

```
> expand(");
```

$$\frac{1}{12}\int(x-1)^{-1}\,dx - \frac{1}{3}\int\frac{x}{x^2+x+1}\,dx - \frac{2}{3}\int(x^2+x+1)^{-1}\,dx + \frac{1}{4}\int\frac{x}{x^2+1}\,dx +$$
$$\frac{1}{4}\int(x^2+1)^{-1}\,dx - \frac{1}{2}\int\frac{x}{\left(x^2+1\right)^2}\,dx + \frac{1}{2}\int(x^2+1)^{-2}\,dx - \int\frac{x}{\left(x^2+1\right)^3}\,dx$$

> value(");

$$\frac{1}{12}\ln(x-1) - \frac{1}{6}\ln(x^2+x+1) - \frac{1}{3}\sqrt{3}\arctan(\frac{1}{3}(2x+1)\sqrt{3}) + \frac{1}{8}\ln(x^2+1) + \frac{1}{2}\arctan(x) + \frac{1}{4(x^2+1)} + \frac{1}{4}\frac{x}{x^2+1} + \frac{1}{4(x^2+1)^2}$$

■

Example 7.34 Evaluate $\int \frac{1-3x+2x^2-x^3}{x(x^2+1)^2} dx$.

Solution The integrand is

> f8 := (1-3*x + 2*x^2 - x^3)/(x*(x^2 + 1)^2);

$$\frac{1-3x+2x^2-x^3}{x(x^2+1)^2}$$

> factor(");

$$-\frac{-1+3x-2x^2+x^3}{x(x^2+1)^2}$$

Thus, the partial fraction decomposition is

> convert(",parfrac,x);

$$x^{-1} - \frac{x+1}{x^2+1} + \frac{x-2}{(x^2+1)^2}$$

The integral is

> Int(",x);

$$\int x^{-1} - \frac{x+1}{x^2+1} + \frac{x-2}{(x^2+1)^2} dx$$

which has the value

> expand(");

$$\int x^{-1} dx - \int \frac{x}{x^2+1} dx - \int (x^2+1)^{-1} dx + \int \frac{x}{(x^2+1)^2} dx - 2\int (x^2+1)^{-2} dx$$

> value(");

$$\ln(x) - \frac{1}{2}\ln(x^2+1) - 2\arctan(x) - \frac{1}{2(x^2+1)} - \frac{x}{x^2+1}$$

■

7.7 Numerical Approximations

Since our very first discussion of the concept of area under a curve, we have used approximations to integrals. In this section, we extend our rather crude approximations based on rectangles to obtain some of the standard numerical integration techniques. One of the chief advantages of knowing in detail the behavior of nice functions such as polynomials is that it gives us a way of approximating functions more accurately with less work.

7.7.1 The Trapezoidal Rule

We wish to compute the area under a curve by using an approximation that averages the partial sums based on rectangles and that overestimates and underestimates the integral. Consider, for example, the function f defined by $f(x) = -x^2+3$ on the interval $[0, 1]$. The function f is decreasing on the interval $[a, b]$ and there are six intervals of width $1/6$, so we can obtain two such estimates using only the rectangles shown in the following graphs

```
> plt1 := leftbox(-x^2+3,x=0..1,6):
> plt1;
```

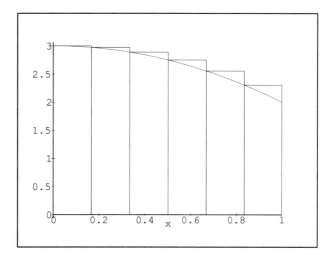

and

```
> plt2 := rightbox(-x^2+3,x=0..1,6):
> plt2;
```

Chapter 7 Integration Techniques

The total area of these sets of rectangles give, respectively, upper and lower estimates of the total area under the curve. For n rectangles from a to b, these bounds are

```
> Upper := leftsum( f(x) , x =a..b , n );
```

$$(b-a)\sum_{i=0}^{n-1} f(a + \frac{i(b-a)}{n})n^{-1}$$

```
> Lower := rightsum( f(x), x =a..b , n );
```

$$(b-a)\sum_{i=1}^{n} f(a + \frac{i(b-a)}{n})n^{-1}$$

The average of these two is clearly a better estimate.

```
> Aver := normal( (value(Lower) + value(Upper))/2 );
```

$$-\frac{1}{2}(-b+a)\left(\sum_{i=1}^{n} f(\frac{an+ib-ia}{n}) + \sum_{i=0}^{n-1} f(\frac{an+ib-ia}{n})\right)n^{-1}$$

In the current example with

```
> f := (x)-> -x^2+3: a := 0: b := 1: n := 6:
```

this evaluates to

```
> evalf(Aver);
```

$$2.662037037$$

and it compares favorably with

```
> evalf( Upper ), evalf( Lower );
```
$$2.745370371, 2.578703704$$

This averaging technique also leads to a general formula. For an unknown procedure g, with n equal-sized intervals from a to b, the formula is

```
> f := 'f': a := 'a': b := 'b': n := 'n':
> trapezoid(g(x),x=a..b,n);
```

$$\frac{1}{2}(b-a)\left(g(a) + 2\sum_{i=1}^{n-1} g(a + \frac{i(b-a)}{n}) + g(b)\right) n^{-1}$$

Basically, all but the first and last terms of the original sums from **leftsum( )** and **rightsum( )** are counted twice.

This formula is called the *trapezoidal rule* because the summands are the areas of the trapezoids taken by using slanted tops (the lines going from the top of the underestimating rectangle to the top of the overestimating rectangle) on each subinterval, as shown by the graph

```
> plots[display]([plt1,plt2]);
```

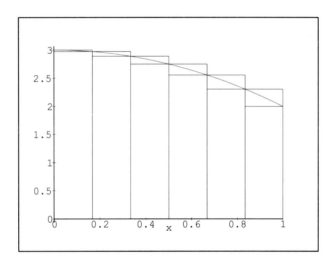

For f over [0, 1], with six intervals, this new estimate is

```
> f := (x) -> -x^2 + 3:
> trapezoid( f(x), x=0..1,6 );
```

$$\frac{5}{12} + \frac{1}{6}\sum_{i=1}^{5}\left(-\frac{1}{36}i^2 + 3\right)$$

```
> evalf(");
```

$$2.662037037$$

The error involved in using this method can be bounded theoretically. It is at most

> tr_error := M*(b-a)^3/(12*n^2);

$$\frac{1}{12}\frac{M(b-a)^3}{n^2}$$

where M is the maximum value assumed by the second derivative of f on the interval $[a, b]$.

Example 7.35 Compute the estimates obtained for the area under the curve

> f := (x)-> 1/x:

for x in the domain $[1, 2]$, by using the trapezoidal rule with five and ten intervals.

Solution For this function and domain, with five intervals, the trapezoidal rule yields the sum

> trapezoid(f(x),x=1..2,5);

$$\frac{3}{20} + \frac{1}{5}\sum_{i=1}^{4}\left(1+\frac{1}{5}i\right)^{-1}$$

which evaluates to

> evalf(");

$$0.6956349206$$

With ten intervals, we get the result

> trapezoid(f(x),x=1..2,10);

$$\frac{3}{40} + \frac{1}{10}\sum_{i=1}^{9}\left(1+\frac{1}{10}i\right)^{-1}$$

> evalf(");

$$0.6937714032$$

∎

Example 7.36 Use the error bound for the trapezoidal rule to discover how many intervals should be used with the trapezoidal rule to guarantee that the answer is within .001 of the correct answer for the function in Example 7.35.

Solution The error bound on this interval is

> err1 := subs(b=2,a=1, tr_error);

$$\frac{1}{12}\frac{M}{n^2}$$

The following graph of f'' over the interval [1, 2], shows that f'' is monotonically decreasing.

```
> plot( D(D(f)), 1..2 );
```

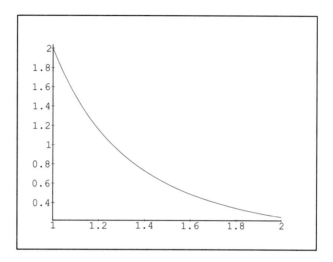

The maximum value that M could have is

```
> D(D(f))(1);
```

$$2$$

so the error bound for using the trapezoidal method for this function and interval is

```
> err1 := subs(M=2,err1);
```

$$\frac{1}{6n^2}$$

It remains to find a suitable value for n. The solution

```
> fsolve( err1=1/1000,n,1..100);
```

$$12.90994449$$

indicates that $n = 13$ is a sufficiently large number of intervals. ∎

We can see how the trapezoidal rule converges to the value of the integral in Example 7.35 as n increases, as follows. For n intervals, the trapezoidal rule is

```
> trapezoid( f(x),x=1..2,n);
```

$$\frac{1}{2}\left(\frac{3}{2}+2\sum_{i=1}^{n-1}\left(1+\frac{i}{n}\right)^{-1}\right)n^{-1}$$

The value of this sum expressed in terms of a special function, Ψ (also known as the *digamma* function), is

```
> value(");
```

$$\frac{1}{2}\left(\frac{3}{2} + 2n\Psi(2n) - 2n\Psi(n+1)\right)n^{-1}$$

A study of this function goes beyond this course, but it can be plotted by *Maple*. The following graph

```
> plot( " , n=1..20 , 0.69..0.72 );
```

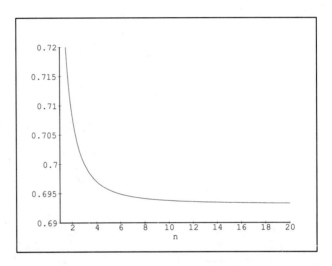

suggests the existence of a horizontal asymptote at the true value

```
> int(f(x),x=1..2);
```

$$\ln(2)$$

which is approximately

```
> evalf(");
```

$$0.6931471806$$

7.7.2 Simpson's Rule

An even better estimate (under most circumstances) than the trapezoidal rule can be obtained by using polynomials of degree two to estimate the area under a pair of adjacent rectangles.

Example 7.37 Use a polynomial of degree two to estimate the area under the curve defined by

```
> f := (x) -> sin(x^2) + 2:
```

on the interval [0, 2]. Compare this estimate with that obtained by using two equal-width rectangles.

7.7 Numerical Approximations

Solution The curve we wish to approximate is shown along with two rectangles by the command

> leftbox(f(x),x=0..2,2);

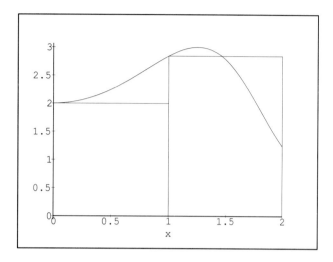

We name this plot *p3*.

> p3 := ":

First we construct a curve defined by a polynomial of degree two that meets this curve at the beginning, the middle, and the end of the curve. If the function is

> g := (x) -> A*x^2 + B*x + C;

$$x \mapsto Ax^2 + Bx + C$$

then suitable values for *A*, *B*, and *C* can be found by solving the system of equations

> { f(0)=g(0), f(1) = g(1) , f(2) = g(2) };

$$\{2 = C, \sin(1) + 2 = A + B + C, \sin(4) + 2 = 4A + 2B + C\}$$

An approximate solution for *A*, *B*, and *C* is

> fsolve(",{A,B,C});

$$\{C = 2.0, A = -1.219872233, B = 2.061343218\}$$

so that the actual function is

> g := makeproc(subs(",g(x)) , x);

$$x \mapsto -1.219872233\, x^2 + 2.061343218\, x + 2.0$$

The following graph shows both curves together.

```
> p4 := plot(g(x),x=0..2):
> plots[display]([p3,p4]);
```

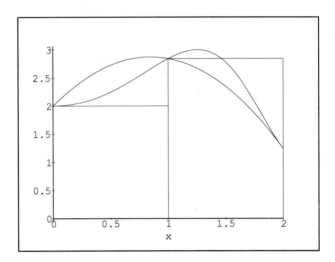

The three estimates for the value of this integral given by the method of rectangles, the trapezoidal method, and Simpson's rule all using two intervals are

```
> evalf( leftsum( f(x),x=0..2,2) ),
> evalf( trapezoid( f(x),x=0..2,2) ),
> evalf( Int( g(x),x=0..2) );
```

$$4.841470985, 4.463069737, 4.869693815$$

The exact answer is closer to

```
> evalf( Int( f(x),x=0..2) );
```

$$4.804776489$$

So in this specific example, the two rectangles do better than the other two methods. If we use six intervals, the power of Simpson's method starts to show, as it is much more accurate than the other two.

```
> evalf( leftsum( f(x),x=0..2,6) ),
> evalf( trapezoid( f(x),x=0..2,6) ),
> evalf( simpson( f(x),x=0..2,6) );
```

$$4.905602556, 4.779468808, 4.810356006$$

■

The method used to construct the polynomial $g(x)$ in the preceding example occurs frequently enough that there is a command in *Maple* to construct this in one step. Given the points

```
> pts := [0,1,2];
```

$$[0, 1, 2]$$

7.7 Numerical Approximations

The function values at these points are

```
> fvals := map(f,");
```

$$[2, \sin(1) + 2, \sin(4) + 2]$$

and the polynomial in x is

```
> interp( pts , fvals ,x );
```

$$\frac{1}{2}\sin(4)x^2 - \frac{1}{2}\sin(4)x - \sin(1)x^2 + 2\sin(1)x + 2$$

Simpson's rule is obtained by using all the subintervals defined by the partition, two at a time, to construct degree two polynomials on the respective subintervals. The result of integrating each polynomial over its domain can be expressed in terms of a sum that is similar in structure to the sum for the trapezoidal rule.

To derive this sum, we repeat the calculations of Example 7.37 in the general setting. Suppose that the curve defining the area is f, and that the two adjacent intervals are of width h. Then the boundaries of the rectangles are at the points

```
> restart: with(student):
> pts := [a,a+h,a+2*h];
```

$$[a, a + h, a + 2h]$$

and the corresponding function values at these points are

```
> fvals := map(f,pts);
```

$$[f(a), f(a + h), f(a + 2h)]$$

We can construct the polynomial $g(x)$ by using **interp( )**.

```
> p := interp(pts,fvals,x):
```

The full formula for $g(x)$ is quite messy, but a simplified version of it in the special case of $a = 0$ simplifies to

```
> map(simplify,subs(a=0,p));
```

$$\frac{1}{2}\frac{(f(2h) - 2f(h) + f(0))x^2}{h^2} - \frac{1}{2}\frac{(3f(0) + f(2h) - 4f(h))x}{h} + f(0)$$

The corresponding formula for the area under the polynomial over the interval $[a, a+2h]$ is suprisingly simple. It is the value of integral

```
> int(p,x=a.. a + 2*h):
```

and simplifies to

```
> simplify(");
```

$$\frac{1}{3}h(f(a) + 4f(a + h) + f(a + 2h))$$

Continuing, the formula for area under the polynomial for the next two intervals (i.e., from $x = a + 2h$ to $x = a + 4h$) is

```
> subs( a=a+2*h,");
```

$$\frac{1}{3}h(f(a+2h)+f(a+4h)+4f(a+3h))$$

and so on. (Notice how a term for $x = a + 2h$ occurs in the formula for the area under both the first and the second polynomial.)

The general formula for the area under the polynomials, using n equal-sized intervals, pairwise over the range $[a, b]$ becomes

```
> simpson( f(x),x=a..b,n);
```

$$\frac{1}{3}(b-a)\left(f(a)+f(b)+4\sum_{i=1}^{\frac{1}{2}n}f(a+\frac{(2i-1)(b-a)}{n})+2\sum_{i=1}^{\frac{1}{2}n-1}f(a+\frac{2i(b-a)}{n})\right)n^{-1}$$

(We assume n is even.) Just as for the trapezoidal rule, there is an absolute bound for the error that occurs in using this method. It is

```
> sp_error := M*(b-a)^5/(180*n^4);
```

$$\frac{1}{180}\frac{M(b-a)^5}{n^4}$$

where M is the maximum value assumed by the fourth derivative of f on the interval $[a, b]$.

Example 7.38 Estimate the area under the curve

```
> f := (x)-> 1/x;
```

$$x \mapsto x^{-1}$$

with Simpson's rule and investigate the size of the error in your answer.

Solution For ten intervals the formula produces the value

```
> simpson(f(x),x=1..2,10);
```

$$\frac{1}{20}+\frac{2}{15}\sum_{i=1}^{5}\left(\frac{9}{10}+\frac{1}{5}i\right)^{-1}+\frac{1}{15}\sum_{i=1}^{4}\left(1+\frac{1}{5}i\right)^{-1}$$

```
> evalf(");
```

$$0.6931502307$$

The error bound for Simpson's rule is:

```
> sp_error := M*(b-a)^5/(180*n^4);
```

$$\frac{1}{180}\frac{M(b-a)^5}{n^4}$$

where M is the maximum value assumed by the fourth derivative of f. The first few derivatives of f are

```
> D(f), (D@@2)(f), (D@@3)(f),` (D@@4)(f);
```

$$x \mapsto -x^{-2},\ x \mapsto \frac{2}{x^3},\ x \mapsto -\frac{6}{x^4},\ x \mapsto \frac{24}{x^5}$$

and from the graph

```
> plot( (D@@4)(f), 1 .. 2);
```

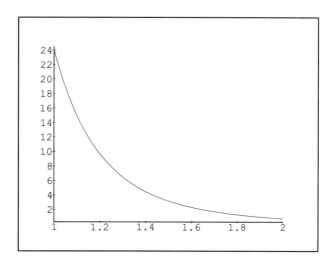

we see that the maximum on this interval occurs at $x = 1$.

```
> ((D@@4)(f))(1);
```

$$24$$

```
> subs(M=",n=10,a=1,b=2,sp_error);
```

$$\frac{1}{75000}$$

The exact answer is:

```
> int(1/x,x=1..2) ;
```

$$\ln(2)$$

```
> evalf(ln(2));
```

$$0.6931471806$$

■

Command	Description
((D@@4)(f))(1);	The fourth derivative of f at 1
(f=g)(t);	A function application producing an equation
(lhs-rhs)(eq1);	A function application
(sin^3)(x);	$\sin(x)$ all cubed
D(D(f))(1);	Second derivative of f at 1
D(f),(D@@2)(f),(D@@3)(f);	The first, second, and third derivatives of f
Int(f(x),x);	Represent an indefinite integral
Int(f(x),x=a..b);	Represent a definite integral
Sum(1/i,i=1..1000):	An unevaluated sum
[[10,0],[10,f2(10)]];	A list of points
assume(a>0);	Introduce a special property
changevar(3+2*x=u,I2,x);	Perform a change of variables
collect(",x);	Regroup the terms of a polynomial
combine((cos^5)(x),trig);	Regroup trig expressions
combine(I1,trig);	Regroup trig expressions
completesquare(I2,x);	Complete the square
constants:=constants,a;	Define a new system constant
convert(f(x),'@'):	Rewrite as a composition $f \circ x$
convert(g,parfrac,x);	Perform a partial fraction decomposition
convert(sin@x,nested);	Rewrite as $sin(x)$
denom(f1);	The denominator of a quotient
diff(sec(x),x);	Compute a derivative
display([p3,p4]);	Display two plots simultaneously
evalf(Int(f(x),x=0..2));	Approximate an integral
evalf(leftsum(f(x),x=0..2,2)),	Approximate a sum
evalf(simpson(f(x),x=0..2,6));	Construct and evaluate Simpson's approximation
evalf(trapezoid(f(x),x=0..2,6));	Construct and evaluate the Trapezoidal approximation
expand(",cos);	Expand, but freeze expressions involving cos
expand(cos(2*x)):	Apply basic trig identities
expand(sin(a+b));	Apply basic trig identities
f@g:	Represent a composition of two functions
factor(");	Factor an algebraic expression

Table 7.1 Some of the Commands Used in Chapter 7

EXERCISE SET 7

Some of the important commands that have been used in this chapter are shown in Table 7.1. Keep them in mind as you explore the examples from the chapter and the exercises listed here.

1. Using fixed precision floating-point arithmetic limits our ability to measure the effect of adding or subtracting a very small number from a very large number. What is the maximum value that will be computed by machine for the sum

    ```
    > Digits := 2; total := 0;
    > for i to 1000 do total := total + 5.0/i od;
    ```

    What is the last term that has any measurable effect on the total? What happens at three digits precision?

2. By using an even higher setting for **Digits**, or by asking *Maple* to find the value of

    ```
    > i := 'i'; Sum( 5/i, i=1..1000);
    ```

    and using **evalf( )**, find the correct value for this sum.

3. Recompute the sum in Exercise 2 by using the following approach.

    ```
    > Digits := 2; total := 0;
    > for i from 1000 by -1 to 1 do
    >     total := total + 5.0/i;
    > od;
    ```

    Why do we get a different answer from the first attempt? If we take into account the limitations of two digit arithmetic, both answers are essentially correct.

4. Use the change of variables technique to express each of the following integrals in terms of $u = x + 5$, $u = 2 * x$, and $u = x^3$.

    ```
    > i1 := Int( x - 5 , x );
    ```

    $$\int x - 5 \, dx$$

    ```
    > i2 := Int( 2*x , x );
    ```

    $$\int 2x \, dx$$

    ```
    > i3 := Int( x^3 , x );
    ```

    $$\int x^3 \, dx$$

    In each case, identify how the equation relating $D(u)$ and $D(x)$ has led to the indicated result.

5. In each case of the Exercise 4, use *Maple* to evaluate the new integral (expressed in terms of u) and transform the solution, by using the "reverse" transformation, back into a solution of the original problem. Compare these solutions with those used by using *Maple* to evaluate the integrals directly.

6. Compare the result of doing a change of variables defined by $u = x^6$ with the change of variables defined by $u = x^6 + 3$ on the integral

```
> Int(6*x^5*sin(x^6-3),x);
```

$$\int 6x^5 \sin(x^6 - 3)\, dx$$

Both of these are relatively good choices for defining a change of variables because they result in integrals of a simpler form that we have already considered.

7. Compare the effect of the two changes of variables in Exercise 6 with those defined by $u = x^2$ and $u = x^3$ or $u = \sin(x)$. Although these are valid changes of variables, in this case they are not at all helpful in simplifying the integration problem.

8. For each of the following integrals, identify a change of variables that would be effective in simplifying each integral.

```
> Int((3*x^2 + 2*x)/sqrt(x^2 + x^2),x);
```

$$\int \frac{1}{2} \frac{(3x^2 + 2x)\sqrt{2}}{x}\, dx$$

```
> Int(x^2*sin(x^3),x);
```

$$\int x^2 \sin(x^3)\, dx$$

```
> Int( x^2*sqrt(x^3 + 3),x);
```

$$\int x^2 \sqrt{x^3 + 3}\, dx$$

Use the transformed integral to construct an antiderivative to the original integration problem. By making the assignment **infolevel[int] := 5;** you can obtain information about how *Maple* obtains a particular integral. How does your answer, based on your change of variables, compare with the answer given by using *Maple* directly?

9. Consider the value of

```
> i1 := Int((3*x^2 + 2*x)*sqrt(x^3 + x^2),x);
```

$$\int (3x^2 + 2x)\sqrt{x^3 + x^2}\, dx$$

You can use differentiation to verify that the value of $i1$ is correct, without knowing anything about how its value was computed. Do so.

Compare the value of $i1$ with the value of

```
> with(student):
> i2 := expand(i1);
```

$$3\int \sqrt{x^3 + x^2}\, x^2\, dx + 2\int \sqrt{x^3 + x^2}\, x\, dx$$

Assuming the value of $i1$ that is reported by *Maple* is correct, how can you verify that the value of $i2$ is correct without differentiating?

10. Compute a value for

    ```
    > with(student):
    > i3 := Int( (10*x+1)*(5*x^2 + x + 3)^3,x);
    ```

    $$\int (10x+1)\left(5x^2 + x + 3\right)^3 dx$$

    directly by using *Maple*'s **value( )** command, and indirectly by first doing the change of variables defined by $u = 5x^2 + x + 3$. Show that either both of these answers are correct or verify that both are correct. (Hint: Compute and plot the difference between the two answers.)

11. The area under the curve

    ```
    > f1 := (x) -> (x^3 + 1)*3*x^2;
    ```

    $$x \mapsto 3\left(x^3 + 1\right)x^2$$

    corresponding to

    ```
    > i1 := Int(f1(x),x=1..2);
    ```

    $$\int_1^2 3\left(x^3 + 1\right)x^2 \, dx$$

    is bounded by the *x*-axis, the two lines

    ```
    > l1 := [ [1,0],[1,f1(1)]]:
    > l2 := [ [2,0],[2,f1(2)]]:
    ```

    and the curve $f1$.

    ```
    > plot({l1,l2,f1(x)},x=1..2,y=0..120);
    ```

    Draw graphs for the integrals corresponding to each of

    ```
    > i2 := changevar(u=x^2,i1,u);
    > i3 := changevar(u=x^3,i1,u);
    > i4 := changevar( u=sqrt(x) , I1 , u );
    ```

    Verify that each of these definite integrals has the same value both by computing exact values for the integrals and by approximating these integral values directly using either **leftbox( )**, or **evalf( )**, or their equivalents.

12. Transform the integral

    ```
    > i5 := Int(u*cos(u),u);
    ```

    $$\int u \cos(u) \, du$$

    into an integral that does not involve a product by using integration by parts. Verify that the value of the resulting integral is a valid antiderivative for $u \cos(u)$.

13. Integration by parts can be used to reexpress the integral

    ```
    > i6 := Int( x^2*ln(x) , x );
    ```

    $$\int x^2 \ln(x) \, dx$$

in many different forms. Examine the transformations

```
> intparts(i6,e);
> expand(");
```

where e is chosen from $\{x, x^2, x^3, \ln(x), x \ln(x)\}$. In each case, identify the integrals and derivatives that need to be computed to carry out the transformation. Can you identify a pattern?

14. Complete the construction of an antiderivative to $x^2 \ln(x)$ for each of the transformations attempted in Exercise 13. This will involve evaluating directly any new integrals that are produced and isolating all the terms involving $i2$ on one side of the equation.

15. Examine the pattern of the antiderivatives generated by

```
> for i to 10 do  i , int(x^i*ln(x),x); od;
```

Use these results to conjecture a suitable value for

```
> i7 := Int(x^k*ln(x),x);
```

Use integration by parts to reexpress $\int x^k \ln(x)\, dx$ in terms of the (known) value of $\int x^{k-1} \ln(x)\, dx$. (Hint: The commands **simplify**( ), **expand**( ), and **isolate**( ) can be used to good effect.)

16. By choosing different parts (e.g., x, x^2, ...), use integration by parts to reformulate

```
> i8 := Int(x^3*exp(x),x);
```

$$\int x^3 e^x\, dx$$

in several different ways. (You may need to use the command **combine(...,power)** to simplify some of the results.) Which of these lead to a successful integration technique?

17. Consider the effect of the commands

```
> i8 := Int(sec(x)^5,x);
> intparts(i8,sec^3(x));
> convert(",sincos);
```

Use a change of variables to rewrite this integral in terms of an integral of a quotient of polynomials. You can use a partial fraction decomposition on that quotient to complete the evaluation of this integral.

18. Use integration by parts and change of variables to reformulate

```
> Int(x^2*arctan(x),x=0..1);
```

$$\int_0^1 x^2 \arctan(x)\, dx$$

in terms of an integral of a single logarithmic term. Compare the value of the resulting integral with that obtained by computing the numerical value of the integral directly (use **evalf**( )). Which answers make sense in light of the graph of the curve over the indicated interval?

19. Identify and make use of standard trigonometric identities[5] to carry out the transformation represented by

    ```
    > combine(sin(x)^3*cos(x)^3,trig);
    ```

    $$-\frac{1}{32}\sin(6x) + \frac{3}{32}\sin(2x)$$

20. Use graphical or numerical computations to verify that the two expressions compared in Exercise 19 are indeed the same. Also verify the equivalence of

    ```
    > expand(");
    ```

    $$-\sin(x)\cos(x)^5 + \sin(x)\cos(x)^3$$

    The **plot( )** and **testeq( )** commands may be helpful.

21. Use the commands **powsubs( )** and **changevar( )** to reexpress the integral

    ```
    > Int(sin(x)^3*cos(x)^3,x);
    ```

    $$\int \sin(x)^3 \cos(x)^3 \, dx$$

    as an integral of a polynomial. Transform the integral of this polynomial into an antiderivative for the original integrand. Verify your result by comparing it with that generated by *Maple*.

22. Results such as in Exercise 21 can also be verified by differentiation. Use differentiation to confirm that *Maple*'s result is an antiderivative. Be warned that the derivative will be in a very different form from the original integrand. An effective strategy for comparing two complicated trigonometric expressions includes: plotting their difference, checking numerical values (see **testeq( )**), or converting both expressions into the standard form produced by **combine(...,trig)**.

23. Use the command **combine(...,trig)** to convert the integrals

    ```
    > Int(sin(x)^8,x);
    ```

    $$\int \sin(x)^8 \, dx$$

    and

    ```
    > Int(cos(x)^8,x);
    ```

    $$\int \cos(x)^8 \, dx$$

    into a form that can easily be integrated knowing only a value for $\int \cos(u) \, du$ and the change of variables technique.

24. Use substitutions and a change of variables to reformulate each of

    ```
    > Int( tan(x)^4*sec(x)^6,x);
    ```

    $$\int \tan(x)^4 \sec(x)^6 \, dx$$

---

[5] You will need to use the command **powsubs( )** to complete mathematical substitutions such as $\sin(x)^2 = 1 - \cos(x)^2$. *Maple*'s command **subs( )** is primarily used to replace names by values.

as an integral of a polynomial.

25. Reformulate the integral

    > Int(sqrt(16-x^2)/x^2,x);

    $$\int \sqrt{16 - x^2}\, x^{-2}\, dx$$

    as an integral involving $\sqrt{1 - u^2}$ for some suitable choice of u.

26. Describe how to use a change of variables to reexpress this integral as an integral of $\sec^2(v)$.

    > I3 := Int(1/x^2/sqrt(x^2 + 4) , x);

    $$\int x^{-2} \frac{1}{\sqrt{x^2 + 4}}\, dx$$

27. Use completing the square and a change of variables to reformulate

    > Int(3*x/sqrt(8 - 2*x - x^2), x);

    $$\int 3x \frac{1}{\sqrt{8 - 2x - x^2}}\, dx$$

    as an integral involving a denominator of the form $\sqrt{1 - v^2}$.

28. Given

    > e1 := (x^3 - 5*x^2 + 8*x - 2)/(x+1)/(x-2)/(x+3)/(x-4);

    $$\frac{x^3 - 5x^2 + 8x - 2}{(x + 1)(x - 2)(x + 3)(x - 4)}$$

    and

    > e2 := a/(x+1) + b/(x-2) + c/(x+3) + d/(x-4);

    $$\frac{a}{x + 1} + \frac{b}{x - 2} + \frac{c}{x + 3} + \frac{d}{x - 4}$$

    develop a set of linear equations that can be solve to find values for a, b, c, and d so that $e1 - e2$. Solve for these values and verify your solution assigning those values and subtracting $e1$ from $e2$. (Also compare your solution with that obtained by using **convert(f,parfrac,x);**.)

29. Use a partial fraction decomposition to reformulate

    > e1 := (x^3 - 5*x^2 + 8*x - 2)/(x+1)/(x-2)/(x+3)/(x-4);

    $$\frac{x^3 - 5x^2 + 8x - 2}{(x + 1)(x - 2)(x + 3)(x - 4)}$$

```
> Int(e1,x);
```

$$\int \frac{x^3 - 5x^2 + 8x - 2}{(x+1)(x-2)(x+3)(x-4)} \, dx$$

Compute each of the integrals corresponding to the terms of this sum to find a value for the integral.

30. Find values of a such that the value of the integral

```
> Int((x^3-5*x^2+a*x-2)/(x^4-2*x^3-13*x^2+14*x+24),x);
```

$$\int \frac{x^3 - 5x^2 + ax - 2}{x^4 - 2x^3 - 13x^2 + 14x + 24} \, dx$$

does not have a logarithmic term of the form $\ln(a+1)$.

31. Compute the estimates obtained for the area under the curve

```
> f := (x) -> (x+1)/(x^2 + x):
```

for x in the domain $[1, 2]$ by using the trapezoidal rule with five and ten intervals.

32. Use the error bound for the trapezoidal rule to discover how many intervals should be used with the trapezoidal rule to guarantee that the answer is within .001 of the correct answer for the function in Exercise 31.

33. Use a polynomial of degree two to estimate the area under the curve defined by

```
> f := (x) -> sin(x^3)/x + 2:
```

on the interval $[1, 3]$. Compare this estimate with that obtained by using two equal-width rectangles.

34. Higher degree polynomials can be used to approximate a curve. Use the same techniques as outlined in Section 7.7 to develop a degree three polynomial that approximates the curve of Exercise 33 over three equal-sized intervals spanning the interval $[1, 3]$. (You can use the command **interp( )** to construct a suitable polyonmial once you have a list of x values and a list of the function values at those points.)

35. By constructing a suitable polynomial on the points

```
> pts := [a,a+h,a+2*h,a+3*h];
```

$$[a, a + h, a + 2h, a + 3h]$$

for an undefined function g, and then integrating that polynomial over the interval $[a, a+3h]$, develop an integration rule similar to Simpson's rule, but based on degree three polynomials.

36. Compare using your rule (as developed in the previous exercise) on two consecutive groups of three intervals, the trapezoidal rule used on all six intervals, and Simpson's rule on three consecutive groups of two intervals (i.e., use **simpson(...,x=0..3,6);**) as a means of evaluating the integral

```
> Int( 2*x^4 -x^2 + 3 , x=0..3 );
```

$$\int_0^3 2x^4 - x^2 + 3 \, dx$$

How do they compare with the correct answer?

8 More Applications of Integration

8.1 Volumes Through Integration

Exact integration allows us to find formulas for the volume of many standard geometric figures. Once an appropriate orientation has been chosen, a formula for the volume follows almost routinely from the algebraic equations describing the relative position of the various corners and faces of the objects. In this section we review some common examples.

The main challenge in each case is to choose a position and orientation that simplifies the algebraic description of the problem. These same descriptions and computations often can be used to help display the surface of the object via the plot command.

Example 8.1 Find the volume of a pyramid whose base is a square with side L and whose height is h.

Solution Because use of symmetry can help keep the algebraic description of the object simple, center the base of the pyramid at the origin of the x-y plane and align it with the axis.

Now think of this pyramid as being constructed of horizontal layers, each of which is a square. The horizontal layer of the pyramid at height z is a square of dimensions $2x$ by $2x$ for some value of x. To find x as a function of z, look at the cross section in the x-z plane (corresponding to $y = 0$). This is a triangle connecting the points $p1$, $p2$, and $-p2$.

```
> p1 := [0,h]: p2 := [L/2,0];
```

$$[\frac{1}{2}L, 0]$$

The line defined by the points $p1$ and $p2$ satisfies the equation

```
> z = slope(p1,p2)*x + h;
```

$$z = -\frac{2hx}{L} + h$$

To express x on this line as a function of z, first isolate x, as in

```
> isolate(",x);
```

$$x = \frac{1}{2}\frac{(-z+h)L}{h}$$

and then construct a suitable procedure:

```
> xval := makeproc( rhs("),z);
```

$$z \mapsto \frac{1}{2}\frac{(-z+h)L}{h}$$

8.1 Volumes Through Integration

The four corners of the layer defined by a specific choice of *z* are

```
> [ [ xval(z),  xval(z), z],
>   [-xval(z),  xval(z), z],
>   [-xval(z),-xval(z), z],
>   [ xval(z),-xval(z), z]
> ];

> [[ xval(z),  xval(z), z],
>   [-xval(z),  xval(z), z],
>   [-xval(z),-xval(z), z],
>   [ xval(z),-xval(z), z]
> ]:
> op(");
```

$$[\frac{1}{2}\frac{(-z+h)L}{h}, \frac{1}{2}\frac{(-z+h)L}{h}, z], [-\frac{1}{2}\frac{(-z+h)L}{h}, \frac{1}{2}\frac{(-z+h)L}{h}, z],$$
$$[-\frac{1}{2}\frac{(-z+h)L}{h}, -\frac{1}{2}\frac{(-z+h)L}{h}, z], [\frac{1}{2}\frac{(-z+h)L}{h}, -\frac{1}{2}\frac{(-z+h)L}{h}, z]$$

The following function constructs such a lists of corner points.

```
> layer := makeproc( ",z):
```

Thus, a specific layer at, say $z = 3$, with $L = 10$ and $h = 12$, can be drawn as

```
> L := 10: h := 12:
> with(plots):

> polygonplot3d( layer(3),style=patch,axes=boxed);
```

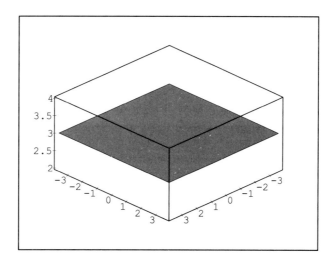

To help visualize such an object, we first construct several layers, as in

```
> n := 21:
> for i from 0 to n do
>     plt[i] := polygonplot3d( layer(i*h/n))
> od:
```

Each layer can be displayed by referring to it directly, as in **plt[3]**; or they can all be displayed simultaneously, as in

```
> display3d(convert(plt,list), style=patch, axes=frame);
```

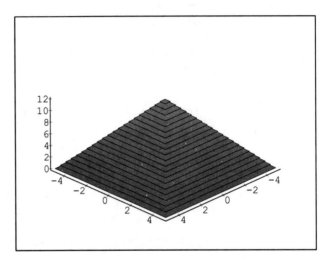

To compute the total volume, first compute the area of each layer as a function of z. This is done by using the function

```
> A := makeproc( (2*xval(z))^2 , z );
```

$$z \mapsto 4\left(-\frac{5}{12}z+5\right)^2$$

The volume is then

```
> V := Int( A(z), z=0..h);
```

$$\int_0^{12} 4\left(-\frac{5}{12}z+5\right)^2 dz$$

which evaluates to

```
> value(");
```

$$400$$

For unspecified L and h,

```
> L := 'L': h := 'h':
```

the area is still computed by the function

```
> A := makeproc( (2*xval(z))^2 , z );
```

$$z \mapsto \frac{(-z+h)^2 L^2}{h^2}$$

and the volume is

```
> V := Int( A(z),z=0..h);
```

$$\int_0^h \frac{(-z+h)^2 L^2}{h^2} \, dz$$

This evaluates to

```
> value(");
```

$$\frac{1}{3}hL^2$$

∎

Example 8.2 Show that the volume of a sphere of radius r is

$$V = \frac{4}{3}\pi r^3$$

Solution When centered at the origin, a sphere is represented by the equation

```
> eq := x^2 + y^2 + z^2 = r^2;
```

$$x^2 + y^2 + z^2 = r^2$$

For the specific value of the radius $r = 2$, a sphere is generated by the commands

```
> with(plots):
> opts := x=-2..2,y=-2..2,z=-2..2,axes=framed:
```

```
> implicitplot3d( subs(r=2,eq) , opts );
```

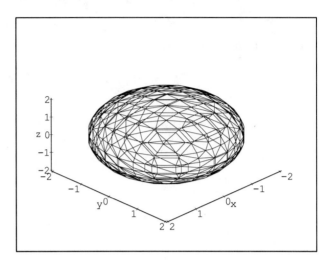

We can compute this volume by using layers orthogonal to the *x*-axis. For a given *x* value, the corresponding layer is a circle of radius at most *r*. The exact radius can be determined from the point of intersection of that circle with the *z* = 0 plane. In terms of *x*, the radius is *y*, where *y* is given by

```
> subs(z=0,eq);
```
$$x^2 + y^2 = r^2$$

```
> isolate(",y^2):
> map(sqrt,");
```
$$y = \sqrt{r^2 - x^2}$$

As a function of *x*, the radius of the circular cross section is

```
> xradius := makeproc( rhs(") , x ):
```

and the area is

```
> Area := makeproc( xradius(x)^2*Pi,x);
```
$$x \mapsto (r^2 - x^2)\,\pi$$

The volume of the sphere is just

```
> Int( Area(x),x=-r..r);
```
$$\int_{-r}^{r} (r^2 - x^2)\,\pi\,dx$$

which evaluates to

```
> value(");
```
$$\frac{4}{3}\pi r^3$$

∎

8.1.1 Visualizing Stacks of Disks

The layers corresponding to each value of x in the previous example are disks. The radius changes in length from 0 to 2 to 0 as x changes from -2 to 2, as shown in the following graph of the radius.

```
> r := 2;
```

$$2$$

```
> plot( xradius , -2..2, scaling=constrained);
```

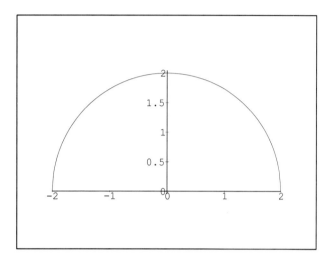

To generate a three-dimensional graph showing a sequence of these disks, we simply need to automate the process of generating the corresponding circles.

How do we draw a circle orthogonal to the x-axis? In the absence of a graphing primitive that draws such circles, we simply approximate the circle by a many-sided polygon. Given the center and the radius of the circle, a sequence of points on the perimeter of the circle can be generated by allowing a "radius" of length r to pivot around the center of the circle in the correct plane. When the radius is at an angle t with the y-axis, the y value is $r\cos(t)$ and the z value is $r\sin(t)$.

A *Maple* procedure that divides such a circle into 30 equal segments computes the coordinates of each of the 31 segment boundary points and displays them as a polygon can be constructed as follows.

First we automate the construction of points on the perimeter of the circle. For example, if the circle is of radius $r1$ at center $[.5, 0, 0]$, the procedure

```
> pt := makeproc( [.5, 0 + r1*cos(t), 0 + r1*sin(t)], t );
```

$$t \mapsto [0.5, r1\ \cos(t), r1\ \sin(t)]$$

builds such points as a function of the angle t. Thus, at $x = 0.5$ the radius of the circular cross section is

```
> r1 := xradius(0.5);
```
$$1.936491673$$

and the points

```
> pt(0),pt(Pi/2),pt(Pi),pt(3*Pi/2);
```
$$[0.5, 1.936491673, 0], [0.5, 0, 1.936491673], [0.5, -1.936491673, 0],$$
$$[0.5, 0, -1.936491673]$$

all lie on the boundary of the disk.

The procedure[1]

```
> xcircle := proc(xval,yval,zval,r)
>    local x,y,t,n,angle,pt;
>    n := 31;
>    angle := 2*Pi/n ;
>    pt := makeproc(
>        [xval, yval + r*cos(t), zval + r*sin(t)]
>        ,t );
>    PLOT3D( POLYGONS( [seq( evalf(pt(t*angle)),t=0..n)] ));
> end:
```

uses this idea to construct a sequence of 31 equally spaced points on the perimeter of a circle and to display[2] a many-sided polygon approximating that circle.

```
> xcircle(.5,0,0,xradius(.5)):
```

```
> display3d(",style=patch,axes=frame);
```

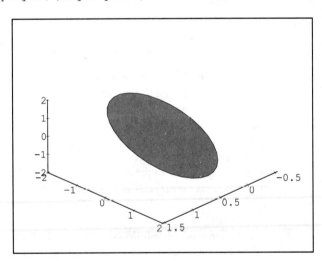

The sphere can now be simulated by a sequence of circles at a distance of $\dfrac{2r}{n}$ apart. Twenty-one such circles are generated by

---

[1] The **makeproc( )** procedure is not defined until the student package has been loaded.

[2] All the *Maple* plot commands generate function calls of the form **PLOT( )** or **PLOT3D( )**. These can be named, saved, and manipulated just like ordinary algebraic objects. The only difference is that they "print" to the screen by drawing a graph. The ':' after the call to **xcircle( )** suppresses that printing.

```
> size := 4/21;
```

$$\frac{4}{21}$$

```
> for i from 0 to 21 do
> p[i] := xcircle(-2+i*size,0,0,xradius(-2+i*size)):
> od:
```

and are shown by the commands

```
> opts := scaling=constrained, axes=frame:

> display3d(convert(p,list), opts);
```

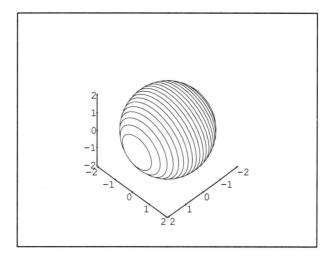

Example 8.3 Find the volume of the solid obtained by rotating about the x-axis the region under the curve $y = \sqrt{x}$ from 0 to 1.

Solution For each value of x, the cross section is a circle of radius y.

```
> xradius := (x) -> sqrt(x):
```

The area of this cross section is given by the function

```
> A := makeproc( xradius(x)^2*Pi , x ):
```

and the volume is given by

```
> Int( A(x),x=0..1);
```

$$\int_0^1 x\pi \, dx$$

which evaluates to

> value(");

$$\frac{1}{2}\pi$$

To view the object as a collection of, say, 16 cross sections, we first construct the cross sections by the commands

```
> size := 1/15: p := table():
> for i from 0 to 15 do
> p[i] := xcircle(i*size,0,0,xradius(i*size)):
> od:
```

and then display them together as

```
> with(plots):

> display3d(convert(p,list),scaling=constrained,axes=frame);
```

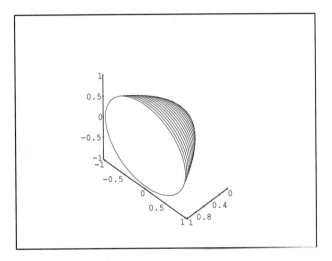

■

8.1.2 Variations on Volume

We conclude this section by looking at several variations on finding volumes. Where it is easy to do so, we also generate an image directly from the plot tools available to us, rather than layer by layer.

Example 8.4 Find the volume of the solid obtained by rotating about the y-axis the region bounded by $y = x^3$, $y = 8$, and $x = 0$.

Solution From the equation

```
> eq := (y = x^3);
```

$$y = x^3$$

we obtain the radius x as a function of y as

> isolate(",x);

$$x = \sqrt[3]{y}$$

This can be used to construct a procedure for computing the radius of each layer.

> rad := makeproc(rhs("),y);

$$y \mapsto \sqrt[3]{y}$$

The outer surface (excluding ends) of the object is displayed by the command

> opts := radius=rad(y),tubepoints=13,axes=frame,
> scaling=constrained:

> tubeplot([0,y,0,y=0..8], opts);

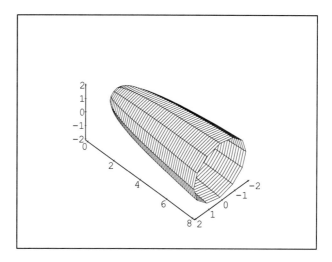

The volume can be approximated by circular disks orthogonal to the y-axis.

> A := makeproc(Pi*rad(y)^2,y);

$$y \mapsto \pi y^{\frac{2}{3}}$$

> Int(A(y),y=0..8);

$$\int_0^8 \pi y^{\frac{2}{3}} \, dy$$

> value(");

$$\frac{24}{5} 8^{\frac{2}{3}} \pi$$

This answer simplifies to

> simplify(");

$$\frac{96}{5}\pi$$

∎

Example 8.5 The region R bounded by the curves $y = x$ and $y = x^2$ is rotated about the x-axis. Find the volume of the resulting solid.

Solution To use disks orthogonal to the x-axis, we must find upper and lower bounds on x. Given

> eqns := { y=x,y=x^2};

$$\{y = x^2, y = x\}$$

we first find two solutions for the intersection of these curves.

> sol := [solve(")];

$$[\{y = 0, x = 0\}, \{y = 1, x = 1\}]$$

The lower and upper x values are

> [subs(sol[1],x), subs(sol[2],x)]:
> xbnd := sort(");

$$[0, 1]$$

The lower and upper y values are

> [subs(sol[1],y), subs(sol[2],y)]:
> ybnd := sort(");

$$[0, 1]$$

The region that will be rotated about the x-axis is displayed as

> implicitplot(eqns,x=xbnd[1]..xbnd[2],y=ybnd[1]..ybnd[2]);

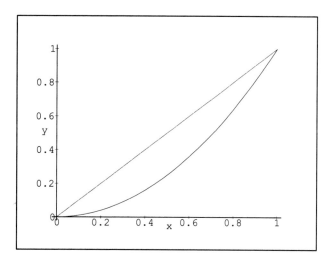

If we use cross sections orthogonal to the x-axis, then we can subtract the volume represented by the inside surface from the volume represented by the outside surface.

```
> rad1 := (x)-> x^2:
> A1 := (x) -> Pi*rad1(x)^2:
> rad2 := (x)-> x^4:
> A2 := (x) -> Pi*rad2(x)^2:
```

The total volume is

```
> Int(A1(x),x=xbnd[1]..xbnd[2])
> - Int(A2(x),x=xbnd[1]..xbnd[2]);
```

$$\int_0^1 \pi x^4 \, dx - \int_0^1 \pi x^8 \, dx$$

This type of volume is sometimes computed as

```
> combine(");
```

$$\int_0^1 \pi x^4 - \pi x^8 \, dx$$

for which the integrand is the area of an annulus formed by a cross section of the entire object. The actual volume is

```
> value(");
```

$$\frac{4}{45}\pi$$

∎

Example 8.6 Find the volume of the solid obtained by rotating about the line $y = 2$ the region bound by the set of curves defined by $\{y = x, y = x^2\}$.

Solution The main change from the version of this problem studied in Example 8.5 involves the radii. We now have

```
> rad1 := makeproc( 2-rad1(x),x):
> rad2 := makeproc( 2-rad2(x),x):
```

so that the two cross-sectional areas for a specific x value are

```
> A1(x), A2(x);
```

$$\pi\left(2-x^{2}\right)^{2},\ \pi\left(2-x^{4}\right)^{2}$$

This time the roles of inside and outside radii have changed. The volume is given by

```
> Int(A2(x),x=xbnd[1]..xbnd[2])
> - Int(A1(x),x=xbnd[1]..xbnd[2]);
```

$$\int_{0}^{1}\pi\left(2-x^{4}\right)^{2}dx - \int_{0}^{1}\pi\left(2-x^{2}\right)^{2}dx$$

which has a value of

```
> value(");
```

$$\frac{4}{9}\pi$$

■

Example 8.7 Find the volume of the solid obtained by rotating about the y-axis the region defined by the curves

```
> eqns;
```

$$\{y = x^{2}, y = x\}$$

Solution In this example, it is convenient to use a cylindrical decomposition. The height of a cylinder with radius x is given by

```
> h1 := (x) -> x^2:
> h2 := (x) -> x^4:
> (h1-h2)(x);
```

$$x^{2} - x^{4}$$

The area of a cylinder with radius x is given by

```
> A := (x) -> 2*Pi*(h1(x)-h2(x)):
```

The volume of this solid of revolution is

```
> V := Int( A(x),x=xbnd[1]..xbnd[2]);
```

$$\int_{0}^{1} 2\pi\left(x^{2} - x^{4}\right) dx$$

which evaluates to

> value(V);

$$\frac{4}{15}\pi$$

∎

Example 8.8 A solid has a circular base of radius 1. Parallel cross sections perpendicular to the base are equilateral triangles. Find the volume of the solid.

Solution Orient the object so that it is centered at the origin with the base at $z = 0$. The boundary of the base is then given by

> x^2 + y^2 = 1;

$$x^2 + y^2 = 1$$

For a specific x, the y value on this boundary is

> isolate(",y^2);

$$y^2 = 1 - x^2$$

> map(sqrt,");

$$y = \sqrt{1 - x^2}$$

The function

> yval := makeproc(rhs("),x):

can be used to compute y in terms of x.

The length of one side of an equilateral triangle cross section positioned at a specific value of x and orthogonal to the x-axis is

> 2*yval(x);

$$2\sqrt{1 - x^2}$$

The height h of the equilateral triangle must satisfy the relation

> h/y=tan(Pi/3);

$$\frac{h}{y} = \sqrt{3}$$

so that as a function of y, h is

> isolate(",h):
> h := makeproc('evalf'(rhs(")),y):

The triangle at x can be displayed as a polygon with three vertices,[3] as in

---

[3] We use **evalf( )** here since the vertices must be expressed in floating-point numbers.

```
> tr := (x) -> PLOT3D( POLYGONS( evalf( [
> [x,yval(x),0],
> [x,-yval(x),0],
> [x,0, h(yval(x))]
> ]))):
```

As *x* ranges from -1 to 1 in steps of size $1/10$, we obtain the triangles

```
> p := table():
> for i from 0 to 20 do p[i] := tr(-1+i*1/10): od:
```

These are displayed simultaneously by the command

```
> with(plots):
> opts := scaling=constrained,style=patch,axes=frame:

> display3d(convert(p,list),opts);
```

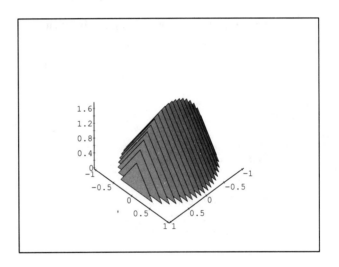

The area of a triangle at *x* is given by

```
> A := makeproc( 1/2*h(x)*2*yval(x) , x );
```

$$x \mapsto 1.732050808\, x\sqrt{1-x^2}$$

In this discussion, we have assumed that *x* and $h(x)$ are nonnegative. If we allow *x* to be negative, then we can still correct these formulas by making careful use of absolute values. Or, we can avoid the introduction of absolute values by relying on symmetry. One half of the volume is given by

```
> Int( A(y),y=0..1);
```

$$\int_0^1 1.732050808\, y\sqrt{1-y^2}\, dy$$

so the total volume is

```
> 2*value(");
```

$$1.154700539$$

∎

Example 8.9 A wedge is cut from a circular cylinder of radius 4 by two planes. One plane is perpendicular to the axis of the cylinder. The other intersects the first at an angle of 30 degrees along a diameter of the cylinder. Find the volume of the wedge.

Solution Align the cylinder about the *z*-axis and take cross sections orthogonal to the *x*-axis. An equation for a circle at $z = 0$ corresponding to a cross section orthogonal to the *z*-axis (the base of the wedge) is then

```
> eq := x^2 + y^2 = 16;
```

$$x^2 + y^2 = 16$$

As a function of *x* the boundary points of this circle are [*x*, *yval*(*x*)], where *yval* is defined by

```
> isolate(",y^2);
```

$$y^2 = 16 - x^2$$

```
> map(sqrt,");
```

$$y = \sqrt{16 - x^2}$$

```
> yval := makeproc(rhs("),x):
```

If the object is oriented so that the *x*-axis intersects the sloping plane forming the top of the wedge, then the cross sections orthogonal to the *x*-axis are all triangles with corners as constructed by **pts( )**.

```
> pts := (x) -> PLOT3D( POLYGONS( evalf( [
> [x,0,0],
> [x,yval(x),0],
> [x,yval(x),tan(Pi/6)*yval(x)]
> ]))):
```

A sequence of 41 cross sections regularly spaced along the *x*-axis can be displayed simultaneously by the commands

```
>   n := 40: width:= 8/n: p := table():
> for i from 0 to n do p[i] := pts(-4 + i*width) od:
> opts:= axes=frame, scaling=constrained:
```

```
> display3d(convert(p,list) , opts);
```

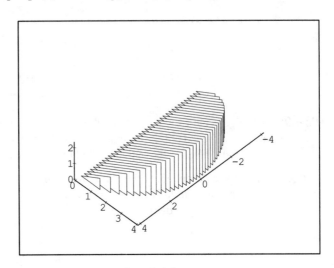

The area of each triangle is $\dfrac{\text{base height}}{2}$. The height as a function of x is defined by

```
> zval := (x)-> tan(Pi/6)*yval(x):
```

and the area is

```
> A := makeproc( 1/2*zval(x)*yval(x),x);
```

$$x \mapsto \frac{1}{6}\sqrt{3}\,(16 - x^2)$$

Thus, the volume of the wedge is

```
> Int( A(x),x=-4..4);
```

$$\int_{-4}^{4} \frac{1}{6}\sqrt{3}\,(16 - x^2)\,dx$$

```
> value(");
```

$$\frac{128}{9}\sqrt{3}$$

∎

8.2 Cylindrical Shells

An effective way of finding volumes of many solids of revolution is by using cylindrical shells. Actual cylinders are easily viewed with **tubeplot( )**.

Example 8.10 Use cylindrical shells to find the volume of the solid obtained by rotating about the x-axis the region under the curve $y = \sqrt{x}$ from 0 to 1. (See Example 8.3.)

Solution Each cylindrical shell is of radius y, for some y in the range $[0, 1]$. For a given y, the corresponding cylinder runs from the curve $y = \sqrt{x}$ to $x = 1$. The equation $y = \sqrt{x}$ can be rearranged to express x as a function of y.

```
> with(plots): with(student):
> isolate( y=sqrt(x),x);
```

$$x = y^2$$

A function to compute the lower x value in terms of y is

```
> lower := makeproc(rhs("),y);
```

$$y \mapsto y^2$$

The surface area of the cylinder of radius y and height $1 - lower(y)$ as a function of y is

```
> A := makeproc( (1-lower(y))*2*Pi*y,y);
```

$$y \mapsto 2\left(1 - y^2\right)\pi y$$

The total volume is

```
> Int( A(y),y=0..1);
```

$$\int_0^1 2\left(1 - y^2\right)\pi y \, dy$$

which evaluates to

```
> value(");
```

$$\frac{1}{2}\pi$$

A representative image based on five cylinders is generated as follows. The central axis of each cylinder runs along the line defined by $[x, y, z]$ for some range of values of x, y and z. To follow the x-axis, both y and z must be 0, while the height of the cylinder, x, ranges from $lower(y)$ to 1.

The individual cylinders are generated by the commands[4]

```
> p := table():   size := 1/9:
> for i from 1 to 8 do
>     r := i*size;
>     p[i] := tubeplot( [x,0,0,x=lower(r)..1],radius=r);
> od:
```

and collectively displayed by

---

[4]We have not included the tube corresponding to a 0 radius.

```
> display3d( convert(p,list),axes=boxed);
```

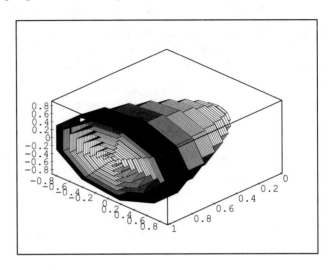

8.3 Arc Length

From the "right point of view," determining the length of a curved line segment, *arc length*, can be very easy. For example, if the function s already computes arc length so that $s(a) - s(0)$ is the distance traveled along the arc as the argument to s changes from 0 to a, this total distance is

```
> I1 := Int( 1 , u = s(0)..s(a) );
```

$$\int_{s(0)}^{s(a)} 1 \, du$$

This simple integral is the area under the graph of a horizontal line. The horizontal axis measures the actual distance along the original arc, and the vertical axis measures the actual velocity as you move along that arc. Movement along the arc is at a steady velocity of 1.

By changing variables, we can change our point of view. For example, if the position u along the curve is a more general function of time, as in $u = s(t)$, then $t = s^{(-1)}(u)$ and this integral can be reexpressed as

```
> I2 := changevar( t = (s@@(-1))(u),I1,t);
```

$$\int_0^a D(s)(t) \, dt$$

In this form $D(s)(t)$ need no longer be constant. The horizontal axis measures time instead of distance.

8.3 Arc Length

This last formula for the length of a curve (i.e., arc length) can also be deduced from the following diagram. Let *dx*, *dy*, and *ds* measure the change in position of the object in the *x* direction, the *y* direction, and along the curve *s*, over some small interval of time *t*. These quantities are related as shown in the graph

```
> with(plots):
> p1 := textplot( [2,-.1 ,'dx'], align=BELOW):
> p2 := textplot( [4.2,1.5 ,'dy'], align=RIGHT):
> p3 := textplot( [2,1.7 ,'ds'], align=ABOVE):
> p4 := plot([[0,0],[4,0],[4,3],[0,0]]):
> display([p.(1..4)]);
```

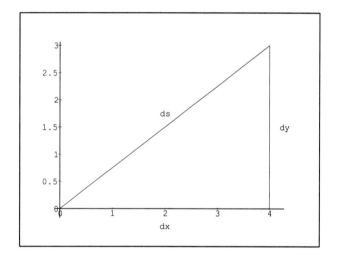

The integral *I2* is the limit of the sum of all these small changes *ds* as the width of the time intervals tend to 0.

We can also reexpress this integral in terms of *x* or *y*. If it is reexpressed as an integral with respect to *x*, the corresponding physical interpretation is that we are monitoring the progress along the arc while proceeding along the *x*-axis at a steady rate.

To benefit from this last point of view we must be able to measure how *s* changes relative to the value of *x(t)* instead of to the value *t*. Pythagoras's theorem and the previous graph give us a way to do it. The relation

```
> eq := D(s)^2 = D(x)^2 + D(y)^2;
```

$$D(s)^2 = D(x)^2 + D(y)^2$$

must hold where all of *s*, *x*, and *y* are unspecified functions of time. By specifying that, for example, $x(t) = t$, we can relate $D(s)$ directly to $D(y)$. From the relation $x(t) = t$ we have $D(x) = 1$ and

```
> subs(D(x)=1,");
```

$$D(s)^2 = 1 + D(y)^2$$

```
> map(sqrt,");
```
$$D(s) = \sqrt{1 + D(y)^2}$$

Thus, with respect to $x(t)$, our original integral for arc length becomes

```
> I3 := Int( rhs(") , x=0..xval(a) );
```
$$\int_0^{xval(a)} \sqrt{1 + D(y)^2} \, dx$$

where $xval(a)$ is the x value of the corresponding point on the arc at $t = a$.

Example 8.11 Find the length of the arc of the semicubical parabola $y^2 = x^3$ between the points $[1, 1]$ and $[4, 8]$.

Solution The curve corresponding to

```
> eq := y^2 = x^3;
```
$$y^2 = x^3$$

is shown by the plot command

```
> with(plots):

> implicitplot(eq,x=1..4,y=1..8);
```

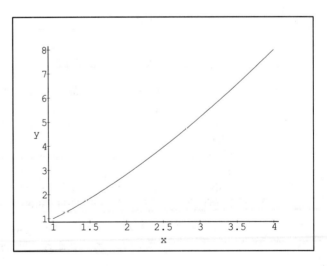

To compute the length of this arc, we regard both x and y as unspecified functions of time. Thus, we have

```
> map(D,eq);
```
$$2D(y)y = 3D(x)x^2$$

and

```
> isolate(",D(y));
```

8.3 Arc Length

$$D(y) = \frac{3}{2}\frac{D(x)x^2}{y}$$

If we define x by $x(t) = t$ (i.e., if we choose to proceed along the x-axis at a constant rate), then $D(x) = 1$ and $D(x)(t) = 1$ for all t. Thus, the previous equation simplifies to

```
> e1 := subs( {D(x)=1,y=sqrt(x^3)},rhs(") );
```

$$\frac{3}{2}\sqrt{x}$$

thereby expressing $D(y)$ entirely in terms of x. The integrand for arc length becomes

```
> sqrt( 1 + (D(y)^2)(t) ) = sqrt(1 + ("^2)(t));
```

$$\sqrt{1 + D(y)(t)^2} = \sqrt{1 + \frac{9}{4}x(t)}$$

With x defined by $x(t) = t$, the arc length can be computed as

```
> arclength := Int( subs(x(t)=t,rhs(")) ,t=1..4);
```

$$\int_1^4 \sqrt{1 + \frac{9}{4}t}\, dt$$

(note that x and t range from 1 to 4), which is approximately

```
> evalf(");
```

$$7.633705416$$

■

When the function x is defined by $x(t) = t$ for all t, we can also obtain the preceding solution directly from $e1$ by integrating with respect to x immediately.

```
> e1;
```

$$\frac{3}{2}\sqrt{x}$$

```
> Int(sqrt(1+"^2),x=1..2);
```

$$\int_1^2 \frac{1}{2}\sqrt{4 + 9x}\, dx$$

The advantage of writing the solution explicitly in terms of t, as in the previous solution, is that the process remains valid for other definitions of x as a function of time t.

REMARK: By isolating $D(x)$ and defining y by the identity function $y(t) = t$, we could have computed arc length with respect to y as follows. We have

```
> isolate(eq,x):
```

and

```
> isolate(D(eq),D(x)):
```

so on using the first equation and $D(y) = 1$ to simplify the right-hand side of the second, we obtain

```
> subs( {"",D(y)=1}, rhs("));
```

$$\frac{2}{3}\frac{1}{\sqrt[3]{y}}$$

The arc length is the integral

```
> Int( sqrt(1 + "^2),y=1..8);
```

$$\int_1^8 \sqrt{1 + \frac{4}{9}y^{-\frac{2}{3}}}\, dy$$

Note that y must range over its full range.

REMARK: The form of the integral obtained by integrating with respect to x in this example is considerably simpler than this one and has the advantage that it can be evaluated exactly.

Example 8.12 Find the length of the arc of the parabola $y^2 = x$ from $[0, 0]$ to $[1, 1]$.

Solution We are given the equation

```
> eq := y^2 = x;
```

$$y^2 = x$$

By treating x and y as functions, we obtain the equation

```
> map(D,");
```

$$2D(y)y = D(x)$$

so that their derivatives are related by

```
> isolate(",D(y));
```

$$D(y) = \frac{1}{2}\frac{D(x)}{y}$$

To integrate with respect to x, we define x by $x(t) = t$ so that $D(x) = 1$ (i.e., $D(x)(t) = 1$ for all t). The previous equation then simplifies to

```
> subs( {D(x)=1},");
```

$$D(y) = \frac{1}{2y}$$

With $y = \sqrt{x}$, the integral for this arc length is

```
> Int( sqrt( 1 + subs(y=sqrt(x),rhs("))^2 ), x=0..1);
```

$$\int_0^1 \sqrt{1 + \frac{1}{4x}}\, dx$$

which is approximately

```
> evalf(");
```

$$1.478942858$$

∎

REMARK: Here it is easier to integrate with respect to y than with respect to x primarily because it is easier to solve $y^2 = x$ for x than for y. This kind of consideration is particularly important when looking for an exact answer.

Example 8.13 Find the length of the arc of the hyperbola $xy = 1$ from the point $(1, 1)$ to the point $(2, \frac{1}{2})$.

Solution From the equation

```
> eq := x*y=1;
```

$$xy = 1$$

we have

```
> eq1 := isolate(",y);
```

$$y = x^{-1}$$

and

```
> eq2 := map(D,");
```

$$D(y) = -\frac{D(x)}{x^2}$$

The specified arc length is given by

```
> Int( sqrt( 1 + subs({D(x)=1,eq1},rhs(eq2))^2) , x=1..2);
```

$$\int_1^2 \sqrt{1 + x^{-4}}\, dx$$

which is approximately

```
> evalf(");
```

$$1.132090393$$

∎

Example 8.14 Find the arc length from $[0, 0]$ as a function of x for the curve $y = x^2 - (\ln x)/8$.

Solution Consider the functional relationship

```
> eq := y = x^2 - ln(x)/8;
```

$$y = x^2 - \frac{1}{8}\ln(x)$$

We use this relationship to express $D(y)$ in terms of $D(x)$. To accomplish this, we regard $\ln(x)$ as a composition, as in

> convert(",'@');

$$y = x^2 - \frac{1}{8} \ln \circ x$$

From this we obtain

> map(D,");

$$D(y) = 2D(x)x - \frac{1}{8} a \mapsto a^{-1} \circ x D(x)$$

> isolate(",D(y));

$$D(y) = 2D(x)x - \frac{1}{8} a \mapsto a^{-1} \circ x D(x)$$

As a relation between expressions when x is the identity function, this is

> eq2 := subs(D(x)=1,convert(",nested));

$$D(y) = 2x - \frac{1}{8x}$$

The arc length from $x = 0$ to $x = t$ is given by

> Int(sqrt(1 + subs({eq},rhs(eq2))^2) , x=1..t);

$$\int_1^t \sqrt{1 + \left(2x - \frac{1}{8x}\right)^2}\, dx$$

It is convenient to have a closed form solution this time. The value of this integral is

> value(");

$$t^2 + \frac{1}{8}\ln(t) - 1$$

so the length of the arc as a function of t is

> arclength := makeproc(",t);

$$t \mapsto t^2 + \frac{1}{8}\ln(t) - 1$$

A graph of this function is shown here.

```
> plot(arclength,1..5);
```

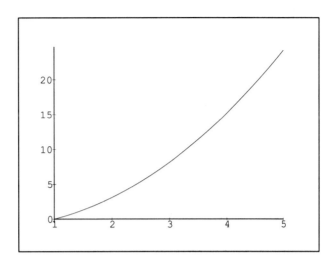

∎

8.4 Surface Area

The surface area of a solid of revolution can be obtained by carefully comparing the surface area of a cylinder to that of a cone. We computed the former by "unwrapping" the surface. The result was a rectangle of dimension h by $2\pi r$, where h was the height of the cylinder and r was the radius.

Now align the cone and the cylinder so that the axis of symmetry for both is the x-axis and the top of both is at the origin, as in Figure 8.1, and compare their cross sections in the x-y plane.

In any given interval, the arc corresponding to the upper edge of this cross section is longer for the cone than for the cylinder. In fact, the ratio of the length of this arc for the cone to this arc for the cylinder is $\sqrt{1 + (D(y)(x))^2}$, where $y(x) = mx$ is the radius of the cone at x and $D(y)(x) = m$ is a constant. (Recall how we computed arc length.)

The "unwrapped" surface of a small part of the cone is bigger than the unwrapped surface of a cylinder of comparable average radius in essentially this same ratio. The surface area of a cylinder of height h is

```
> Int( 2*Pi*y(x),x=0..h);
```

$$\int_0^h 2\pi\, y(x)\, dx$$

(for a cylinder, $y(x)$ is a constant), and by analogy the surface area of the cone of height h is

```
> Int( 2*Pi*y(x)*sqrt(1 + D(y)(x)^2),x=0..h);
```

406   Chapter 8   More Applications of Integration

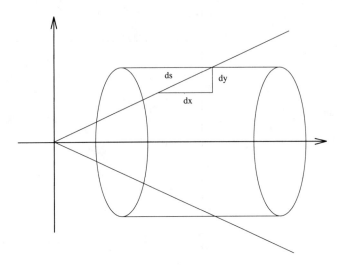

Figure 8.1 Stretching cones into cylinders

$$\int_0^h 2\pi\, y(x) \sqrt{1 + D(y)(x)^2}\, dx$$

This remains true when the graph of the radius with respect to x does not form a straight line.

Example 8.15   The curve $y = \sqrt{4 - x^2}$, $-1 \leq x \leq 1$, is an arc of the circle $x^2 + y^2 = 4$. Find the area of the surface obtained by rotating this arc about the x-axis. (The surface is a portion of a sphere of radius 2.)

Solution   The surface in question is

```
> f := (x) -> sqrt(4-x^2);
```

$$x \mapsto \sqrt{4 - x^2}$$

```
> with(plots):
> opts := axes =framed:

> tubeplot( [x,0,0,x=-1..1],radius=f(x),opts);
```

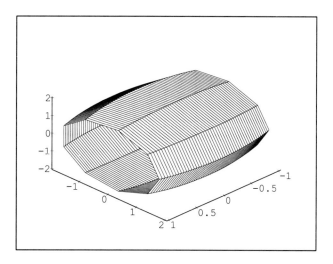

The surface area is given by the integral

```
> Int( f(x)*sqrt( 1 + D(f)(x)^2),x=-1..1);
```

$$\int_{-1}^{1} \sqrt{4 - x^2} \sqrt{1 + \frac{x^2}{4 - x^2}}\, dx$$

which evaluates to

```
> value(");
```

$$4$$

∎

Example 8.16 The arc of the parabola $y = x^2$ from $(1, 1)$ to $(2, 4)$ is rotated about the y-axis. Find the area of the resulting surface.

Solution This time the radius is the x value, so we need to express x as a function of y.

```
> y = x^2;
```

$$y = x^2$$

```
> map(sqrt,"): isolate(",x);
```

$$x = \sqrt{y}$$

```
> f := makeproc(rhs("),y):
```

The surface in question is

```
> opts := axes=framed:
```

```
> tubeplot( [x,0,0,x=-1..1],radius=f(x),opts);
```

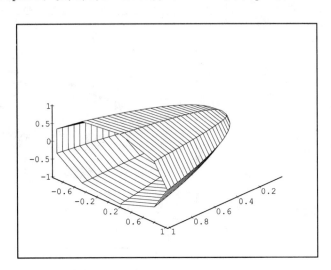

The surface area is given by the integral

```
> Int( f(y)*sqrt( 1 + D(f)(y)^2),y=1..4);
```

$$\int_1^4 \sqrt{y}\sqrt{1+\frac{1}{4y}}\,dy$$

which evaluates to

```
> value(");
```

$$\frac{17}{12}\sqrt{17} - \frac{5}{12}\sqrt{5}$$

■

Example 8.17 Find the area of the surface generated by rotating the curve $y = e^x$, $1 \leq x \leq 2$, about the line $y = 1$.

Solution The radius as a function of x is

```
> f := (x) -> exp(x)-1;
```

$$x \mapsto e^x - 1$$

The surface is

```
> opts := axes =framed:
```

```
> tubeplot( [x,1,0,x=1..2],radius=f(x),opts);
```

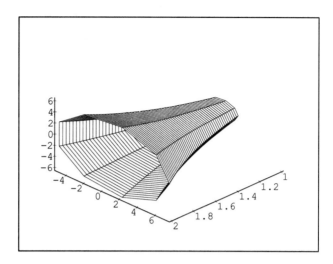

The surface area is given by the integral

```
> Int( f(x)*sqrt( 1 + D(f)(x)^2),x=1..2);
```

$$\int_1^2 (e^x - 1) \sqrt{1 + e^{x2}} \, dx$$

which is approximately

```
> evalf(");
```

$$19.31234505$$

∎

EXERCISE SET 8

Some of the important commands that have been used in this chapter are shown in Table 8.1. Keep them in mind as you explore the examples from the chapter and the exercises listed here.

1. By constructing a sequence of horizontal cross sections, generate a graph of a pyramid whose base is a rectangle that is twice as long as it is wide (h by $2h$) and whose height is h. Find the volume of the pyramid.
2. Generate a graph of the pyramid described in Exercise 1 by constructing a sequence of cross sections orthogonal to the longest horizontal dimension. Find the volume of the pyramid by integrating with respect to that dimension. Note that this gives you a way of checking that the answer you previously obtained is correct.
3. Generate a graph of the pyramid described in the first two exercises using only four polygons.

Command	Description
(x) -> sqrt(x);	A function definition
Int(A(z),z=0..h);	Representation of a definite integral
PLOT3D(...);	A 3D plot data structure
POLYGONS([...]);	A data structure used internally by plot
[x(z), x(z), z];	A point on a 3D curve
axes=boxed;	A plot option
axes=frame;	A plot option
combine(");	Undo an expand
convert(f(x),'@');	Rewrite $f(x)$ as $f \circ x$
convert(f@x,nested);	Rewrite $f \circ x$ as $f(x)$
convert(plt,list);	Convert to a list
display3d([p1,p2],insequence=true);	Animate a sequence of plots
evalf(");	Write numbers in floating-point
for i from 0 to 21 do ... od;	Repeat a sequence of commands
implicitplot(eq,x=1..4,y=1..8);	Generate an implicit plot
implicitplot3d(eq,...);	Generate an implicit plot
isolate(eq,x);	Solve for x
isolate(eq,y^2);	Solve for y^2
makeproc(e1,z);	Construct a procedure
map(sqrt,[a,b]);	Construct a list of square roots
p := table():	Define a table
p1 := [0,h];	Define a point
polygonplot3d(...);	Generate a surface from data points
scaling=constrained;	A plot option
simplify(e1);	Perform algebraic simplifications
slope(p1,p2);	Slope of a line defined by two points
sort([c,a,b]);	Sort a list
style=patch;	A plot option
subs({D(x)=1,y=sqrt(x^3)},");	A simultaneous substitution
subs(D(x)=1,");	A substitution
tubeplot([x,1,0,x=1..2],radius=f(x));	A solid of revolution
tubepoints=13;	A tube plot option
value(e1);	Evaluate an inert representation
with(plots);	Load special plot routines
xcircle(.5,0,0,xradius(.5));	A user-defined procedure

Table 8.1 Some of the Commands from Chapter 8

4. Use **implicit3d( )** to generate a three-dimensional plot of the surface described by the equation

   > eq1 := 2*x^2 + y^2 + z^2 = 9;
   $$2x^2 + y^2 + z^2 = 9$$

   Find the volume of this object by using cross sections orthogonal to the x-axis. Depending on how you compute the area of each cross section, you may need to decide which sign to use for any expressions with absolute values to complete an exact integration.

5. Find the volume of the object described by the equation

   > eq2 := a*x^2 + y^2 + z^2 = r^2;
   $$ax^2 + y^2 + z^2 = r^2$$

6. Consider the curve defined by

   > eq := y = 3*sqrt(x+3) + 2;
   $$y = 3\sqrt{x+3} + 2$$

   Use **tubeplot( )** to generate a graph of the solid of revolution obtained by rotating the region below this curve from $x = 0$ to $x = 1$. Find the volume of that solid.

7. Repeat Exercise 6 for the solid obtained by rotating the region between the curve *eq* and the line $y = 1$ about the line $y = 1$.

8. Use **tubeplot( )** to generate a graph of the solid obtained by rotating about the y-axis the region bounded by the two curves

   > eq := {y-2 = 2*x^2-8*x+8, y-2 = 2*x-4};
   $$\{y - 2 = 2x^2 - 8x + 8, y - 2 = 2x - 4\}$$

   Find the volume of the resulting solid.

9. Generate a graph of the solid obtained by rotating about the line $y = 1$ the region from Exercise 8. Find the volume of the resulting solid.

10. Generate a sequence of cross sections illustrating the method of integration you used in Exercise 9.

11. A wedge is cut from a cylinder of radius 5 by two planes. Viewed from a plane p perpendicular to the axis of the cylinder, one plane intersects p at an angle of $\pi/5$ radians along a diameter of the cylinder and the other plane intersects p along that same diameter at an angle of $-\pi/4$. Generate a graph of the wedge, and find its volume.

12. Generate a graph of the cylindrical shells that would be used to find the volume of the solid obtained by rotating about the y-axis the region between the curve $y = \sqrt{x}$ and the y-axis and from $x = 0$ to $x = 1$. Find the volume of the resulting solid.

13. Verify your volume computation from Exercise 12. by computing the same volume using circular disks orthogonal to the y-axis.

14. Consider the curve defined implicitly by

    > eq := x^3+x^2*y = y^2;
    $$x^3 + x^2 y = y^2$$

with x ranging from $x = 1$ to $x = 3$. Generate a graph of the portion of this curve with $view = [1..3, 0..10]$. Construct an integral for the arclength of this curve, and find an approximate value to this length by using a numerical integration technique or **evalf( )**. Ensure that your answer is sensible by comparing it with the apparent length of curve as seen on the graph.

15. Find the length of the arc of the curve defined by

    ```
    > eq := y^2 + y = x;
    ```

    $$y^2 + y = x$$

    from $x = 2$ to $x = 5$ with positive y values, by setting up and evaluating an integral with respect to y. Remember to verify your answer graphically.

16. Verify the computation in Exercise 15 by defining and evaluating (using **evalf( )**) an appropriate integral with respect to x.

17. Reformulate the integral for arclength from Exercise 16 as an integral in terms of t, where $x = 2 * t$. Verify your answer numerically.

18. The curve $y = \sqrt{9 - 2x^2}$, $-1 \le x \le 1$, is an arc of the curve defined by $2x^2 + y^2 = 9$. Generate a graph of the object obtained by rotating the region beneath this arc about the x-axis. Construct an integral that can be used to compute the surface area of this object. Evaluate this integral. Verify that the answer is reasonable by comparing it with the surface area of a similar portion of a perfect sphere.

19. (*Adapted from the Maple Calculus Workbook, by Geddes, Marshman, McGee, Ponzo, and Char.*) Deanna and Craig have been asked to construct a large number of beads as part of a school art project. Because of budget restrictions, the beads must consume the smallest total amount of material possible, but must have a previously specified length (l) and hole size (radius r). Assist them by completing the following investigation.

    a. Consider the sphere defined by

       ```
       > eq := x^2 + y^2 + z^2 = 9;
       ```

       $$x^2 + y^2 + z^2 = 9$$

       Find the volume V of the cap of the sphere defined by slicing the sphere at the plane $x = k$, where k is a constant satisfying $k > 0$.

    b. After removing two caps of the same size, one defined by the plane $x = k$ and the other by the plane $x = -k$, a hole is bored through the sphere connecting the two faces left by the caps. Find the volume of the portion of the sphere that remains.

    c. If the hole through the sphere is of radius r, and the length of the bead is l, find the radius of the sphere that minimizes the volume of material needed to make a bead.

9 Parametric Equations

9.1 Introduction

We have seen numerous examples of curves defined by relations such as $y = f(x)$ or $x = g(y)$ or even $f(x, y) = c$. We will call such an equation a *Cartesian* representation of the curve. Given a Cartesian representation of the curve, we can display the curve by using commands such as

```
> plot(f(x),x);
> implicitplot(x^2 + y^2 = 4,x=-2..2,y=-2..2);
```

Such representations extend to surfaces by using equations such as $z = f(x, y)$ or $x^2 + y^2 + z^2 = 9$. These are displayed by commands such as

```
> plot3d(f(x,y),x=-2..2,y=-3..3);
> implicitplot3d(x^2+y^2+z^2=1,x=-1..1,y=-1..1,z=-1..1);
```

and

```
> tubeplot( [t,0,0,t=-3..3],radius=sqrt(9-t^2));
```

To define a tube plot, we must specify a position, such as $[t, 0, 0]$, along the axis of symmetry, and the radius of the object at that point. The point and the radius are specified in terms of a parameter t. In general, as t varies, the current point on the curve is given by $[x(t), y(t), z(t)]$. In this chapter, we examine such *parametric* representations in greater detail. For the most part, we confine our discussion to two dimensions.

9.2 Parametric Curves

When we introduced the concept of arc length in Section 8.3, we treated x and y as functions of a third variable t.

Example 9.1 Sketch the curve $[x(t), y(t)]$ as t ranges from 0 to 5, where x and y are given by

```
> x := (t) -> t:
> y := (t) -> t^2 -3*t + 2:
```

and compute the total arc length.

Solution To draw such a curve, we use the command **plot( )** directly.

```
> plot( [x(t),y(t),t=1..5] );
```

Chapter 9 Parametric Equations

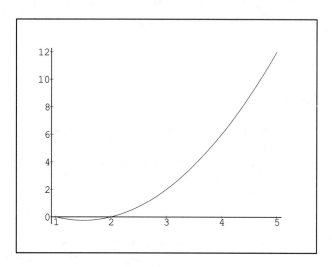

Each value of t corresponds to a specific point $[x(t), y(t)]$ on the curve.

Because $x(t) = t$, this curve can also be described directly by the equation $y = x^2 - 3x + 2$. Using the arc length formula from Chapter 8, the length of this arc is

```
> Int( sqrt( 1 + D(y)(x)^2),x=0..5);
```

$$\int_0^5 \sqrt{10 + 4x^2 - 12x}\, dx$$

which is approximately

```
> evalf(");
```

$$15.86171873$$

■

In general, if t denotes time, then the pair $[x(t), y(t)]$ is the Cartesian position of an object at time t. As t varies, the object moves. There are many different ways of moving along the path. For example, we may want to traverse the path at a constant rate, quickly, or slowly, or even vary our speed as we move along the path. Each choice results in a different a way of defining the functions x and y.

Consider a curve defined by the equation $y(t) = f(x(t))$. If we control our rate of progress along the curve so that our progress in the direction of the x-axis is at a constant rate of 1 (i.e., x is just the identity function $(t) \to t$), then $y(t) = f(t)$. Similarly, if we control our rate of progress along the curve so that our progress in the direction of the y-axis is at a constant rate of 1 (i.e., y is the identity function), then $x(t) = g(t)$, where $g = f^{(-1)}$ is the functional inverse of f satisfying $g(f(t)) = t$.

There are many different parameterizations of the same curve. For example, each of the following commands plots exactly the same curve segment.

9.2 Parametric Curves

```
> plot( [x(t),y(t),t=1..9] );
> plot( [x(t^2),y(t^2),t=1..3]);
> plot( [x(t^2+1),y(t^2+1),t=0..sqrt(8)]);
> plot( [x(1/t),y(1/t),t=1..1/9]);
```

REMARK: A curve remains the same, independent of the parameterization. Different parameterizations correspond to traversing the same curve in different manners, or at different rates.

We are not limited to having one of x or y to be the identity function. The following parametric plot is a portion of the parabola $y = x^2$.

```
> plot( [sin(t),sin(t)^2,t=0..2*Pi],labels=[x,y]);
```

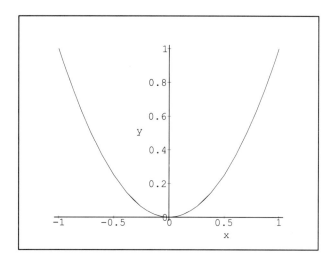

This parametric representation traces out the path by starting at $x = 0$, proceeding to $x = 1$, moving back to $x = -1$, and finally finishing at $x = 0$, so that we actually traverse the curve segment twice.

9.2.1 Finding Cartesian Representations

Because there are so many different ways of parameterizing the same curve, it can be difficult to recognize familiar curves such as circles, parabolas, and ellipses from just their parametric descriptions. In such cases, it is useful to be able to obtain the Cartesian representation. To find a Cartesian representation from a given parametric representation of a curve, we proceed as in the following example.

Example 9.2 Find a Cartesian representation of the parametric curve $[f(t), g(t), t = 1..9]$ defined by

```
> f := (t) -> t^2-3*t +2: g := (t) -> t + 2:
```

Solution A graph of this curve is

```
> plot([f(t),g(t),t=1..9],labels=[x,y]);
```

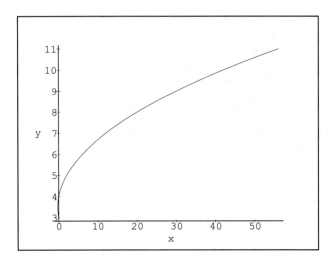

Each of the two equations

```
> eq1 := x=f(t);  eq2 := y=g(t);
```

$$x = t^2 - 3t + 2, \; y = t + 2$$

can be solved for *t*. There are two solutions to *eq1*, as given by

```
> sol1 := [solve(eq1,{t})];
```

$$[\left\{t = \frac{3}{2} - \frac{1}{2}\sqrt{1+4x}\right\}, \left\{t = \frac{3}{2} + \frac{1}{2}\sqrt{1+4x}\right\}]$$

For *eq2* we have the solution

```
> sol2 := solve(eq2,{t});
```

$$\{t = y - 2\}$$

The Cartesian representations corresponding to the two solutions to *eq1* are

```
> sols := subs(sol2,sol1);
```

$$[\left\{y - 2 = \frac{3}{2} - \frac{1}{2}\sqrt{1+4x}\right\}, \left\{y - 2 = \frac{3}{2} + \frac{1}{2}\sqrt{1+4x}\right\}]$$

It remains to be determined which of these solutions corresponds to the parametric curve. From the graph of the parametric curve, we see that the correct Cartesian representation of this curve must result in *y* being positive at, say, $x = f(50)$. At $x = f(50)$, the two solutions to *eq1* lead to *y* values of

```
> map( fsolve , subs(x=50,sols) );
```

$$[\{y = -3.588723440\}, \{y = 10.58872344\}]$$

so the equation

```
> y-2 = 3/2+1/2*(1+4*x)^(1/2);
```

$$y - 2 = \frac{3}{2} + \frac{1}{2}\sqrt{1 + 4x}$$

is the correct Cartesian representation of this curve. The x values must lie between the minimum and the maximum value of f on the domain $[1, 9]$, and the y values lie between the minimum and the maximum of g on that same domain. ∎

Example 9.3 Find a Cartesian representation of the parametric curve

```
> [cos(t),sin(t),t=0..2*Pi];
```

$$[\cos(t), \sin(t), t = 0 \ldots 2\pi]$$

Solution This curve is shown by the plot command

```
> plot(",scaling=constrained);
```

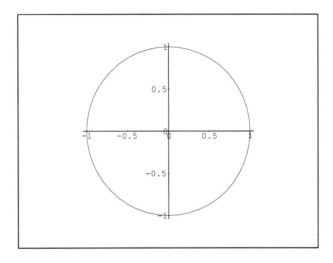

Again, we eliminate t from the equations that define x and y in terms of t.

```
> eq1 := x = cos(t): eq2 := y = sin(t):
```

A slightly different way of using *Maple* to do this is by using the following commands.

```
> sol1 := solve(eq1,t):
> sol2 := solve(eq2,t):
> sol1 = sol2;
```

$$\arccos(x) = \arcsin(y)$$

This solution can be further simplified by applying **sin( )** to both sides, as in

```
> map(sin,");
```

$$\sqrt{1 - x^2} = y$$

The x values range from

```
> minimize( cos(t) );
```
$$-1$$

to

```
> maximize( cos(t) );
```
$$1$$

∎

REMARK: To carry out the preceding simplifications, we could have started with the identity $\sin(t)^2 + \cos(t)^2 = 1$. Given *eq1* and *eq2*, this leads directly to the equation

```
> x^2 + y^2 = 1;
```
$$x^2 + y^2 = 1$$

and provides us with an approach to simplifying $\sin(\arccos(x))$.

Example 9.4 A *cycloid* is the curve traced by a point P fixed to the perimeter of a circle as the circle rolls along the x-axis. A parametric description of such a curve is

$$[rt - r\sin(t), r - r\cos(t), t = 0..4\pi]$$

Sketch this curve for $r = 2$ and find a Cartesian representation of the curve.

Solution For a circle of radius 2, the curve traced by P is drawn by the commands

```
> r := 2:
> opts := labels= [x,y] , scaling=constrained:
> plot( [r*t - r*sin(t), r - r*cos(t),t=0..4*Pi],opts);
```

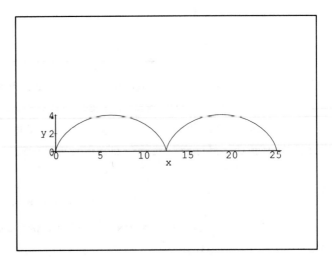

Each arch of the cycloid corresponds to one complete roll of the circle along the x-axis.

9.2 Parametric Curves

To describe the curve in terms of the Cartesian coordinates x and y, we begin by trying to eliminate t. The original curve is defined by the parametric equations

```
> eqns := {x=r*t - r*sin(t),y=r - r*cos(t)};
```

$$\{y = 2 - 2\cos(t), x = 2t - 2\sin(t)\}$$

When we attempt to isolate t in each of these equations, as in

```
> map(isolate,eqns,t);
```

$$\left\{t = \pi - \arccos(\frac{1}{2}y - 1), -2t + 2\sin(t) = -x\right\}$$

Maple succeeds with only one of the equations. This partial success is enough, however, for it allows us to eliminate t from the second equation. That is, by using the equation

```
> isolate(y = r - r*cos(t),t);
```

$$t = \pi - \arccos(\frac{1}{2}y - 1)$$

we can simplify the remaining equation to

```
> subs(",x=r*t - r*sin(t)): ";
```

$$x = 2\pi - 2\arccos(\frac{1}{2}y - 1) - \sqrt{-y^2 + 4y}$$

∎

REMARK: The parametric representation is much easier to work with than the Cartesian representation in this example.

Recall that each parameterization represents a different "point of view." For the preceding parameterization of the cycloid, the parameter t measures the angle between the line from the center of the circle to the point P and the y-axis. The parametric descriptions

```
> [ r*(2*t) - r*sin(2*t) , r - r*cos(2*t) , t = 0..2*Pi ];
```

$$[4t - 2\sin(2t), 2 - 2\cos(2t), t = 0\ldots 2\pi]$$

and

```
> expand(");
```

$$[4t - 4\sin(t)\cos(t), 4 - 4\cos(t)^2, t = 0\ldots 2\pi]$$

represent the same curve but they traverse it twice as fast as the parameterization in Example 9.4.

9.3 Tangents and Areas Revisited

When a curve is described by an equation of the form $y = f(x)$, we construct a tangent line at $x = a$ by first finding the slope of the tangent line at $x = a$. This is

$$\left.\frac{dy}{dx}\right|_{x=a} = f'(a)$$

When the same curve is defined parametrically, as in $[x(t), y(t), t = a \cdots b]$, both x and y are functions of some parameter t, say, time, and $y(t) = f(x(t))$. In this case, the expression $f(x)$ is really a composition of functions, as in

```
> y = f@x;
```

$$y = f \circ x$$

and when these "functions" are applied to t, we obtain the equation

```
> "(t);
```

$$y(t) = f(x(t))$$

The (Cartesian) slope $\dfrac{dy}{dx}$ can be computed directly in terms of t by the function $\dfrac{D(y)}{D(x)}$. This is obtained from $y = f \circ x$ by differentiating both sides and dividing through by $D(x)$, as in

```
> map(D, y=f@x );
```

$$D(y) = D(f) \circ x \, D(x)$$

```
> slpe := "/D(x);
```

$$\frac{D(y)}{D(x)} = D(f) \circ x$$

Thus, at time t, the object will be at position $[x(t), y(t)]$ and the Cartesian slope of this parametrically defined curve will be

```
> slpe(t);
```

$$\frac{D(y)(t)}{D(x)(t)} = D(f)(x(t))$$

As always, different parameterizations correspond to different points of view. When x is the identity function, so that $x(t) = t$ and $D(x) = 1$, then y is essentially a function of x, and its derivative, as a function of x, is defined by

```
> subs({D(x)=1,x(t)=t},"): ";
```

$$D(y)(t) = D(f)(t)$$

Example 9.5 Given a parametric curve $[x(t), y(t), t = a \cdots b]$ satisfying

```
> eq := y = sin(x);
```

$$y = \sin(x)$$

find the Cartesian slope $\dfrac{dy}{dx}$ of the curve as a function of t.

Solution First we rewrite sin(x) as a composition, as in

```
> convert(eq,'@');
```

$$y = \sin \circ x$$

The derivatives of the functions x and y are related by the equation

```
> map(D,");
```

$$D(y) = \cos \circ x \, D(x)$$

and $\dfrac{dy}{dx}$, as a function of t, is given by

```
> "/D(x);
```

$$\frac{D(y)}{D(x)} = \cos \circ x$$

At t, this function evaluates to

```
> "(t);
```

$$\frac{D(y)(t)}{D(x)(t)} = \cos(x(t))$$

∎

When x is the identity function $(t) \to t$, then $D(x)(t) = 1$ and $x(t) = t$ and the slope at time t simplifies to

```
> eq2 := subs(x=((t)->t),"): ";
```

$$D(y)(t) = \cos(t)$$

REMARK: When x is the identity function, y is an explicit function of x and $\dfrac{dy}{dx}$ can also be obtained by differentiating expressions, as in

```
> subs(y=y(x),eq);
```

$$y(x) = \sin(x)$$

```
> map(diff,",x);
```

$$\frac{d}{dx} y(x) = \cos(x)$$

Example 9.6 Given a parametric curve $[x(t), y(t), t = a \cdots b]$, satisfying

```
> eq := y = x^3 + x^2 + x;
```

$$y = x^3 + x^2 + x$$

find the Cartesian slope $\dfrac{dy}{dx}$ of this curve as a function of t, and as a function of x.

Solution Since no function applications occur in *eq*, we do not need to use the command **convert(",'@')** to convert function applications to compositions. The derivatives of the functions x and y are related by the equation

```
> map(D,eq):
```

which can be rerranged as

```
> isolate(",D(y));
```

$$D(y) = 3D(x)x^2 + 2D(x)x + D(x)$$

The Cartesian slope $\dfrac{dy}{dx}$ is

```
> slpe := expand( "/D(x));
```

$$\frac{D(y)}{D(x)} = 3x^2 + 2x + 1$$

and, at time t, the slope is

```
> slpe(t);
```

$$\frac{D(y)(t)}{D(x)(t)} = 3x(t)^2 + 2x(t) + 1$$

To interpret y as a function of x, make x the identity function. We obtain the function

```
> x_slpe := subs(x=(x->x),slpe):
```

which, when evaluated at $x = a$, yields

```
> x_slpe(a);
```

$$D(y)(a) = 3a^2 + 2a + 1$$

■

It is possible that neither x nor y are identity functions.

Example 9.7 Let the curve $f(x, y) = 0$ be defined parametrically by $[x(t), y(t), t = a..b]$ with

```
> x := (t) -> 3*t^2+2:
> y := (t) -> t^2 -3*t + 2:
```

Find the slope dy/dx as a function of t.

Solution The desired slope as a function of t is

```
> D(y)/D(x);
```

$$\frac{t \mapsto 2t - 3}{t \mapsto 6t}$$

but this time $D(x) \neq 1$. We can also define this "function" as

```
> slpe := makeproc("(t),t);
```

$$t \mapsto \frac{1}{6}\frac{2t-3}{t}$$

∎

REMARK: If we are to carry out the division $\frac{D(y)}{D(x)}$ successfully, then $D(x)(t)$ should not be 0 in the interval under consideration.

Example 9.8 For a radius of 3, the cycloid is defined parametrically by

```
> x := (t) -> 3*(t-sin(t)):
> y := (t) -> 3*(1-cos(t)):
```

Find $\frac{dy}{dx}$, the Cartesian slope of the tangent line, as a function of t, and evaluate this at $t = 5$.

Solution The graph of part of this cycloid is

```
> plot( [x(t),y(t),t=0..6*Pi]);
```

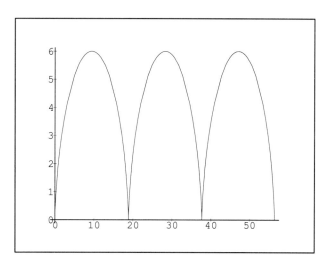

The slope dy/dx is computed by the function

```
> slpe := D(y)/D(x);
```

$$\frac{D(y)}{D(x)}$$

At time $t = 5$, it is

> slpe(5);

$$\frac{3\sin(5)}{3-3\cos(5)}$$

9.3.1 Second Derivatives

To study the concavity of the graph $y = f(x)$, we compute $\frac{d^2y}{dx^2}$. For curves defined parametrically in terms of t, this is computed as

$$\frac{d^2y}{dx^2} = \frac{d}{dx}\left(\frac{dy}{dx}\right) = \frac{\frac{d}{dt}\left(\frac{dy}{dx}\right)}{\frac{dx}{dt}}$$

The corresponding "function" is

> D(D(y)/D(x)) / D(x);

$$\left(-\frac{\left(D^{(2)}\right)(x)\,D(y)}{D(x)^2} + \frac{\left(D^{(2)}\right)(y)}{D(x)}\right) D(x)^{-1}$$

which simplifies to

> concav := normal(");

$$\frac{-\left(D^{(2)}\right)(x)\,D(y) + \left(D^{(2)}\right)(y)\,D(x)}{D(x)^3}$$

When this function is evaluated at t, we obtain the expression

> concav(t);

$$\frac{-\left(D^{(2)}\right)(x)(t)\,D(y)(t) + \left(D^{(2)}\right)(y)(t)\,D(x)(t)}{D(x)(t)^3}$$

Example 9.9 Investigate the concavity of the cycloid $[x(t), y(t), t = a..b]$, when x is defined by $t \to 3(t - \sin(t))$ and y is defined by $t \to 3(1 - \cos(t))$.

Solution To study concavity, we examine the behavior of the second derivative $\frac{d^2y}{dx^2}$. At time t, this evaluates to

> x := (t) -> 3*(t-sin(t)): y := (t) -> 3*(1-cos(t)):
> concav(t);

$$\frac{-9\sin(t)^2 + 3\cos(t)(3 - 3\cos(t))}{(3 - 3\cos(t))^3}$$

which simplifies to

> simplify(");

$$-\frac{1}{3\left(1 - 2\cos(t) + \cos(t)^2\right)}$$

From the general form of this expression, we see that this is always negative except at $\cos(t) = 1$. This observation is confirmed by the graph

```
> opts := t=0..20,y=0..6,labels=[x,y]:
> plot({concav(t),[x(t),y(t),t=0..2*Pi]}, opts);
```

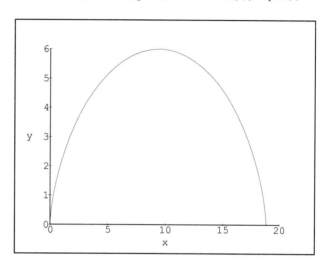

■

Parametric curves may intersect vertical lines more than once, and they may also have two horizontal tangent lines for a given value of x.

Example 9.10 Find the x values corresponding to horizontal tangent lines for the parametric curve defined by $[x(t), y(t), t = 0..\pi]$ with

```
> x := (t) -> 3*cos(2*t):
> y := (t) -> 3*sin(2*t):
```

Solution The formula for the slope of the Cartesian tangent line is

```
> eq1 := D(y)/D(x);
```

$$\frac{t \mapsto 6\cos(2t)}{t \mapsto -6\sin(2t)}$$

To find the horizontal tangent lines, we must solve

```
> eq2 := "(t) = 0;
```

$$-\frac{\cos(2t)}{\sin(2t)} = 0$$

for t.

```
> solve(");
```

$$\frac{1}{4}\pi$$

This shows only one solution, but note that there is another solution at $t = 3\pi/4$, as verified by

```
> value( subs(t=3*Pi/4, eq2) );
```

$$0 = 0$$

The tangent lines at $\frac{\pi}{4}$ and $\frac{3\pi}{4}$ and the parametric curve are displayed simultaneously by the command

```
> plot({y(Pi/4),y(3*Pi/4),[x(t),y(t),t=0..2*Pi]});
```

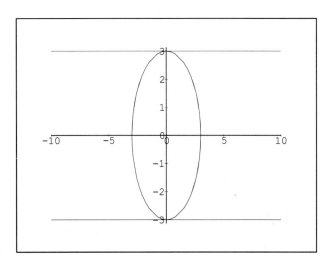

∎

9.3.2 Areas

To compute the area under the curve $y = f(x)$, as x ranges from a to b, we compute the integral

```
> Int( 'y' , x=a..b );
```

$$\int_a^b y\,dx$$

When y and x are defined as parameters of t, this can be rewritten as an integral in terms of t, as in the following example.

Example 9.11 Find the area under one arch of the cycloid $y = f(x)$ when the curve is defined parametrically by

```
> x := (t) -> 3*(t-sin(t)):
> y := (t) -> 3*(1-cos(t)):
```

Solution One arch of the cycloid is generated as *t* varies from 0 to 2π, as in

```
> plot( [x(t),y(t),t=0..2*Pi],labels=[x,y]);
```

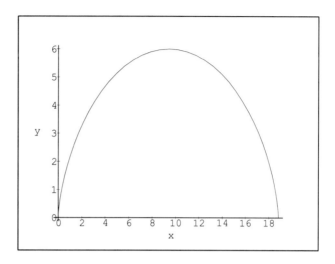

If *y* were a function of *x*, then the area under this arch would be

```
> I1 :=Int( 'y'(x) , x=x(0)..x(2*Pi));
```

$$\int_0^{6\pi} y(x)\, dx$$

In terms of *t*, this becomes, after the change of variables defined by $x = x(t)$,

```
> I2 := Int( y(t)*D(x)(t) , t=0..2*Pi);
```

$$\int_0^{2\pi} (3 - 3\cos(t))^2 \, dt$$

which evaluates to

```
> value(");
```

$$27\pi$$

∎

9.4 Arc Length and Surface Area Revisited

The basic formulas for arc length and surface area can be derived for parametric curves in much the same way we derived our original formulas, but with fewer actual steps since we no longer need to choose a parameterization.

9.4.1 Arc Length

Recall that arc length is given by the integral $\int_0^a D(s)(t)dt$. The function s measures the distance along the curve as a function of time so that the actual distance traveled from $t = 0$ to $t = a$ is

```
> s(a)-s(0) = Int( D(s)(t),t=0..a);
```

$$s(a) - s(0) = \int_0^a D(s)(t)\,dt$$

To carry out the calculation we need a formula for $D(s)(t)$. This is obtained from the fact that

```
> D(s)(t) = sqrt( D(x)(t)^2 + D(y)(t)^2 );
```

$$D(s)(t) = \sqrt{D(x)(t)^2 + D(y)(t)^2}$$

The resulting formula for arc length is

```
> L := subs("","");
```

$$s(a) - s(0) = \int_0^a \sqrt{D(x)(t)^2 + D(y)(t)^2}\,dt$$

Example 9.12 Compute the length of the perimeter of a circle of radius r.

Solution If we position the circle at the origin, then a parametric formulation of the perimeter of the circle is $[x(t), y(t), t = 0..2\pi]$ where

```
> x := (t)-> r*cos(t):
> y := (t)-> r*sin(t):
```

The length of the perimeter s is $s(2\pi) - s(0)$:

```
> subs(a=2*Pi,L);
```

$$s(2\pi) - s(0) = \int_0^{2\pi} \sqrt{r^2 \sin(t)^2 + r^2 \cos(t)^2}\,dt$$

which evaluates to

```
> value(");
```

$$s(2\pi) - s(0) = 2r\pi$$

∎

REMARK: A change of variables of the form $u = 3t$ amounts to constructing a different parameterization of the original circle.

```
> changevar(u=3*t,L,u);
```

$$s(a) - s(0) = \int_0^{3a} \frac{1}{3}\sqrt{r^2 \sin(\frac{1}{3}u)^2 + r^2 \cos(\frac{1}{3}u)^2}\,du$$

9.4.2 Surface Area

The parametric formula for surface area is derived in a similar manner. Recall that we derived the original formula by computing the surface area of each segment as a "distorted" segment of a cylinder of average height (see Figure 8.1). The width of such a segment is magnified by a factor of

```
> D(s)(t) = sqrt( D(x)(t)^2 + D(y)(t)^2 );
```

$$D(s)(t) = \sqrt{D(x)(t)^2 + D(y)(t)^2}$$

which is the same factor that arises when the length of a horizontal line is compared with the length of an arc, as shown on page 399. The corresponding integral for surface area is

```
> S := Int(2*Pi*y(t)*sqrt(D(x)(t)^2 + D(y)(t)^2), t=0..a);
```

$$\int_0^a 2\pi y(t) \sqrt{D(x)(t)^2 + D(y)(t)^2}\, dt$$

assuming that the edge of the cross section is the path traced by $s(t)$ as t varies from 0 to a.

Example 9.13 Find the surface area of a sphere of radius r.

Solution The curve bounding the cross section of the sphere in the $z = 0$ plane is a circle in the x-y plane of radius r. The points on this curve are obtained parametrically as $[x(t), y(t), t = 0..2\pi]$, with x and y defined by

```
> x := (t)-> r*cos(t);
```

$$t \mapsto r\cos(t)$$

```
> y := (t)-> r*sin(t);
```

$$t \mapsto r\sin(t)$$

The entire surface is obtained by rotating the top half of a circular cross section about the x-axis. The surface area is

```
> subs(a=Pi,S);
```

$$\int_0^\pi 2\pi r \sin(t) \sqrt{r^2 \sin(t)^2 + r^2 \cos(t)^2}\, dt$$

which evaluates to

```
> value(");
```

$$4r^2\pi$$

∎

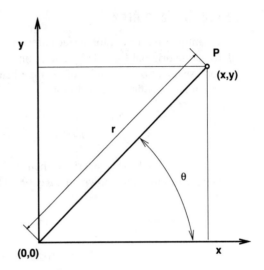

Figure 9.1 Polar Coordinates of P

9.5 Polar Coordinates

Figure 9.1 illustrates two methods of specifying a point in the Cartesian plane. We may specify the x and y coordinates as a pair $[x, y]$ or we may specify an angle θ and a distance r from the origin as a pair $[r, \theta]$. The pair $[r, \theta]$ is referred to as the *polar coordinates* of the point P.

The two representations are related by the equations

```
> eqns := { x = r*cos(t), y = r*sin(t) };
```
$$\{y = r\sin(t), x = r\cos(t)\}$$

These equations can be solved explicitly for r and t to yield

```
> eqns2 := solve(", {r,t});
```
$$\left\{r = \sqrt{x^2 + y^2}, t = \arctan(\frac{y}{x})\right\}$$

It is often useful to keep in mind the related equation

```
> r^2 = subs(", r^2);
```
$$r^2 = x^2 + y^2$$

Example 9.14 Find the polar coordinates of the point with Cartesian coordinates $[1, 1]$.

Solution This is just the point

```
> subs( {x=1,y=1}, eqns2 );
```

$$\left\{ r = \sqrt{2}, t = \frac{1}{4}\pi \right\}$$

The polar coordinates of this point are

```
> subs(",[r,t] );
```

$$[\sqrt{2}, \frac{1}{4}\pi]$$

∎

Example 9.15 Find the Cartesian coordinates of the point with polar coordinates $[2, \pi/3]$.

Solution This is the point

```
> subs( {r=2,t=Pi/3}, eqns );
```

$$\left\{ x = 1, y = \sqrt{3} \right\}$$

The Cartesian coordinates of this point are

```
> subs(",[x,y]);
```

$$[1, \sqrt{3}]$$

∎

9.5.1 Curve Sketching in Polar Coordinates

For some curves, the polar representation is more convenient than the Cartesian representation. The **plot( )** command allows us to specify that polar coordinates are being used for parametric plots.

Example 9.16 A circle of radius 2 can be represented in Cartesian coordinates either implicitly by the equation $x^2 + y^2 = 2^2$ or parametrically in terms of the angle t by $[2\cos(t), 2\sin(t), t = 0 \cdots 2\pi]$. The corresponding plot commands are

```
> with(plots):
> implicitplot( x^2 + y^2 = 4 , x=-2..2,y=-2..2);
```

and

```
> plot( [2*cos(t),2*sin(t),t=0..2*Pi] );
```

Both these plots draw the graph in Cartesian coordinates, but in the parametric form we must compute values of x and y.

This same parametric plot can be defined in polar coordinates as

```
> plot( [2,t,t=0..2*Pi],coords=polar );
```

An implicit equation[1] for this circle in polar coordinates is simply $r = 2$.

---
[1] *Maple* V Release 2 does not generate implicit plots that are defined in polar polar coordinates.

Parametric plots defined in polar coordinates are the natural choice when the equation defining the curve is of the form $r = f(\theta)$.

Example 9.17 Generate the Cartesian plot of the curve defined by the polar equation $r = \cos(2t)$.

Solution This curve in the Cartesian plane is generated by the command

```
> plot( [cos(2*t),t,t=0..2*Pi],coords=polar);
```

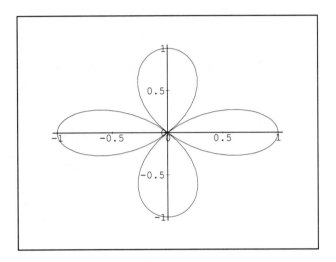

9.5.2 Tangent Lines and Polar Coordinates

All our graphs have been drawn in Cartesian plane even when the curve has been originally defined in polar coordinates. The slope of a tangent line is still $\dfrac{dy}{dx}$. It remains to express $\dfrac{dy}{dx}$ in terms of t and r. To accomplish this we proceed as follows.

If we regard x, y, r, and t as functions of yet another parameter, say, u, we can rewrite the equations

```
> eqns := {x = r*cos(t), y= r*sin(t)};
```

$$\{y = r\sin(t), x = r\cos(t)\}$$

To interpret these as relations between functions, convert them to the compositional form.

```
> with(student):
> convert( eqns , `@` );
```

$$\{y = r\sin \circ t, x = r\cos \circ t\}$$

Now we can use these equations to establish the relations

```
> map(D,");
```

9.5 Polar Coordinates

$$\{D(y) = D(r)\sin \circ t + r \cos \circ t\, D(t),\ D(x) = D(r)\cos \circ t + r - \sin \circ t\, D(t)\}$$

between the derivatives of x, y, r, and t. Just as before, the function defined by

```
> eq := D(y)/D(x) = subs(",D(y)/D(x));
```

$$\frac{D(y)}{D(x)} = \frac{D(r)\sin \circ t + r \cos \circ t\, D(t)}{D(r)\cos \circ t + r - \sin \circ t\, D(t)}$$

computes the desired slope.

If we make one of these functions the identity function, we can reinterpret all the remaining functions as functions of that chosen variable. For example, if t is the identity function, then each of x, y, and r are essentially functions of the angle t, and the slope is

```
> slpe := subs( { D(t)=1, t = ((u) -> u) } , rhs("));
```

$$\frac{D(r)\sin + r\cos}{D(r)\cos - r\sin}$$

Without knowing any more details about the curve, we can still conclude that, for example, at $t = 3\pi/2$ the actual slope $\frac{dy}{dx}$ is

```
> slpe(3*Pi/2);
```

$$-\frac{D(r)(\frac{3}{2}\pi)}{r(\frac{3}{2}\pi)}$$

If r is a constant function (as for a circle), then the slope at $t = \frac{3\pi}{2}$ simplifies to 0, indicating a horizontal tangent line at that point on the curve.

Example 9.18   Given the cardioid defined by the polar equation

```
> eq := r = 1 + sin(t):
```

graph this curve and find the angles t where the slope $\frac{dy}{dx}$ of the tangent line is 0. What is the slope of the (Cartesian) tangent line to this curve at $t = \frac{\pi}{3}$?

Solution   As a function of t, we have

```
> r := unapply(rhs(eq),t);
```

$$t \mapsto 1 + \sin(t)$$

By examining the graph

```
> plot( [r(t),t,t=0..2*Pi] , coords=polar, labels=[x,y]);
```

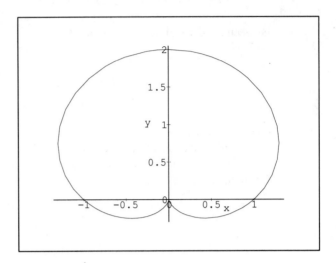

we see that there is at least one time where the tangent line is horizontal.

The Cartesian slope of the tangent line at the point $[r(t), t]$, found at angle t, is given by

```
> slpe(t);
```

$$\frac{\cos(t)\sin(t) + (1 + \sin(t))\cos(t)}{\cos(t)^2 - (1 + \sin(t))\sin(t)}$$

This slope will be 0 when

```
> slpe(t) = 0;
```

$$\frac{\cos(t)\sin(t) + (1 + \sin(t))\cos(t)}{\cos(t)^2 - (1 + \sin(t))\sin(t)} = 0$$

We can solve this for t. Two such solutions are given by

```
> solve(",{t});
```

$$\left\{ t = \frac{1}{2}\pi \right\}, \left\{ t = -\frac{1}{6}\pi \right\}$$

An additional solution also occurs at $t = 2\pi - \pi/6$ as verified by

```
> slpe(2*Pi-Pi/6);
```

$$0$$

At $t = \pi/3$ the slope of the tangent line is

```
> slpe(Pi/3);
```

$$\frac{\frac{1}{2}\sqrt{3} + \frac{1}{2}}{\frac{1}{4} - \frac{1}{2}\left(1 + \frac{1}{2}\sqrt{3}\right)\sqrt{3}}$$

which simplifies to

```
> simplify(");
```

$$-1$$

∎

9.6 Areas in Polar Coordinates

The area of a sector of a circle is directly proportional to the size of the central angle. Given a central angle of $dt = \theta$ (see Figure 9.1 on page 430), and a radius of r, we obtain a sector area of

$$A = \frac{t}{2\pi}\pi r^2 = \frac{r^2 dt}{2}$$

For a general curve described in polar coordinates, we can estimate the area bounded by the curve by sweeping around the origin in small sectors and allowing the radius to adjust as we move from one small sector to the next. The total area as the central angle changes from a to b is computed by summing these estimates for each sector while allowing the maximum angle of the contributing sectors to tend to 0. This yields the formula

$$A = \int_a^b \frac{r(t)^2}{2} dt$$

for the sector-like area swept out from $t = a$ to $t = b$.

Example 9.19   Find the area contained in one petal of the four-leafed rose defined by $r = \cos(2t)$.

Solution   A careful examination of the sketch of this curve generated in the previous example suggests that one petal corresponds to t values ranging from $t = -\pi/4$ to $t = \pi/4$. Indeed, the radii at $t = -\pi/4$ $t = \pi/4$ are 0.

```
> cos(-2*Pi/4), cos(2*Pi/4);
```

$$0, 0$$

The corresponding graph is

```
> plot( [cos(2*t),t,t=0..2*Pi],coords=polar);
```

The area corresponding to this sweep of a ray about the origin is given by

```
> r := (t) -> cos(2*t);
```
$$t \mapsto \cos(2t)$$

```
> A := Int( 1/2*r(t)^2,t=-Pi/4..Pi/4);
```
$$\int_{-\frac{1}{4}\pi}^{\frac{1}{4}\pi} \frac{1}{2} \cos(2t)^2 \, dt$$

```
> value(");
```
$$\frac{1}{8}\pi$$

∎

9.7 Arc Lengths in Polar Coordinates

If s is the function that measures arc length and if each of s, y, x, and r are functions of parameter u, then

```
> eq1 := D(s)^2 = D(x)^2 + D(y)^2;
```
$$D(s)^2 = D(x)^2 + D(y)^2$$

and arc length, as the parameter u varies from a to b, is given by

```
> L := Int( D(s)(u),u=a..b);
```
$$\int_a^b D(s)(u) \, du$$

9.7 Arc Lengths in Polar Coordinates

When polar coordinates are used, regard each of x, y, r, and t as functions of some other variable u. Then, we have

```
> eqns := {x = r*cos@t, y = r*sin@t };
```

$$\{x = r\cos \circ t, y = r\sin \circ t\}$$

Substituting this into equation $eq1$, we obtain

```
> subs(eqns , eq1 ): ";
```

$$D(s)^2 = D(r)\cos - r\sin \circ t^2 \, D(t)^2 + D(r)\sin + r\cos \circ t^2 \, D(t)^2$$

If in addition, we make t the identity function, we obtain the equation

```
> subs( t = ((u) -> u) , " ): ";
```

$$D(s)^2 = (D(r)\cos - r\sin)^2 + (D(r)\sin + r\cos)^2$$

which can be simplified by expanding

```
> expand(");
```

$$D(s)^2 = D(r)^2 \cos^2 + r^2 \sin^2 + D(r)^2 \sin^2 + r^2 \cos^2$$

and factoring

```
> factor(");
```

$$D(s)^2 = \left(\cos^2 + \sin^2\right)\left(D(r)^2 + r^2\right)$$

Also, the "function" $\sin^2 + \cos^2$ simplifies to the constant function $(u) \to 1$ so that $D(s)$ becomes

```
> subs( sin^2 + cos^2=1,");
```

$$D(s)^2 = D(r)^2 + r^2$$

On solving this for $D(x)$, we obtain

```
> map(sqrt,");
```

$$D(s) = \sqrt{D(r)^2 + r^2}$$

The formula for arc length in polar coordinates becomes

```
> L := subs(",L): ";
```

$$\int_a^b \sqrt{D(r)(u)^2 + r(u)^2}\, du$$

Chapter 9 Parametric Equations

Example 9.20 Find the length of the boundary of the cardioid defined by the polar equation $r = 1+\sin(t)$.

Solution The cardioid is shown by the command

```
> plot( [1 + sin(t),t,t=0..2*Pi],coords=polar);
```

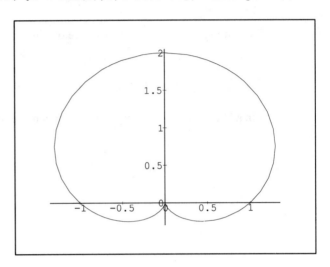

Since

```
> r := (t) -> 1 + sin(t);
```
$$t \mapsto 1 + \sin(t)$$

the arc length is given by

```
> L;
```
$$\int_a^b \sqrt{\cos(u)^2 + (1 + \sin(u))^2}\, du$$

with

```
> a := 0; b := 2*Pi;
```
$$2\pi$$

This is approximately

```
> evalf(L);
```
$$8.0$$

∎

Example 9.21 Find the length of the loop of the "chonchoid" as defined by $r = 4 + 2\sec(t)$.

Solution A natural first step is to try and visualize the loop. The following parametric plot shows this loop

9.7 Arc Lengths in Polar Coordinates

```
> r := (t) -> 4 + 2*sec(t);
```

$$t \mapsto 4 + 2\sec(t)$$

```
> opts := coords=polar,view=[-3..3,-3..3]:

> plot([r(t),t,t=0..2*Pi],opts);
```

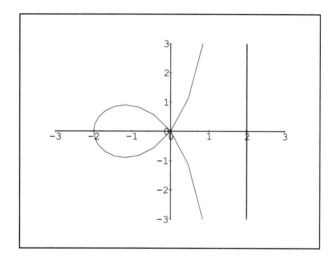

REMARK: If you do not restrict the view on this particular plot, the fact that sec(*t*) is unbounded near $\pi/2$ causes scaling problems.

The loop corresponds to a restricted domain of *t*. To identify which domain, observe that the 0s for *r* will occur only when $4 + 2\sec(t) = 0$ or, equivalently, when $\cos(t) = -1/2$. The plot

```
> plot( {cos(t),-1/2},t=1..5 );
```

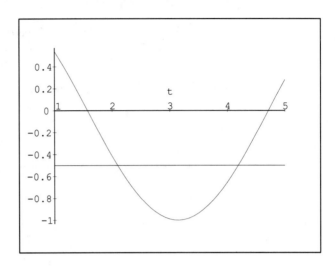

shows us that there should be two solutions, one near 2 and the other near 4. Approximate t values corresponding to these solutions are

```
> a := fsolve( r(t) = 0 , t , 2..2.5);
```
$$2.094395102$$

and

```
> b := fsolve( r(t) = 0 , t , 4..4.5);
```
$$4.188790205$$

The arc length corresponds to the integral

```
> Int( sqrt( r(t)^2 + D(r)(t)^2 ) , t=a..b);
```
$$\int_{2.094395102}^{4.188790205} \sqrt{(4 + 2\sec(t))^2 + 4\sec(t)^2 \tan(t)^2}\, dt$$

which has an approximate value of

```
> evalf(");
```
$$5.812830812$$

∎

EXERCISE SET 9

Some of the important commands that have been used in this chapter are shown in Table 9.1. Keep them in mind as you explore the examples from the chapter and the exercises listed here.

1. Sketch the curve $[f(t), g(t)]$ as t ranges from 0 to 5, for each of the pairs of functions defined on page 442.

Command	Description
{x=r*cos(t), y=r*sin(t)};	Equations defining polar coordinates
solve(eq,{r,t});	Simultaneous solution of a set of equations
Int(y,x=a..b);	Representation of a definite integral
[solve(eq1,{t})];	A list of solutions
changevar(u=3*t,L,u);	A change of variables
convert(f(x),'@');	Change $f(x)$ to $f \circ x$
coords=polar;	A parametric plot option
evalf(L);	Convert to a floating-point number
expand(");	Perform algebraic expansions
factor(");	Factor polynomials
fsolve(r(t)=0,t,2..2.5);	Find an approximate solution
isolate(",y);	Solve for a variable
labels=[x,y];	A plot option
makeproc(f(t),t);	Construct a procedure from an expression
map(D,eq):	Apply D to every element of a set
map(diff,",x);	Differentiate every element of a set with respect to x
maximize(f(t));	Maximize an algebraic expression
minimize(cos(t));	Minimize an algebraic expression
plot([x(t),y(t),t=1..5]);	Generate a parametric plot
simplify(e1);	Simplify an algebraic expression
subs(x=((t)->t),");	Replace x by the identity function
unapply(rhs(eq),t);	Construct a procedure from an expression
value(e1);	Evaluate an inert expression
view=[-3..3,-3..3]:	A plot option
x := (t) -> t+2:	A procedure definition

Table 9.1 Some of the Commands Used in Chapter 9

a. > f := (t) -> t:
   > g := (t) -> (t+3)^2 + 5:

b. > f := (t) -> (t+3)^2 + 5:
   > g := (t) -> (t+3)^2 + 5:

c. > f := (t) -> (t+3)^2 + 5:
   > g := (t) -> t:

d. > f := (t) -> sin(2*t):
   > g := (t) -> cos(3*t):

2. How are the curves in Exercise 1 affected if we trace out $[f(t), g(t), t = 0..5]$ instead of $[f(t), g(t), t = 0..5]$?

3. The tracing of each of the curves in Exercise 1 can be animated. To accomplish this, first generate a list of plots, each with the same view but a different range for t (i.e., $t = 1..2, t = 1..3, t = 1..4$, and so on), and then display them together using **display( )** with the plot option **insequence=true**. Try animating the construction of each of the curves from Exercise 1.

4. Find a cartesian representation of each of the curves

   > [f(t),g(t),t=1..5];

   from Exercise 1.

5. Transform each of the following equations defining curves in terms of the cartesian coordinates, x and y, into curves defined in terms of the polar coordindates r and t.

   a. > y = x^2 + 3;
   $$y = x^2 + 3$$

   b. > y^2 = x^2 + 3;
   $$y^2 = x^2 + 3$$

   c. > x*y = 3;
   $$xy = 3$$

   d. > y = x^2 + 3;
   $$y = x^2 + 3$$

   e. > y^3 - x^2 + 3;
   $$y^3 = x^2 + 3$$

   Recall that

   > implicitplot(f(x,y)=0 , x=-2..2,y=-2..2);

   corresponds to

   > plot([r(t),t,t=a..b],coords=polar);

   Generate two plots of each of the curves, one using each set of coordinates.

6. Transform each of the following equations defining a curve in terms of a radius r and an angle t into an equation in the cartesian coordinates x and y.

a. ```
> r = 3 - 2*sin(t);
```
$$r = 3 - 2\sin(t)$$

b. ```
> r = 2*cos(t);
```
$$r = 2\cos(t)$$

c. ```
> r = sin(t) + cos(t);
```
$$r = \sin(t) + \cos(t)$$

d. ```
> r = 2 + cos(2*t);
```
$$r = 2 + \cos(2t)$$

Generate two plots of each of the curves, one using each set of coordinates.

7. Find the area that lies inside the curve
```
> r = 2 + cos(2*t);
```
$$r = 2 + \cos(2t)$$

but outside the curve
```
> r = 2 + sin(t);
```
$$r = 2 + \sin(t)$$

8. Generate a graph of
```
> c := [3*cos(t)-cos(2*t),3*sin(t)-sin(2*t),t=0..2*Pi];
```
$$[3\cos(t) - \cos(2t), 3\sin(t) - \sin(2t), t = 0\ldots 2\pi]$$

Where do the tangent lines to the curve defined by c become vertical? Where do they become horizontal?

9. Regard both x and y as functions of a parameter t. Find a formula in terms of t for the slope of the tangent line of the (cartesian) curve defined by
```
> eq := y = -x^3 + 2*x^2 - x;
```
$$y = -x^3 + 2x^2 - x$$

10. Let the curve $f(x, y) = 0$ be defined parametrically by
```
> x := (t) -> 5*t^2+2*t:
> y := (t) -> t^3 -2*t + 1;
```
$$t \mapsto t^3 - 2t + 1$$

Plot the slope of the cartesian tangent line as a function of t.

11. For a given radius 5, the cycloid is defined parametrically by
```
> x := (t) -> 5*(t-sin(t));
```
$$t \mapsto 5t - 5\sin(t)$$

```
> y := (t) -> 5*(1-cos(t));
```
$$t \mapsto 5 - 5\cos(t)$$

Plot $\dfrac{dy}{dx}$, the slope of the cartesian tangent line as a function of t. Plot the second derivative of x with respect to y as a function of t.

12. Compute the total length of a perimeter of the curve defined by
```
> x^2 + 3*y^2 = 12;
```
$$x^2 + 3y^2 = 12$$

13. Find the surface area of the solid of revolution obtained by rotating about the x-axis the curve defined parametrically by
```
> x := (t) -> 2*r*cos(t);
```
$$t \mapsto 2r\cos(t)$$
```
> y := (t) -> 3*r*cos(t);
```
$$t \mapsto 3r\cos(t)$$

14. Find the cartesian coordinates of the point with polar coordinates $(3, \pi/4)$.

15. Consider the equations
```
> eq := { x = 2*r*sin(t), y = 3*r*cos(t) + 1 };
```
$$\{y = 3r\cos(t) + 1, x = 2r\sin(t)\}$$

   a. Use the convert command to reinterpret $\sin(t)$ and $\cos(t)$ as the functional compositions $\sin \circ t$ and $\cos \circ t$.
   b. Assuming that each of the functions x, y, r, and t is a function of some parameter u, and that t is defined by the identity function $(u) \to u$, use these equations to construct a new function that computes the rate of change of x with respect to y.
   c. Show that in the special case where r is the constant function that the rate of change of y with respect to x at time t is given by $-3/2 \tan(t)$.

16. Find the points on the cardioid defined by
```
> r = 5*(1 + sin(t));
```
$$r = 5 + 5\sin(t)$$

   at $t = \pi/7$ and find the time where the tangent line is horizontal.

17. Find the area contained by one petal of the rose defined by $r = \cos(5*t)$.

18. Find the length of the boundary of the cardioid defined by
```
> r = 5*(1 + sin(t));
```
$$r = 5 + 5\sin(t)$$

19. The limacon of Pascal is defined by
```
> r = 1 + 2*cos(t);
```
$$r = 1 + 2\cos(t)$$

Find the length of the arc stretching from $t = 0$ to $t = 2$.

20. Consider the curve defined by the polar equation

    ```
    > r = 1/sqrt(t);
    ```

    $$r = \frac{1}{\sqrt{t}}$$

    Find the length of the arc stretching from $t = 1$ to $t = 2$.

10 Sequences and Series

10.1 Introduction

Infinite sequences and series arise in many contexts. A useful way to regard a sequence is as a function f whose domain is the set of positive integers; the first term of the sequence is $f(1)$, the second term is $f(2)$, and so on. A series is simply the sum of the terms of such a sequence. When the nth term of a sequence, or the nth summand of a series, is given by an algebraic formula we can make general observations about the behavior of the terms of the sequence and the value of of the series from the structure of that formula. A special type of series — power series — provides us with an alternative representation of functions that is especially useful for constructing accurate approximations to non-polynomial functions such as exp, sin, and cos.

10.2 Sequences

A sequence $a_1, a_2, a_3, \ldots, a_n, \ldots$ is an ordered listing of the values of a function whose domain is a subset of the nonnegative integers. When the domain is $1 \ldots n$, we can explicitly "list" the individual terms of the sequence, as in

```
> a := [ 1, 1, 2, 3, 5, 8, 5, 3 , 2 , 1 , 1];
```
$$[1, 1, 2, 3, 5, 8, 5, 3, 2, 1, 1]$$

The 4th term of this sequence a is $a_4 = 3$. In *Maple* we refer to a_4 as

```
> a[4];
```
$$3$$

There are important differences between lists and sets. Sets do not allow repeated elements and you have no control over the order of the elements that do occur. For example, the set of elements occurring in a is

```
> convert(a,set);
```
$$\{1, 2, 3, 5, 8\}$$

which is the range of the associated function.

Sequences may be infinite. An infinite sequence $[a_1, a_2, a_3, \ldots]$ is sometimes denoted by $[a_n]$ or $[a_n]_{n=1}^{\infty}$. Just as for functions over the real numbers, functions defined over the positive integers (i.e., sequences) can be defined by formulas. A list of the first few terms of the sequence $[n^2]_{n=1}^{\infty}$ is

```
> a := [ seq( n^2,n=1..10)];
```
$$[1, 4, 9, 16, 25, 36, 49, 64, 81, 100]$$

Similarly, a list of the first few terms of the sequence $[b_n]_{n=1}^{\infty}$, where b_n is defined by

$$b_n = \frac{(-1)^n(n+1)}{4^n}$$

is

```
> b := [ seq( (-1)^n*(n+1)/4^n,n=1..10) ];
```

$$[-\frac{1}{2}, \frac{3}{16}, -\frac{1}{16}, \frac{5}{256}, -\frac{3}{512}, \frac{7}{4096}, -\frac{1}{2048}, \frac{9}{65536}, -\frac{5}{131072}, \frac{11}{1048576}]$$

Maple procedures may be used to compute the *n*th term. For example, the function

```
> f := (n) -> cos(n*Pi/6);
```

$$n \mapsto \cos(\frac{1}{6}n\pi)$$

can be used to generate the first few terms of the sequence

```
> c := [ seq( f(n),n=1..10) ];
```

$$[\frac{1}{2}\sqrt{3}, \frac{1}{2}, 0, -\frac{1}{2}, -\frac{1}{2}\sqrt{3}, -1, -\frac{1}{2}\sqrt{3}, -\frac{1}{2}, 0, \frac{1}{2}]$$

REMARK: The **seq( )** command generates a *Maple* "expression sequence." The expression sequence is then used to construct either a list [ ] or a set { }, depending on the purpose at hand.

10.2.1 Recurrence Relations

A very useful approach to defining the terms of a sequence is through a recurrence relation. A well-known example of a sequence that is generated in this manner is the *Fibonacci sequence f*. It is recursively defined by the three equations

$$f_1 = 1, \ f_2 = 1, \text{ and } f_n = f_{n-1} + f_{n-2} \quad \text{for } (n > 2)$$

The terms of this sequence can also be generated by the *Maple* procedure **fibonacci( )**, which is found in the **combinat** package. To load it, use the command

```
> with(combinat,fibonacci);
```

[*fibonacci*]

The first few terms of the sequence are

```
> [seq( fibonacci(i),i=1..10) ];
```

$$[1, 1, 2, 3, 5, 8, 13, 21, 34, 55]$$

Each successive term is the sum of the two previous ones. The 100th term in this sequence is

```
> fibonacci(100);
```

$$354224848179261915075$$

which has

```
> length(");
```

$$21$$

digits.

Recursively Defined Procedures

It is easy to define such procedures in *Maple*. For example, the procedure

```
> F := proc(n) option remember: F(n-1) + F(n-2) end:
```

also computes the terms of the *Fibonacci* sequence, providing we define the two values

```
> F(1) := 1:  F(2) := 1:
```

These two statements are absolutely essential. To see this, consider what happens when we attempt to compute $F(5)$ using F. The procedure definition requests values for $F(4)$ (and $F(3)$), which in turn trigger requests for the values of $F(3)$ (and $F(2)$). The computation of $F(3)$ in turn triggers a request for the values of $F(2)$ and $F(1)$, and so on. Unless we reach two consecutive known function values, this chain of requests can never end.

The statement **option remember**, appearing in the procedure definition for F, is important for "efficient" computation. Its purpose is to prevent the procedure from computing any particular function value more than once. It works by maintaining a table of previously computed function values. This table is consulted prior to any attempt to compute a function value. The efficiency gain can be quite dramatic. For example, without **option remember**, $F(3)$ would be computed 21 times to compute $F(10)$, 2584 times to compute $F(20)$, and 135301852344706746049 times (a twenty-one digit number) to compute $F(100)$. With option remember, $F(3)$ is computed only once in each case.

Solving Recurrences

There is usually more than one way to define a given sequence. For example, the Fibonacci recurrence relation can be solved to yield an explicit formula for the nth term of the sequence. The techniques necessary to do this are not part of this course, but we can use *Maple* to find such a formula. Recall that the recurrence was defined by

```
> n := 'n': f := 'f':
> eqns := {f(n) = f(n-1) + f(n-2), f(1)=1,f(2)=1};
```

$$\{f(n) = f(n-1) + f(n-2), f(1) = 1, f(2) = 1\}$$

Maple's recurrence solver, **rsolve( )** constructs the solution

```
> rsolve(" , f(n));
```

$$\frac{1}{5}\sqrt{5}\left(\frac{2}{-1+\sqrt{5}}\right)^n - \frac{1}{5}\sqrt{5}\left(-\frac{2}{1+\sqrt{5}}\right)^n$$

10.2 Sequences

This formula for $f(n)$ can be simplified to

```
> simplify(");
```

$$-\frac{1}{5}\sqrt{5}\left(-\left(-\frac{1}{2}+\frac{1}{2}\sqrt{5}\right)^{-n}+\left(-\frac{1}{2}-\frac{1}{2}\sqrt{5}\right)^{-n}\right)$$

and then used to define a procedure f.

```
> f := unapply(",n):
```

It is a pleasant surprise that this rather complicated looking formula always simplifies to an exact integer when evaluated at a positive integer. For example,

```
> f(10);
```

$$-\frac{1}{5}\sqrt{5}\left(-\left(-\frac{1}{2}+\frac{1}{2}\sqrt{5}\right)^{-10}+\left(-\frac{1}{2}-\frac{1}{2}\sqrt{5}\right)^{-10}\right)$$

simplifies to

```
> normal(",expanded);
```

$$55$$

the 10th term of the Fibonacci sequence.

10.2.2 Asymptotic Behavior of Sequences

When the terms of a sequence are defined by a formula, we can investigate how those terms grow as $n \to \infty$.[1] This is usually accomplished by using limits to carry out a term-by-term comparison with the terms of a known sequence.

Example 10.1 Show that the terms of the sequence defined by $f_n = \dfrac{n}{n+1}$ tend to 1 as $n \to \infty$.

Solution This effect can be seen by generating a plot. The following graph shows the points of the sequence corresponding to the function

```
> f := (n) -> n/(n+1):
```

The points of the sequence are listed as pairs, $[n, f(n)]$, as in

```
> pts := [seq( [i,f(i)],i=1..30)]:
```

They are plotted as distinct points by the command

```
> plot( {1,pts} , x=1..30, y=0..1 , style=point);
```

---

[1] Though the method is not used here, such information can also be obtained directly from the form of the recurrence relation itself.

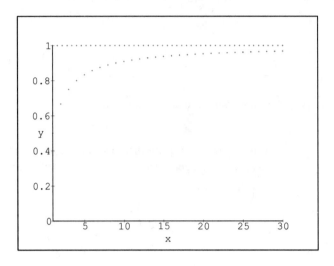

This graph is consistent with the algebraic result

```
> Limit(f(n),n=infinity):
> " = value(");
```

$$\lim_{n \to \infty} \frac{n}{n+1} = 1$$

∎

Example 10.2 Show that the Fibonacci sequence grows approximately the same as $.4472\,(1.618)^n$ as $n \to \infty$.

Solution Recall our formula for computing the nth term of the Fibonacci sequence.

```
> eqns := {f(n) = f(n-1) + f(n-2), f(1)=1,f(2)=1}:
> rsolve(eqns,f):
> f := unapply(",n);
```

$$n \mapsto -\frac{1}{5}\left(-1+\sqrt{5}\right)\sqrt{5}\left(-\frac{2}{1-\sqrt{5}}\right)^n \left(1-\sqrt{5}\right)^{-1} - \frac{1}{5}\sqrt{5}\left(-\frac{2}{1+\sqrt{5}}\right)^n$$

This yields the approximate formula

```
> evalf( f(n), 4 );
```

$$0.4472\,1.618^n - 0.4472\,(-0.6180)^n$$

for the nth term of the sequence. The growth of this expression, as a function of n, is largely determined by the power whose base is 1.618. This is because the other power, whose base is less than 1 in absolute value, gets close to 0 as $n \to \infty$. ∎

10.3 Series

Given an infinite sequence $[a_n]_{n=1}^{\infty}$, we can investigate the sum of the terms of the sequence. This sum

$$a_1 + a_2 + a_3 + \ldots + a_n + \ldots$$

is called an *infinite series* and is also denoted by

$$\sum_{n=1}^{\infty} a_n \quad \text{or} \quad \sum a_n$$

What happens when we attempt to compute such sums? If all the terms of the underlying sequence are large and positive, the sum is clearly ∞, but there are many other examples where the answer is finite.

Example 10.3   Show that the series

$$\sum_{n=1}^{\infty} \frac{1}{2^n}$$

has the value 1.

Solution   Consider the first few terms of this sum. The *partial* sum

```
> S(k) := Sum(1/2^n,n=1..k);
```

$$\sum_{n=1}^{k} (2^n)^{-1}$$

can be shown to evaluate to[2]

```
> value(");
```

$$-2\left(\frac{1}{2}\right)^{k+1} + 1$$

As k approaches infinity, this partial sum approaches

```
> Limit(",k=infinity):
> " = value(");
```

$$\lim_{k \to \infty} \left(-2\left(\frac{1}{2}\right)^{k+1} + 1\right) = 1$$

∎

---

[2] You can use mathematical induction to verify this formula.

DEFINITION 10.1 If the limit

$$\lim_{k \to \infty} \left(\sum_{n=1}^{k} a_n \right) = L$$

is finite, then the corresponding series is *convergent* and it converges to *L*.

Series that are not convergent are *divergent*.

Example 10.4 Show that the *geometric series* defined by

$$1 + r + r^2 + r^3 + \ldots + r^n + \ldots = \sum_{n=0}^{\infty} r^n$$

is convergent if $0 \leq r < 1$.

Solution To determine if such a series is convergent or divergent, we examine the behavior of the partial sums. These are

```
> S := (k) -> Sum( r^n,n=0..k );
```

$$k \mapsto \sum_{n=0}^{k} r^n$$

(Note that it is sometimes convenient to start a sequence such as $[r^n]$ at $n = 0$.)

If $r \geq 1$, then each term of the sum is at least 1 and the partial sum has a value of at least $k + 1$. The corresponding series is divergent because a lower bound on the sum is

```
> Limit(k+1,k=infinity);
```

$$\lim_{k \to \infty} (k + 1)$$

which evaluates to ∞.

The remaining possibility is that $0 \leq r < 1$. This can be indicated to *Maple* by the command

```
> assume(r,RealRange(0,Open(1))):
```

The value of the partial sum is then

```
> value(S(k));
```

$$\frac{r^{k+1}}{r-1} - (r-1)^{-1}$$

and the limit is

```
> limit(",k=infinity);
```
$$-(r-1)^{-1}$$

∎

Example 10.5 Determine if the series $\sum_{n=1}^{\infty} \dfrac{1}{n(n+1)}$ is convergent or divergent.

Solution Again we examine the partial sum

```
> S := (k) -> Sum( 1/n/(n+1),n=0..k );
```
$$k \mapsto \sum_{n=0}^{k} \frac{1}{n(n+1)}$$

When we attempt to evaluate this partial sum, *Maple* reports that it cannot do so.

```
> value(S(k));
```
$$\sum_{n=0}^{k} \frac{1}{n(n+1)}$$

However, we can still make progress by examining this sum from a different point of view. The summand is

```
> op(1,");
```
$$\frac{1}{n(n+1)}$$

It can be rewritten as

```
> convert(",parfrac,n);
```
$$n^{-1} - (n+1)^{-1}$$

so that the partial sum $S(k)$ can be written as

```
> Sum(",n=1..k);
```
$$\sum_{n=1}^{k} \left(n^{-1} - (n+1)^{-1}\right)$$

or

```
> expand(");
```
$$\sum_{n=1}^{k} n^{-1} - \sum_{n=1}^{k} (n+1)^{-1}$$

The second sum is essentially the first one shifted by one index. When these two sums are combined, all but the first term of the first sum and the last term of the second sum cancel. Thus, the partial sum evaluates to

```
> 1 - 1/(k+1);
```

$$1 - (k+1)^{-1}$$

As $k \to \infty$ this becomes

```
> limit(",k=infinity);
```

$$1$$

∎

In order for a series to converge, the summands must become small. Even the rate at which the summands become small is crucial. If they become small too slowly, the series may still diverge, as in the following very important example.

Example 10.6 Show that the *harmonic* series

$$\sum_{n=1}^{\infty} \frac{1}{n} = 1 + \frac{1}{2} + \frac{1}{3} + \ldots$$

is divergent.

Solution The partial sums

```
> S(k) = Sum( 1/n,n=1..k);
```

$$S(k) = \sum_{n=1}^{k} n^{-1}$$

occur frequently enough that a special function $\Psi()$ has been defined to help represent this value. The exact value of this sum is

```
> value(");
```

$$S(k) = \Psi(k+1) + \gamma$$

where the constant γ is approximately

```
> evalf(gamma,30);
```

$$0.577215664901532860606512090082$$

To investigate the behavior of this partial sum as a function of k, we first plot some sample points. The procedure

```
> S := proc(k) local n: Sum( 1/n,n=1..k)
> end:
```

computes the partial sums and can be used to create a list of sample points $[i, S(i)]$.

```
> pts := [ seq( [10*i+1,S(10*i+1)], i=1..40)]:
```

These points are shown by the command

```
> plot(value(pts),style=point);
```

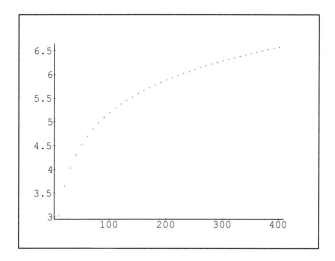

The partial sums do not appear to approach a finite limit as $i \to \infty$, and in fact they never do. For example, even the points

```
> pts := [ seq( [100*i+1,S(100*i+1)], i=1..40)]:
> plot(value(pts),style=point);
```

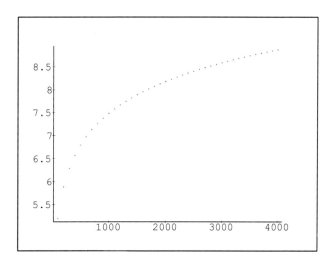

show no indication of approaching a finite limit.

To see why this is the case, consider a partial sum of the form

```
> S(2^n);
```

$$\sum_{n=1}^{2^n} n^{-1}$$

and examine only the last half of the sum. There are 2^{n-1} terms, each of size at least $1/2^n$, so the last half of the partial sum is at least as big as

```
> 2^(n-1)/2^n;
```

$$\frac{2^{n-1}}{2^n}$$

This simplifies to

```
> simplify(");
```

$$\frac{1}{2}$$

The same is true of the last half of the first 2^{n-1} terms, and so on, so that, by including 2^n terms, we can guarantee that the total sum is at least

$$1/2 + 1/2 + \ldots + 1/2 = n/2$$

and $\lim_{n \to \infty} S(2^n) = \infty$. This also shows that $\sum_{n \to \infty} S(n) = \infty$. ∎

10.3.1 Arithmetic on Series

Convergent series can be combined and manipulated in much the same manner as finite sums. That is, if the series $\sum a_n$ and $\sum b_n$ are both convergent and c is a constant, all of the following are true.

1.

$$\sum_{n=1}^{\infty} ca_n = c \sum_{n=1}^{\infty} a_n$$

2.

$$\sum_{n=1}^{\infty} a_n + b_n = \sum_{n=1}^{\infty} a_n + \sum_{n=1}^{\infty} b_n$$

3.

$$\sum_{n=1}^{\infty} a_n - b_n = \sum_{n=1}^{\infty} a_n - \sum_{n=1}^{\infty} b_n$$

Example 10.7 Use the above rules to evaluate

$$\sum_{n=1}^{\infty} \left(\frac{4}{n(n+1)} + \frac{1}{2^n} \right)$$

Solution The summation to be evaluated is

```
> S := Sum(4/n/(n+1)+1/2^n,n=1..infinity);
```

$$\sum_{n=1}^{\infty}\left(\frac{4}{n(n+1)}+(2^n)^{-1}\right)$$

This can be evaluated as

```
> with(student):
> expand(S);
```

$$4\sum_{n=1}^{\infty}\frac{1}{n(n+1)}+\sum_{n=1}^{\infty}(2^n)^{-1}$$

A list of the summations occurring in this expression is

```
> [op(")];
```

$$[4\sum_{n=1}^{\infty}\frac{1}{n(n+1)},\sum_{n=1}^{\infty}(2^n)^{-1}]$$

Each of these can be evaluated using techniques already shown in this section. We obtain

```
> map(value,");
```

$$[4,1]$$

which can be recombined to yield the final result

```
> convert(",`+`);
```

$$5$$

■

10.4 Testing for Convergence and Divergence

There are a variety of tests that indicate whether a particular summation is convergent or divergent. Some of these are investigated in this section.

10.4.1 The Integral Test

Consider the series

```
> s1 := Sum(1/n,n=1..infinity);
```

$$\sum_{n=1}^{\infty}n^{-1}$$

The partial sums for this series are closely related to the integral

```
> Int(1/n,n=1..infinity);
```

$$\int_1^\infty n^{-1}\,dn$$

For example, the partial sum

```
> S := Sum(1/n,n=1..20);
```

$$\sum_{n=1}^{20} n^{-1}$$

corresponds, in some sense, to the integral approximation

```
> with(student):

> rightbox( 1/x, x=1..21,20);
```

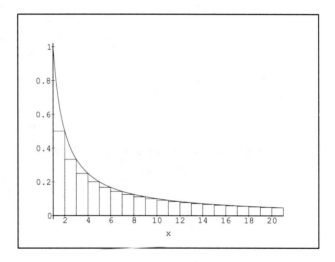

which has the value

```
> rightsum(1/x,x=1..21,20);
```

$$\sum_{i=1}^{20}(1+i)^{-1}$$

Furthermore, the indefinite integral $\int_1^\infty f(x)\,dx$ is computed as the limit

$$\lim_{t\to\infty}\left(\int_1^t f(x)\,dx\right)$$

in a manner exactly analogous to how we computed the value of some infinite series (e.g., Example 10.3).

10.4 Testing for Convergence and Divergence

THEOREM 10.1 Suppose that f is a continuous positive decreasing function on $[0, \infty)$. Then the integral $\int_1^\infty f(x)\,dx$ and the series $\sum_{n=1}^\infty f(n)$ are both convergent or both divergent.

Example 10.8 Use integration to test the series $\sum_{n=1}^\infty \dfrac{1}{1+n^2}$ for convergence.

Solution The function defined by $f : (x) \to \dfrac{1}{1+n^2}$ is continuous, positive, and decreasing, so we can examine the corresponding integral instead of the partial sums. The integral of interest is $\int_1^\infty f(x)\,dx$. We have

```
> Int( 1/(1 + x^2),x=1..t);
```

$$\int_1^t \left(1 + x^2\right)^{-1} dx$$

which evaluates to

```
> value(");
```

$$\arctan(t) - \frac{1}{4}\pi$$

and which as $t \to \infty$ becomes

```
> limit(",t=infinity);
```

$$\frac{1}{4}\pi$$

This is finite, so by the integral test, the summation $\sum_{n=1}^\infty \dfrac{1}{1+n^2}$ is convergent. ∎

Example 10.9 Use integration to show that the series $\sum_{n=1}^\infty \dfrac{1}{n^p}$ is convergent if $p > 1$ and divergent otherwise.

Solution To use the integral test, we examine

```
> I1 := Int( 1/x^p,x=1..infinity);
```

$$\int_1^\infty (x^p)^{-1} dx$$

This evaluates to

```
> v1 := value(");
```

$$\lim_{x \to \infty^-} \left(-\frac{x^{-p+1}}{p-1} + (p-1)^{-1} \right)$$

which is $\frac{1}{p-1}$ providing $p > 1$ and ∞ otherwise.

Basically, the value of the integral approximates the value of the sum, and vice versa.

10.4.2 Comparison Tests

We can always compare a series term by term with another series. Suppose that all the terms of the two series are nonnegative. If the terms from one series are smaller than the terms from the other, pairwise, and the series with the larger terms converges, then both series converge. Similarly, if the terms of a series are bigger, term by term, than the terms of a known divergent series, then both series diverge.

Example 10.10 Compare the series $\sum_{n=1}^{\infty} \frac{5}{3n^2 + 5n + 4}$ and $\sum_{n=1}^{\infty} \frac{5}{3n^2}$

Solution We start by comparing the individual terms of the series. The plots

```
>   plot( {5/3/n^2 , 5/(3*n^2 + 5*n + 2)},n=1..15,y=0..0.3);
```

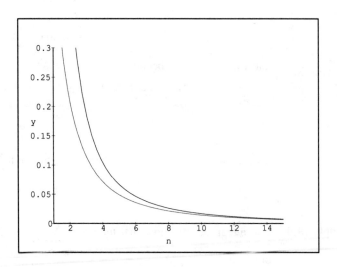

and

```
>   plot( {5/3/n^2 , 5/(3*n^2 + 5*n + 2)},n=160..200);
```

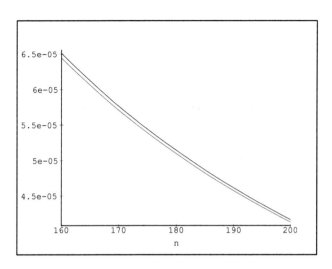

suggest that the nth term $\frac{5}{3n^2+5n+4}$ is bounded above by $\frac{5}{3n^2}$. (Note that the numerators are the same and that the denominator of the first is smaller than that of the second when $n > 0$.) Furthermore, the series

> Sum(5/3/n^2,n=1..infinity);

$$\sum_{n=1}^{\infty} \frac{5}{3n^2}$$

can be rewritten as

> expand(");

$$\frac{5}{3} \sum_{n=1}^{\infty} n^{-2}$$

which converges. Thus, both series converge. ∎

REMARK: By using techniques that go beyond the scope of this course, we can show that this upper bound has an exact value of

> value(");

$$\frac{5}{18}\pi^2$$

Example 10.11 Show that the series $\sum_{i=1}^{\infty} \frac{\ln(n)}{n}$ diverges.

Solution   This can be done by comparing the series term by term with the series $\sum_{i=1}^{\infty} \frac{1}{n}$. Because $\ln(x)$ is an increasing function and

```
> evalf(ln(3));
```
$$1.098612289$$

is already greater than 1, all pairs of terms beyond $n = 3$ satisfy the relation

```
> 1/n < ln(n)/n;
```
$$n^{-1} < \frac{\ln(n)}{n}$$

The value of the series

```
> Sum(1/n,n=3..infinity);
```
$$\sum_{n=3}^{\infty} n^{-1}$$

is

```
> value(");
```
$$\infty$$

(i.e., it diverges). Therefore, both series diverge.   ∎

REMARK: Term-by-term comparisons can be used to guarantee that a sum will be large or small, depending on the direction of the comparison.

Comparisons using Limits

It is really only the comparisons "near" infinity that matter because we can always deal with the first few terms of the series (even thousands of them) separately as a finite sum. Limits help us construct effective term-by-term comparisons by giving us some idea of when two sequences are "converging at the same rate."

Example 10.12   Show that the series

```
> s1 := Sum( (2*n^3 + 5*n)/sqrt(5 + n^10),n=1..infinity);
```
$$\sum_{n=1}^{\infty} (2n^3 + 5n) \frac{1}{\sqrt{5 + n^{10}}}$$

converges.

Solution   We can show this by comparing the size of a general term of this series with an overly simplified version of the same term. In this case, the numerator is of degree 3 and the denominator is of degree 5 so we try to compare the growth of the original terms to those of the convergent series

```
> s2 := Sum( n^3/n^5,n=1..infinity);
```

$$\sum_{n=1}^{\infty} n^{-2}$$

We can compare the two series term by term. The ratio of the formulas for nth terms is

```
> op(1,s1)/op(1,s2);
```

$$(2n^3 + 5n)\, n^2 \frac{1}{\sqrt{5+n^{10}}}$$

Near ∞, this becomes

```
> limit(",n=infinity);
```

$$2$$

The fact that this is a constant means that the two series are converging at essentially the same rate. When this happens we can easily turn one series into a term-by-term upper bound for the other. For example, the terms of $s2$ can be multiplied by a constant factor. In this case we only need to make the terms of the series $s2 = \sum_{n=1}^{\infty} \frac{1}{n^2}$ bigger by a factor of 2 to successfully construct a term-by-term upper bound, so any series, such as

```
> s3 := 3*s2 = combine(3*s2);
```

$$3\sum_{n=1}^{\infty} n^{-2} = \sum_{n=1}^{\infty} \frac{3}{n^2}$$

can be used to complete the comparison test and prove convergence. ∎

10.4.3 Ratio Tests

We can also use the ratio of the size of two successive terms to get an indication of how quickly the terms are going to zero.

THEOREM 10.2 A series $a = \sum_{n=1}^{\infty} a_n$ is *absolutely convergent* if

$$\lim_{n \to \infty} \left| \frac{a_{n+1}}{a_n} \right| = L < 1$$

Absolutely convergent series are convergent. If the limit of the indicated ratio is $L = 1$, then we obtain no information. If the limit of the ratio satisfies $L > 1$, then the series diverges.

Example 10.13 Show that the series

```
>   s4 := Sum( (-1)^n*n^3/3^n,n=1..infinity);
```

$$\sum_{n=1}^{\infty} \frac{(-1)^n n^3}{3^n}$$

is absolutely convergent.

Solution We use the ratio test. The nth term of the sequence is

```
>   op(1,s4);
```

$$\frac{(-1)^n n^3}{3^n}$$

```
>   f := makeproc(",n);
```

$$n \mapsto \frac{(-1)^n n^3}{3^n}$$

The ratio we need to examine is essentially

```
>   f(n+1)/f(n);
```

$$\frac{(-1)^{n+1} (n+1)^3 3^n}{3^{n+1} (-1)^n n^3}$$

```
>   simplify(");
```

$$-\frac{1}{3} \frac{(n+1)^3}{n^3}$$

However, we are interested in the absolute value of this expression.

```
>   abs(");
```

$$\frac{1}{3} \frac{|n+1|^3}{|n|^3}$$

As $n \to \infty$, this becomes

```
>   limit(",n=infinity);
```

$$\frac{1}{3}$$

so the series is absolutely convergent. ∎

10.4.4 The Root Test

The nth term is often given in terms of powers. In such cases, we can often obtain a measure of how fast such terms are tending to zero by examining the nth root of a_n. Again, if this value is small enough, then the series converges.

THEOREM 10.3 If $\lim\limits_{n\to\infty} \sqrt[n]{|a_n|} = L < 1$, then the series $\sum\limits_{n=1}^{\infty} a_n$ is absolutely convergent.

If $L > 1$ the series diverges, whereas if $L = 1$ we do not have sufficient information available to decide one way or the other.

Example 10.14 Test for convergence of the series $\sum\limits_{n=1}^{\infty} \left(\dfrac{7n+3}{9n+2}\right)^n$

Solution The nth term of the series is

```
> ((7*n + 3)/(9*n + 2))^n;
```

$$\left(\frac{7n+3}{9n+2}\right)^n$$

The nth root is

```
> "^(1/n);
```

$$\left(\left(\frac{7n+3}{9n+2}\right)^n\right)^{n^{-1}}$$

which in the limit is

```
> limit(",n=infinity);
```

$$\frac{7}{9}$$

Thus, this series converges absolutely. ∎

The ratio test and the root test help us identify series that are *absolutely convergent*. There are series that are convergent, but not absolutely convergent, but they all involve a mixture of positive and negative terms.

10.5 Alternating Series

The series $\sum\limits_{n=1}^{\infty} \dfrac{1}{n}$ is divergent and yet, when the signs of the successive terms alternate, as in $\sum\limits_{n=1}^{\infty} \dfrac{(-1)^n}{n}$, the series converges. In general, alternating series converge much more easily than series with all nonnegative terms.

THEOREM 10.4 If the alternating series

$$\sum_{n=1}^{\infty}(-1)^n a_n = a_1 - a_2 + a_3 - a_4 + \ldots$$

is such that

1. $a_{n+1} \leq a_n$ for all n
2. $\lim_{n\to\infty} a_n = 0$

then the series converges.

REMARK: In effect, any alternating series is convergent providing that the successive terms decrease in absolute value to 0.

We conclude this section with an example of how to find the sum of the alternating harmonic series by relating the partial sums to those of the harmonic series.

Example 10.15 Use the harmonic series to find a value for the sum of the alternating harmonic series.

Solution The partial sums of the harmonic series H, and this alternating series S, are computed by the procedures

```
> H := proc(k) local n; Sum( 1/n,n=1..k);
> end:
> S := proc(k) local n; Sum( -(-1)^n/n,n=1..k);
> end:
```

For example, the kth partial sums of these two series are

```
> k := 'k':
> H(k),S(k);
```

$$\sum_{n=1}^{k} n^{-1}, \quad \sum_{n=1}^{k} -\frac{(-1)^n}{n}$$

It turns out that we can reexpress the partial sum

```
> p1 := S(2*k);
```

$$\sum_{n=1}^{2k} -\frac{(-1)^n}{n}$$

as

```
> p2 := H(2*k)-H(k);
```

$$\sum_{n=1}^{2k} n^{-1} - \sum_{n=1}^{k} n^{-1}$$

For example, the first five such terms are

```
> for k to 5 do   k , value( p1 = p2) od;
```

$$1, \frac{1}{2} = \frac{1}{2}$$

$$2, \frac{7}{12} = \frac{7}{12}$$

$$3, \frac{37}{60} = \frac{37}{60}$$

$$4, \frac{533}{840} = \frac{533}{840}$$

$$5, \frac{1627}{2520} = \frac{1627}{2520}$$

and the 40th term is

$$\frac{4050907877358880643668631284365 01}{5897065319053451438476867691932 80} = \frac{4050907877358880643668631284365 01}{5897065319053451438476867691932 80}$$

To see that this formula is true for all "bigger values" of k we examine in detail what happens as we move from k to $k+1$. As we move from k to $k+1$, the term

```
> k := 'k':   p1;
```

$$\sum_{n=1}^{2k} -\frac{(-1)^n}{n}$$

increases by the amount

```
> d1 := 1/(2*k+1)-1/(2*k+2);
```

$$(2k+1)^{-1} - (2k+2)^{-1}$$

Similarly, the term

```
> p2;
```

$$\sum_{n=1}^{2k} n^{-1} - \sum_{n=1}^{k} n^{-1}$$

increases by the amount

```
> d2 := 1/(2*k+1)+1/(2*k+2) - 1/(k+1);
```

$$(2k+1)^{-1} + (2k+2)^{-1} - (k+1)^{-1}$$

Furthermore, $d1$ and $d2$ both simplify to the same thing, as in

```
> normal(d1) = normal(d2);
```

$$\frac{1}{2(2k+1)(k+1)} = \frac{1}{2(2k+1)(k+1)}$$

so the two series have the same value. We can use this observation to compute a value for the series $S(n)$. We have

```
> Limit(S(2*k),k=infinity) = Limit(H(2*k)-H(k),k=infinity);
```

$$\lim_{k\to\infty}\sum_{n=1}^{2k}-\frac{(-1)^n}{n} = \lim_{k\to\infty}\left(\sum_{n=1}^{2k}n^{-1} - \sum_{n=1}^{k}n^{-1}\right)$$

and the right-hand side is just

```
> Sum( 1/n,n=k..2*k);
```

$$\sum_{n=k}^{2k} n^{-1}$$

This (recall the integral test of Section 10.4.1) is

```
> ln(2*k) - ln(k);
```

$$\ln(2k) - \ln(k)$$

which simplifies to

```
> expand(");
```

$$\ln(2)$$

∎

10.6 Power Series

A *power series* in x is a series of the form

$$\sum a_n = \sum_{n=0}^{\infty} c_n x^n = c_0 + c_1 x + c_2 x^2 + c_3 x^3 + \ldots$$

The nth term of the series $\sum a_n$ is the product of a coefficient c_n that does not involve x, and the power x^n. Each value for x results in a series. This may or may not converge, depending on the value of x and the sequence of coefficients $[c_n]$.

Power series can be used to define functions of x, as in

```
> f := (x) -> sum( a[i]*x^i,i=0..infinity);
```

$$x \mapsto \sum_{i=0}^{\infty} a_i x^i$$

Example 10.16 Show that the function
```
> f := (x) -> 1/(1-x);
```
$$x \mapsto (1-x)^{-1}$$
corresponds to the function g defined by the power series $\sum x^i$, as in
```
> g := (x) -> Sum(x^n,n=0..infinity);
```
$$x \mapsto \sum_{n=0}^{\infty} x^n$$

Solution Consider the partial sum
```
> i := 'i': s1 := 1 + Sum(x^i,i=1..k);
```
$$1 + \sum_{i=1}^{k} x^i$$

Such a sum can be expressed in terms of the power series $g(x)$ as
```
> s1 = g(x) - x^(k+1)*g(x);
```
$$1 + \sum_{i=1}^{k} x^i = \sum_{n=0}^{\infty} x^n - x^{k+1} \sum_{n=0}^{\infty} x^n$$

```
> factor(");
```
$$1 + \sum_{i=1}^{k} x^i = -\sum_{n=0}^{\infty} x^n \left(-1 + x^{k+1}\right)$$

This equation can be rearranged as
```
> isolate(",g(x));
```
$$\sum_{n=0}^{\infty} x^n = \frac{-1 - \sum_{i=1}^{k} x^i}{-1 + x^{k+1}}$$

In the special case where $k = 0$, this simplifies to
```
> value( subs(k=0,rhs(")));
```
$$-(-1+x)^{-1}$$

∎

The identity $\dfrac{1}{1-x} = \sum_{n=0}^{\infty} x^n$ is valid for those values of x for which the power series converges. In this case, we can identify the x values for which this converges by using the ratio test. The nth term of this power series is generated by
```
> f := (n) -> x^n;
```

$$n \mapsto x^n$$

The ratio of the $(n+1)$st to the nth term is

```
> f(n+1)/f(n);
```

$$\frac{x^{n+1}}{x^n}$$

The absolute value of this ratio is

```
> abs( simplify(") );
```

$$|x|$$

When $|x| < 1$, the series converges and when $|x| > 1$, the series diverges.[3] This power series has a *radius of convergence* of 1. Inside its radius of convergence the power series converges to $\dfrac{1}{1-x}$.

Approximations Based on Power Series

The great advantage of a power series representation is that we can estimate function values by using only the first few terms of a convergent power series. Such polynomials are easy to evaluate. Furthermore, the error we introduce by leaving off the infinite "tail" of the series is generally proportional to the size of the first term not included in our approximation. For example, the function

```
> f := (x) -> 1/(1-x);
```

$$x \mapsto (1-x)^{-1}$$

is approximated by

```
> h := (x) -> 1 + sum( x^i,i=1..10);
```

$$x \mapsto 1 + x + x^2 + x^3 + x^4 + x^5 + x^6 + x^7 + x^8 + x^9 + x^{10}$$

as in

```
> plot( {f(x),h(x)},x=-1.2..1.2,y=-2..15);
```

---

[3]For the moment we will ignore what happens when $|x| = 1$.

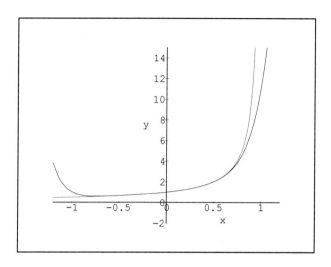

The error in the approximation typically increases near the radius of convergence. For example, we have

> f(.9), h(.9);

10.0, 6.861894039

and

> f(-.9),h(-.9);

0.5263157895, 0.6914792611

while near $x = 0$ the difference between the two representations is insignificant.

> f(0.1),h(0.1);

1.111111111, 1.111111111

This is consistent with our observation that the total error is proportional to the first omitted term, in this case x^{11}. Compare

> .9^11;

0.3138105961

with

> .1^11;

1.0×10^{-11}

10.7 Constructing Power Series

Because power series provide us with a convenient way of approximating many functions, it is important that we be able to construct them. There are two basic approaches. One is to construct a power series by modifying an existing series, or by combining two or more known series. The other is to construct a formula for the *n*th term of the series.

10.7.1 Algebraic Manipulations of Power Series

Series can be combined or manipulated in a variety of ways. They may be added term by term, multiplied as if they were polynomials, differentiated term by term or integrated term by term, and, under the right circumstances, transformed by substitutions. All the operations presented in this section are valid for series independent of questions about convergence, though, in some instances, the radius of convergence may change. Generally speaking, if we begin with absolutely convergent series, the result will also be absolutely convergent.

Example 10.17 Let f be defined by

```
> f := (x) -> 1/(1-x);
```

$$x \mapsto (1-x)^{-1}$$

Find a power series representation for $f(3x^2)$.

Solution   In Section 10.6 we constructed the power series representation

```
> Sum(x^n,n=0..infinity);
```

$$\sum_{n=0}^{\infty} x^n$$

for $f(x)$. Transformations of the form $x = cu^k$, where c and $k > 0$ are constant and k is an integer, can be carried out term by term. The series for $f(3u^2)$ is

```
> subs( x=3*u^2, f(x) = Sum( x^n,n=0..infinity));
```

$$(1 - 3u^2)^{-1} = \sum_{n=0}^{\infty} (3u^2)^n$$

The resulting series is still absolutely convergent, though in this particular case the new radius of convergence is only $\frac{1}{3}$ that of the original series.[4] ∎

If we are interested in only the first few terms (say 10) of the series, then we can use the **series( )** command to construct them, as in

```
> series( f(3*x^2) , x , 10);
```

$$(1 + 3x^2 + 9x^4 + 27x^6 + 81x^8 + O(x^{10}))$$

The term $O(x^{11})$ indicates that the terms that are not shown all contain a factor of x^{11}. If no order is specified, then only the terms up to degree 6 are generated, as in

```
> series( f(x) , x );
```

$$(1 + x + x^2 + x^3 + x^4 + x^5 + O(x^6))$$

The terms represented by $O(x^6)$ can be eliminated by converting the series to a polynomial, as in

---

[4]The radius of convergence can be computed using either the root test or the ratio test.

```
> convert(",polynom);
```
$$1 + x + x^2 + x^3 + x^4 + x^5$$

Example 10.18 Find a power series representing
```
> g := (x) -> 1/(1-x)^2;
```
$$x \mapsto (1-x)^{-2}$$

Solution The algebraic formula used to define g corresponds to
```
> diff(f(x),x);
```
$$(1-x)^{-2}$$

where f is as defined in Example 10.17. Thus, the series can be obtained from the series for $f(x)$ by differentiating term by term. The resulting series is $\sum_{n=1}^{\infty} nx^{n-1}$. The first few terms of this series are
```
> series(g(x),x);
```
$$(1 + 2x + 3x^2 + 4x^3 + 5x^4 + 6x^5 + O(x^6))$$

The radius of convergence is unaffected by differentiation. ■

Example 10.19 Obtain the first few terms of the power series for $\int \frac{1}{1-x} dx$ by integrating the series for $\frac{1}{1-x}$ term by term.

Solution We have
```
> f(x) = series(f(x),x);
```
$$(1-x)^{-1} = (1 + x + x^2 + x^3 + x^4 + x^5 + O(x^6))$$

Recall that, when integrating both sides of an equation, the resulting equation need only be in agreement to within a constant. Thus, this implies that
```
> Int(lhs("),x) + C = Int(rhs("),x);
```
$$\int (1-x)^{-1} dx + C = \int (1 + x + x^2 + x^3 + x^4 + x^5 + O(x^6)) dx$$

which evaluates to
```
> value(");
```
$$-\ln(1-x) + C = (x + \frac{1}{2}x^2 + \frac{1}{3}x^3 + \frac{1}{4}x^4 + \frac{1}{5}x^5 + \frac{1}{6}x^6 + O(x^7))$$

To determine a value for C, evaluate this equation at some value of x where the series converges. In this case, $x = 0$ is a particularly convenient choice because all the terms represented by $O(x^7)$ then disappear. The equation at $x = 0$ is

474   Chapter 10 Sequences and Series

```
> subs({O(x^7)=0,x=0},");
```
$$C = 0$$

and so the value of C is 0.   ∎

The power series for $-\ln(1-x)$ is

$$-\ln(1-x) = \sum_{n=1}^{\infty} \frac{x^n}{n}$$

Its radius of convergence is the same as for the series representing $\frac{1}{1-x}$.

The following three power series are very important. Each of them converges for all values of x.

Example 10.20   A power series representation for $\sin(x)$ is

$$\sin(x) = \sum_{n=0}^{\infty} \frac{(-1)^n x^{2n+1}}{(2n+1)!}$$

The first few terms are

```
> series(sin(x),x,10);
```
$$\left(x - \frac{1}{6}x^3 + \frac{1}{120}x^5 - \frac{1}{5040}x^7 + \frac{1}{362880}x^9 + O\left(x^{10}\right)\right)$$

∎

Example 10.21   A power series representation for $\cos(x)$ is

$$\cos(x) = \sum_{n=0}^{\infty} \frac{(-1)^n x^{2n}}{2n!}$$

The first few terms are

```
> series(cos(x),x,10);
```
$$\left(1 - \frac{1}{2}x^2 + \frac{1}{24}x^4 - \frac{1}{720}x^6 + \frac{1}{40320}x^8 + O\left(x^{10}\right)\right)$$

∎

Example 10.22   A power series representation for $\exp(x)$ is

$$\exp(x) = \sum_{n=0}^{\infty} \frac{x^n}{n!}$$

The first few terms are

```
> series(exp(x),x);
```
$$(1 + x + \frac{1}{2}x^2 + \frac{1}{6}x^3 + \frac{1}{24}x^4 + \frac{1}{120}x^5 + O(x^6))$$

■

Power series may be multiplied together like polynomials.

Example 10.23 Find a power series representation for $\sin(x)\cos(2x)$.

Solution The series representations for $\sin(x)$ and $\cos(2x)$ are

```
> series(sin(x),x,8);
```
$$(x - \frac{1}{6}x^3 + \frac{1}{120}x^5 - \frac{1}{5040}x^7 + O(x^9))$$

and

```
> series(cos(2*x),x,8);
```
$$(1 - 2x^2 + \frac{2}{3}x^4 - \frac{4}{45}x^6 + O(x^8))$$

The series we seek is the product of these two series.

```
> "*"";
```
$$\left((1 - 2x^2 + \frac{2}{3}x^4 - \frac{4}{45}x^6 + O(x^8))\right) \left((x - \frac{1}{6}x^3 + \frac{1}{120}x^5 - \frac{1}{5040}x^7 + O(x^9))\right)$$

It can be obtained by computing this product in much the same way that we multiply out the product of two polynomials. The only difference is that we systematically group together any terms involving degree 8 or more as part of the term $O(x^8)$. The result is

```
> series(",x,8);
```
$$(x - \frac{13}{6}x^3 + \frac{121}{120}x^5 - \frac{1093}{5040}x^7 + O(x^9))$$

■

We may divide one power series by another providing that the denominator has a constant term. Again the operation basically involves multiplying polynomials.

Example 10.24 Use polynomial arithmetic to show that the power series for $\frac{\sin(x)}{\cos(x)}$ begins as

```
> series( sin(x)/cos(x),x);
```
$$(x + \frac{1}{3}x^3 + \frac{2}{15}x^5 + O(x^7))$$

Solution We wish to have values for $a_0, a_1, \ldots$ so that

```
> sin(x)/cos(x)=sum(a[i]*x^i,i=0..6) + O(x^7);
```

$$\frac{\sin(x)}{\cos(x)} = a_0 + a_1 x + a_2 x^2 + a_3 x^3 + a_4 x^4 + a_5 x^5 + a_6 x^6 + O(x^7)$$

This can be be rephrased as a statement about products by multiplying through by $\cos(x)$, as in

```
> "*cos(x);
```

$$\sin(x) = \cos(x)\left(a_0 + a_1 x + a_2 x^2 + a_3 x^3 + a_4 x^4 + a_5 x^5 + a_6 x^6 + O(x^7)\right)$$

For power series, this equation implies that

```
> series(sin(x),x)
> = series(cos(x),x)*(sum(a[i]*x^i,i=0..6) + O(x^7));
```

$$(x - \frac{1}{6}x^3 + \frac{1}{120}x^5 + O(x^6)) =$$

$$\left((1 - \frac{1}{2}x^2 + \frac{1}{24}x^4 + O(x^6))\right)\left(a_0 + a_1 x + a_2 x^2 + a_3 x^3 + a_4 x^4 + a_5 x^5 + a_6 x^6 + O(x^7)\right)$$

The right-hand side can be multiplied out in full to give

```
> series(lhs("),x) = series(rhs("),x);
```

$$(x - \frac{1}{6}x^3 + \frac{1}{120}x^5 + O(x^6)) = (a_0 + a_1 x + \left(a_2 - \frac{1}{2}a_0\right)x^2 + \left(-\frac{1}{2}a_1 + a_3\right)x^3 +$$

$$\left(-\frac{1}{2}a_2 + a_4 + \frac{1}{24}a_0\right)x^4 + \left(-\frac{1}{2}a_3 + a_5 + \frac{1}{24}a_1\right)x^5 + O(x^6))$$

This must be true for all values of x. Thus, the coefficients of the various powers of x must be the same on both sides of this equation. Equivalently, all the coefficients of the powers of x in

```
> series(lhs(")-rhs("),x);
```

$$(-a_0 + (-a_1 + 1)x + \left(-a_2 + \frac{1}{2}a_0\right)x^2 + \left(\frac{1}{2}a_1 - a_3 - \frac{1}{6}\right)x^3 +$$

$$\left(\frac{1}{2}a_2 - a_4 - \frac{1}{24}a_0\right)x^4 + \left(\frac{1}{120} + \frac{1}{2}a_3 - a_5 - \frac{1}{24}a_1\right)x^5 + O(x^6))$$

must be 0. The ones shown here are

```
> convert(",polynom):
> coeffs(",x);
```

$$-a_0, \; -a_1 + 1, \; \frac{1}{120} + \frac{1}{2}a_3 - a_5 - \frac{1}{24}a_1, \; \frac{1}{2}a_1 - a_3 - \frac{1}{6}, \; -a_2 + \frac{1}{2}a_0, \; \frac{1}{2}a_2 - a_4 - \frac{1}{24}a_0$$

and on setting them to 0 we obtain the coefficients of the unknown series.

```
> solve({"});
```

$$\left\{a_4 = 0, a_2 = 0, a_1 = 1, a_5 = \frac{2}{15}, a_3 = \frac{1}{3}, a_0 = 0\right\}$$

∎

10.7 Constructing Power Series

Not all power series have an infinite number of summands. A polynomial in x is simply a power series in which the coefficients of the powers of x beyond the highest degree term are all 0.

Example 10.25 Find a power series for $(1 + x)^8$.

Solution The first 20 terms of this power series are generated by

> series((1+x)^8,x,20);

$$(1 + 8x + 28x^2 + 56x^3 + 70x^4 + 56x^5 + 28x^6 + 8x^7 + x^8)$$

The coefficients beyond x^8 are all 0 so there are no terms left over to be represented by $O(x^{21})$. ∎

10.7.2 Constructing Coefficients

Consider the series

> eq := g(x) = series(g(x),x);

$$(1 - x)^{-2} = (1 + 2x + 3x^2 + 4x^3 + 5x^4 + 6x^5 + O(x^6))$$

and note what always happens when we evaluate the series at $x = 0$.

> subs(x=0,rhs("));

$$1$$

The result is always exactly equal to the constant term.

Also note what happens when we differentiate this series term by term, with respect to x.

> diff(series(g(x),x),x);

$$(2 + 6x + 12x^2 + 20x^3 + 30x^4 + O(x^5))$$

The coefficient of x in the original series becomes the constant term of the new series, and so

> subs(x=0,");

$$2$$

is the coefficient of x in the original series.

By repeating this process we can compute any coefficient. For example, to get the coefficient of x^4 in the original series we first differentiate four times.

> diff(series(g(x),x),x$4);

$$(120 + 720x + O(x^2))$$

The resulting constant term

> subs(x=0,");

$$120$$

is the coefficient of x^4 in the original series, multiplied by 4!. Thus, the original coefficient of x^4 was

```
> "/4!;
```

$$5$$

DEFINITION 10.2 The *Maclaurin series* for $f(x)$ is

```
> f := 'f':
> series(f(x),x);
```

$$(f(0) + D(f)(0)x + \frac{1}{2}\left(D^{(2)}\right)(f)(0)x^2 + \frac{1}{6}\left(D^{(3)}\right)(f)(0)x^3 + \frac{1}{24}\left(D^{(4)}\right)(f)(0)x^4 +$$
$$\frac{1}{120}\left(D^{(5)}\right)(f)(0)x^5 + O\left(x^6\right))$$

Example 10.26 Obtain the Maclaurin series for $\sin(x)$ from the Maclaurin series of an arbitrary function f.

Solution The general Maclaurin series is

```
> series(f(x),x);
```

$$(f(0) + D(f)(0)x + \frac{1}{2}\left(D^{(2)}\right)(f)(0)x^2 + \frac{1}{6}\left(D^{(3)}\right)(f)(0)x^3 + \frac{1}{24}\left(D^{(4)}\right)(f)(0)x^4 +$$
$$\frac{1}{120}\left(D^{(5)}\right)(f)(0)x^5 + O\left(x^6\right))$$

When $f = \sin$, this becomes

```
> subs( f=sin,"): ";
```

$$(x - \frac{1}{6}x^3 + \frac{1}{120}x^5 + O\left(x^6\right))$$

which is exactly the same power series that is generated by the command **series(sin(x),x)**.   ∎

Example 10.27 Use derivatives to compute the first few terms of the Maclaurin series for

```
> e1 := x^2*sin(x)/cos(x);
```

$$\frac{x^2 \sin(x)}{\cos(x)}$$

Solution The first few terms of the requested series are

```
> series(e1,x);
```

$$(x^3 + \frac{1}{3}x^5 + O\left(x^7\right))$$

To compute these using derivatives, proceed as follows. Consider the coefficient of x^3. This coefficient is obtained by evaluating the derivative

```
> diff(x^2*sin(x)/cos(x),x$3)/3!;
```

$$1 + \frac{\sin(x)^2}{\cos(x)^2} + \frac{2x\sin(x)}{\cos(x)} + \frac{2x\sin(x)^3}{\cos(x)^3} + \frac{1}{3}x^2 + \frac{4}{3}\frac{x^2\sin(x)^2}{\cos(x)^2} + \frac{x^2\sin(x)^4}{\cos(x)^4}$$

at $x = 0$. The result is

```
> subs(x=0,"): ";
```

$$1$$

Similarly, the coefficient of x^4 is

```
> subs(x=0,diff(e1,x$4)/4!);
```

$$0$$

Example 10.28 Use derivatives to find the first few terms of the power series for $(1+x)^k$.

Solution The first few terms are

```
> series((1+x)^k,x);
```

$$(1 + kx + \frac{1}{2}k(k-1)x^2 + \frac{1}{6}k(k-1)(k-2)x^3 + \frac{1}{24}k(k-1)(k-2)(k-3)x^4 +$$
$$\frac{1}{120}k(k-1)(k-2)(k-3)(k-4)x^5 + O(x^6))$$

For example, the coefficient of x^3 is

```
> subs(x=0,diff((1+x)^k,x$3)/3!);
```

$$\frac{1}{6}k^3 - \frac{1}{2}k^2 + \frac{1}{3}k$$

which can be factored as

```
> factor(");
```

$$\frac{1}{6}k(k-1)(k-2)$$

The computation of the coefficients in the previous example did not depend on k being a positive integer. Two important examples where k is not an integer are

```
> series(sqrt(1+x),x);
```

$$(1 + \frac{1}{2}x - \frac{1}{8}x^2 + \frac{1}{16}x^3 - \frac{5}{128}x^4 + \frac{7}{256}x^5 + O(x^6))$$

and

```
> series((1+x)^(-1),x);
```

$$(1 - x + x^2 - x^3 + x^4 - x^5 + O(x^6))$$

If k is not a positive integer, then the series has an infinite number of nonzero coefficients.

10.7.3 Taylor Series

Sometimes we will need to use a power series built on powers of the form $(x-c)^n$, where c is a constant. A method, similar to the method outlined in the previous section, of taking derivatives and setting all the powers $(x-c)^n$ to 0 leads to the following type of series.

DEFINITION 10.3 The *Taylor series* for $f(x)$ expanded about $x = c$ is the series

$$f(x) = \sum_{n=0}^{\infty} \frac{D^{(n)}(f)(c)}{n!}(x-c)^n = f(c) + \frac{D(f)(c)}{1!}(x-c) + \frac{D^{(2)}(f)(c)}{2!}(x-c)^2 + \ldots$$

The first four terms of such a series are computed by the command

```
> taylor( f(x),x=c,4);
```

$$(f(c) + D(f)(c)(x-c) + \frac{1}{2}\left(D^{(2)}\right)(f)(c)(x-c)^2 + \frac{1}{6}\left(D^{(3)}\right)(f)(c)(x-c)^3 + O\left((x-c)^4\right))$$

The third argument specifies the number of terms to include in the result. If this degree is not specified, then six terms are used.[5]

Example 10.29 The Taylor series of $\sin(x)$ about $x = 0$ is the same as the power series we generated earlier for $\sin(x)$.

```
> taylor(sin(x),x=0);
```

$$(x - \frac{1}{6}x^3 + \frac{1}{120}x^5 + O\left(x^6\right))$$

∎

Example 10.30 The Taylor series of $\sin(x)$ about $x = 1$ is

```
> taylor(sin(x),x=1);
```

$$(\sin(1) + \cos(1)(x-1) - \frac{1}{2}\sin(1)(x-1)^2 - \frac{1}{6}\cos(1)(x-1)^3 + \frac{1}{24}\sin(1)(x-1)^4 + \frac{1}{120}\cos(1)(x-1)^5 + O\left((x-1)^6\right))$$

∎

---

[5]The first six terms includes terms up to degree 5.

10.8 Approximations

The real strength of a power series representation is that it allows us to approximate a wide range of functions (including the trigonometric and exponential functions) by polynomials.

For example, the sine function can be approximated near $x = 0$ by the the polynomial

```
> taylor(sin(x),x);
```

$$(x - \frac{1}{6}x^3 + \frac{1}{120}x^5 + O(x^6))$$

```
> p5 := convert(",polynom);
```

$$x - \frac{1}{6}x^3 + \frac{1}{120}x^5$$

Consider the plot

```
> plot( { sin(x), p5 } ,x=-5..5,y=-2..2);
```

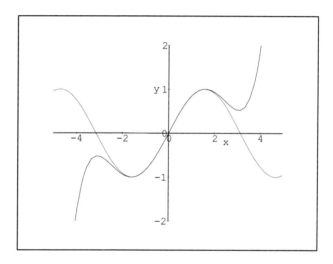

For x in the range $-1.5 \ldots 1.5$, the two curves are almost indistinguishable.

The polynomials obtained by throwing away the "Big OH" terms (i.e., $O(x^{10})$) are called *Taylor polynomials*, and they all form useful approximations. It turns out that Taylor polynomials of degree 1 correspond to tangent lines.

Example 10.31 Use a Taylor polynomial to construct a tangent line to the curve f at $x = a$.

Solution For the function f, the Taylor polynomial of degree 1, at $x = a$, is

```
> f := 'f':
> taylor(f(x),x=a,2);
```

$$(f(a) + D(f)(a)(x - a) + O((x - a)^2))$$

```
> p := convert(",polynom);
```
$$f(a) + D(f)(a)(x - a)$$

The equation
```
> y = p;
```
$$y = f(a) + D(f)(a)(x - a)$$

is an equation for the tangent line to f at that point. ∎

Thus, the tangent line to $\sin(x)$ at $x = 0$ is given by
```
> taylor(sin(x),x,2);
```
$$(x + O(x^2))$$
```
> p1 := convert(",polynom);
```
$$x$$

and the tangent line to $\sin(x)$ at $x = 2$ is given by
```
> taylor( sin(x),x=2,2);
```
$$(\sin(2) + \cos(2)(x - 2) + O((x - 2)^2))$$
```
> q1 := convert(",polynom);
```
$$\sin(2) + \cos(2)(x - 2)$$

this is verified by the plot
```
> plot( {sin(x),p1,q1},x=-5..5,y=-2..2);
```

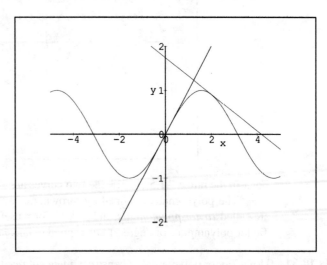

The higher the degree of the polynomial, the better approximation we have.

Example 10.32 Graph the function $\sin(x)$ and the degree 9 Taylor polynomial about $x = 0$ in a manner that indicates where the accuracy of this approximation starts to fail.

Solution The degree 9 Taylor polynomial for $\sin(x)$ about $x = 0$ is
```
> taylor(sin(x),x,10);
```

$$(x - \frac{1}{6}x^3 + \frac{1}{120}x^5 - \frac{1}{5040}x^7 + \frac{1}{362880}x^9 + O(x^{10}))$$

```
> p9 := convert(",polynom);
```

$$x - \frac{1}{6}x^3 + \frac{1}{120}x^5 - \frac{1}{5040}x^7 + \frac{1}{362880}x^9$$

Through trial and error we are able to come up with the following graph, indicating that the two curves are in close agreement over a surprisingly wide range.

```
> plot( {sin(x),p9},x=-5..5,y=-2..2);
```

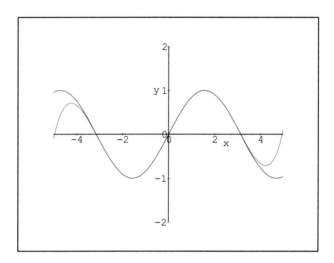

■

10.8.1 Error Analysis

Suppose we are working with a polynomial $p(x)$ of degree n in x that was obtained by truncating the Taylor series of $f(x)$ about $x = c$. How big is the portion of the series we ignored (i.e., the portion represented by the term $O(x^{n+1})$)? The *remainder* — the error incurred by leaving off terms of the series — is essentially the size of the first missing term. To be more precise, it is exactly

$$R_n(x) = \frac{D^{(n+1)}}{(n+1)!}(z)(x-c)^{n+1}$$

for some z in between the current x value and c.

This information can be used to estimate the error incurred by using a Taylor polynomial instead of the original function.

Example 10.33 Determine the maximum error that could occur by using a degree 5 Taylor polynomial expanded about $x = 0$ to approximate $f(x)$ for

484   Chapter 10   Sequences and Series

```
> f := (x) -> cos(x)*sin(2*x):
```
for values of x in the range $-1 \leq x \leq 1$.

Solution   The degree 4 Taylor polynomial expanded about $x = 0$ is $p5$ defined by

```
> taylor(f(x),x=0,5):
> p4 := convert(",polynom);
```

$$2x - \frac{7}{3}x^3$$

A graph of $p4$ and $f(x)$ is shown below

```
> plot({f(x),p4},x=-1..1,y=-1.5..1.5);
```

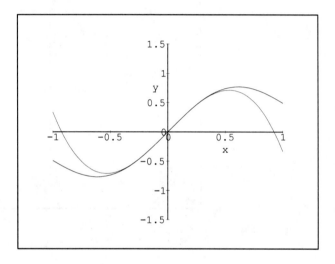

The exact error in computing $f(x)$ is given by the formula

```
> e4 := diff(cos(z) + sin(2*z),z,z,z,z,z)*(x)^5/5!;
```

$$\frac{1}{120}(-\sin(z) + 32\cos(2z))x^5$$

However, it is only exact for some unknown value of z that depends on x.

We do know that $-1 \leq x \leq 1$, but we do not know which z in $-x \leq z \leq x$ corresponds to the exact error in computing $f(x)$ for any specific x. Still, it is enough to obtain an upper estimate on the error by examining the value of this error formula for all possible z. We simply examine all possible pairs $[x, z]$ such that $-1 \leq x \leq 1$ and $-1 \leq z \leq 1$ and then take the worst case that we can find as our upper bound on the error that can occur.

The easiest way to estimate this worst case bound on the error is to plot the error formula as a function of x and z. The result is a surface, as in

```
> plot3d( abs(e4) , x=-1..1,z=-1..1,axes=framed);
```

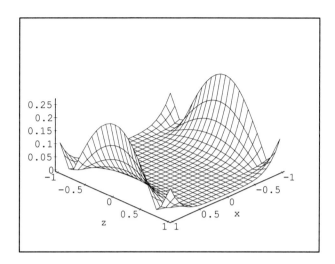

Here, the worst case (the highest point on the surface) would appear to occur at $|x| = 1$ and $z = 0$. Thus, an upper bound on the error is

```
> subs(x=1,z=0,e4);
```

$$\frac{4}{15}$$

which is approximately

```
> evalf(");
```

$$0.2666666667$$

The difference between $p4$ and $f(x)$ will never be greater than this bound on the prescribed interval.

∎

Example 10.34 What degree polynomial would be required in estimating $f(x)$ to ensure that the error was never greater than 0.005 on the same interval as in the previous example.

Solution We must look at error terms of increasing degree until we find one for which the worst case estimate is sufficiently small. The error term as a function of degree is

```
> err := (n) -> diff(f(z),z$n+1)*(x)^(n+1)/(n+1)!;
```

$$n \mapsto \frac{\frac{d^n}{dz^{n+1}}(\cos(z)\sin(2z)) x^{n+1}}{(n+1)!}$$

An examination of the plot

```
> plot3d(abs(err(10)),x=-1..1,z=-1..1,axes=framed);
```

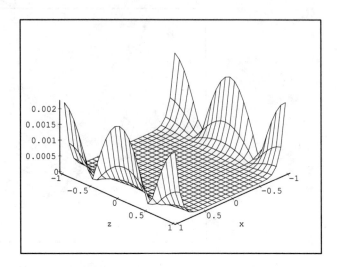

suggests that again the maximum error occurs at $z = 0$ and $x = 1$. At degree 10 the error is at most

```
> abs( err(10) );
```

$$\frac{1}{39916800} |88573 \sin(z) \sin(2z) - 88574 \cos(z) \cos(2z)| \, |x|^{11}$$

which, at $x = 1$ and $z = 0$, is approximately

```
> evalf( subs(x=1,z=0,") );
```

$$0.002218965449$$

This is smaller than 0.005.

Can we use a smaller degree? At degree 9, the automatic scaling of the plot

```
> plot3d(abs(err(9)),x=-1..1,z=-1..1,axes=framed);
```

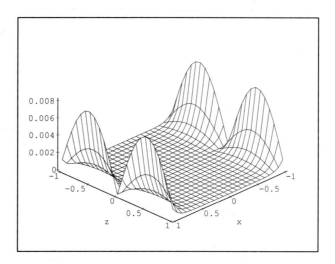

suggests that the best bound we can construct is still a bit greater than 0.005. The worst case bound would appear to occur on the cross section at $x = 1$. To find a more accurate estimate of the maximum error, we examine this two-dimensional curve in more detail. For

> x := 1;

$$1$$

we have

> err(9);

$$-\frac{1181}{145152}\cos(z)\sin(2z) - \frac{7381}{907200}\sin(z)\cos(2z)$$

The relative maximum of this curve occurs where there is a horizontal tangent line. We can find this algebraically as

> diff(",z);

$$\frac{88573}{3628800}\sin(z)\sin(2z) - \frac{44287}{1814400}\cos(z)\cos(2z)$$

> s := solve("=0,z);

$$\frac{1}{2}\pi, \; \arcsin(\frac{1}{729}\sqrt{132861}), \; -\arcsin(\frac{1}{729}\sqrt{132861})$$

This worst case analysis for the degree 9 polynomial gives bounds of

> for t in s do evalf(subs(z=t,[z,err(9)])); od;

$$[1.570796327, 0.008136022928]$$

$$[0.5236004050, -0.008136229608]$$

Command	Description
Int(f(x),x=1..t);	Represent a definite integral
Limit(f(n),n=infinity):	Represent a limit
Sum(f(n),n=1..infinity);	Represent a sum
[seq([i,f(i)],i=1..9)];	A list of points in 2D
[seq(n,n=1..10)];	A list of numbers
assume(r,RealRange(0,Open(1))):	Specify an assumption
[coeffs(e1,x)];	A list of the coefficients of a polynomial
combine(3*s2);	Opposite of expand
convert(",parfrac,n);	Represent a quotient as a partial fraction
convert(",polynom);	Convert a series to a polynomial
convert(lst,'+');	Convert a list to a sum
diff(series(g(x),x),x$4);	Differentiate a series four times
evalf(ln(3));	Evaluate to an approximate number
expand(S);	Perform algebraic expansions
factor(e);	Perform an algebraic factorization
fibonacci(100);	Compute the 100th Fibonacci number
isolate(",g(x));	Solve for $g(x)$
k := 'k':	Unassign the value of k

Table 10.1 Some of the Commands Used in Chapter 10

$$[-0.5236004050, 0.008136229608]$$

the worst of which is for

```
> z := arcsin( sqrt(132681) / 729 );
```

$$\arcsin(\frac{1}{729}\sqrt{132681})$$

This bound is still bigger than 0.005 so more than nine terms are required in the series. ∎

EXERCISE SET 10

Some of the important commands that have been used in this chapter are shown in Tables 10.1 and 10.2. Keep them in mind as you explore the examples from the chapter and the exercises listed here.

1. Show that the terms of the sequence defined by $f_n = \dfrac{n}{n+1}$ tend to 1 as $n \to \infty$.
2. Use the command **rsolve( )** to find a formula for the nth term of the sequence defined by

```
> eqs := {f(n) = f(n-1) + 2*f(n-2), f(0)=1,f(1)=3};
```

$$\{f(n) = f(n-1) + 2f(n-2), f(0) = 1, f(1) = 3\}$$

Command	Description
length(n);	Find the number of digits in a number
limit(",k=infinity);	Compute a limit
makeproc(",n);	Construct a procedure from an expression
normal(d1);	Remove common factors from quotients
normal(e,expanded);	Remove common factors and multiply out powers
op(1,Int(...));	Choose the first argument
O(x^7);	An error term involving powers of x of size at least 8
option remember:	Create a remember table for a procedure
F(1):=1:	Assign to the remember table of a procedure
plot(f(x),x=1..10);	Plot an expression
plot3d(abs(e4),opts);	Generate a 3D plot of an expression
rightsum(1/x,x=1..21,20);	Construct the sum for approximating rectangles
rsolve(eqs,f(n));	Solve a recurrence equation
series(f(x),x,10);	Generate a power series
simplify(");	Simplify an expression
style=point;	A plot option
subs(O(x^7)=0,");	Remove the error term from a series
taylor(f(x),x=1);	Generate a taylor series about $x = 1$
unapply(",n):	Construct a procedure from an expression
value(e);	Evaluate an inert expression

Table 10.2 Some of the Commands Used in Chapter 10

Verify that this formula is correct for $n = 1..10$. Use limits to compare the rate of growth of $f(n)$, as reported by this formula, with the rate of growth of 2^n, as $n \to \infty$. Which grows faster? Use $f(n)$ to invent your own sequence whose terms grow at exactly the same rate as 2^n.

3. Use *Maple* to generate values for the sums

    ```
    > Sum(1/k^i,i=1..infinity);
    ```

    $$\sum_{i=1}^{\infty} (k^i)^{-1}$$

    for $k = 1, 2, 3, 4, 5$. Create a formula for sum as a function of k. Does your formula work for nonpositive values of k?

4. Assume that k is a positive integer and then compute a formula for the partial sum

    ```
    > Sum(1/k^i,i=1..n);
    ```

    $$\sum_{i=1}^{n} (k^i)^{-1}$$

    for an indeterminant value of n. Carry out the following investigation
    a. Verify that your formula is correct for at least one value of n.
    b. Compare the result of taking the limit as $n \to \infty$ with the results from Exercise 3.
    c. How much does the the value of the sum increase by if we include $n + 1$ terms instead of n terms?

5. Use the information gathered in Exercise 4 to prove, by mathematical induction, that the formula is correct.

6. Determine if the series $\sum_{n=1}^{\infty} \dfrac{1}{n(n-1)}$ is convergent or divergent by constructing an appropriate partial sum and then taking the limit of its value as the number of terms in the partial sum tends to infinity.

7. Assume that the value given by *Maple* for

    ```
    > Sum(1/i^2,i=1..infinity);
    ```

    $$\sum_{i=1}^{\infty} i^{-2}$$

    is correct. Show how to use this value, together with the fact that

    ```
    > a := 1/(i)^2/(i+1)^2;
    ```

    $$\frac{1}{i^2 (i+1)^2}$$

    can be written as

    ```
    > convert(",parfrac,i);
    ```

    $$i^{-2} - \frac{2}{i} + (i+1)^{-2} + \frac{2}{i+1}$$

to derive a formula for

```
> Sum(a,i=1..infinity);
```

$$\sum_{i=1}^{\infty} \frac{1}{i^2(i+1)^2}$$

8. By comparing the sum

```
> Sum(1/(i+5),i=1..n);
```

$$\sum_{i=1}^{n} (i+5)^{-1}$$

with the integral

```
> Int(1/(x+5),x=1..n);
```

$$\int_{1}^{n} (x+5)^{-1}\, dx$$

show that the sum is divergent.

9. Compute the limit

```
> Limit(ln(x)/x^(1/2),x=infinity);
```

$$\lim_{x \to \infty} \ln(x) \frac{1}{\sqrt{x}}$$

In general, ln(x) grows slower than any positive power of x. Use this fact to construct a series that can be compared with

```
> s := Sum(ln(x)/x^2,x=1..infinity);
```

$$\sum_{x=1}^{\infty} \frac{\ln(x)}{x^2}$$

term by term to show that the series x converges. Use the comparison test to show that this series converges.

10. Try to use each of the ratio test, the root test, and the comparison test to determine the convergence or divergence of the following series.

   a. `> Sum( (-1)^n*n^5/2^n,n=1..infinity);`

   $$\sum_{n=1}^{\infty} \frac{(-1)^n n^5}{2^n}$$

   b. `> Sum( (3*n+13)/(11*n - 2)^n,n=1..infinity);`

   $$\sum_{n=1}^{\infty} \frac{3n+13}{(11n-2)^n}$$

   c. `> Sum( (ln(n)/(3*n))^2,n=1..infinity);`

   $$\sum_{n=1}^{\infty} \frac{1}{9} \frac{\ln(n)^2}{n^2}$$

d. > Sum(exp(-n)*n!,n=1..infinity);

$$\sum_{n=1}^{\infty} e^{-n} n!$$

e. > Sum(sin(2*n)/n^3,n=1..infinity);

$$\sum_{n=1}^{\infty} \frac{\sin(2n)}{n^3}$$

f. > Sum((n+2)!/n!/10^n,n=1..infinity);

$$\sum_{n=1}^{\infty} \frac{(n+2)!}{n! \, 10^n}$$

(Hint: In instances involving $n!$, the commands **convert(...,GAMMA)** and **simplify(...,GAMMA)** may be of some help. The GAMMA function provides a representation of $n!$ that can be differentiated.)

11. The first few terms of two particular power series are given as

> f := 1 + x + x^2 + 2*x^3 + 3*x^4 + O(x^5);

$$1 + x + x^2 + 2x^3 + 3x^4 + O(x^5)$$

and

> g := 3 + 5*x + 7*x^2 + 9*x^3 + 11*x^4 + O(x^5);

$$3 + 5x + 7x^2 + 9x^3 + 11x^4 + O(x^5)$$

Find as many terms as possible of the power series corresponding to the product $a\, b$.

12. Find values of a, b, and c so that the first few terms of the series

> series((1+3*x)/(1+b*x + c*x^2),x=0,5);

$$(1 + (3-b)x + (-c - -(-3+b)b)x^2 + ((-3+b)c + (c+3b-b^2)b)x^3 +$$
$$((c+3b-b^2)c + (3c-2cb-3b^2+b^3)b)x^4 + O(x^5))$$

are given by

> 1+5*x+5*x^2-15*x^3-55*x^4 + O(x^5);

$$1 + 5x + 5x^2 - 15x^3 - 55x^4 + O(x^5)$$

13. Find a power series representation for

> sin(5*x^2 + x);

$$\sin(5x^2 + x)$$

14. Obtain the series representation to order 20 of

> sin(x^2);

$$\sin(x^2)$$

expanded about $x = 0$. Use this series representation to obtain a series representation of

> Int(sin(x^2),x);

$$\int \sin(x^2)\, dx$$

(Note: This particular antiderivative cannot be represented in terms of polynomials, logarithms, and exponentials and the standard trigonometric functions we have studied in this course.)

15. The power series representation for exp(x) can be used to approximate $exp(1)$. The commands

    ```
    > taylor(exp(x),x,4);
    > convert(",polynom);
    > f :=unapply(",x);
    ```

    construct a *Maple* function that uses the first few terms of the series to construct a suitable function. Generate a simultaneous plot of both f and *exp* over the interval $[0, 3]$. How many terms are required before the value of $exp(1)$ is accurate to 4 decimal places? How many terms are required before the value of $exp(2)$ is accurate to 4 decimal places? Is the approximation more accurate at $x = 2$ if we do the series expansion about $x = 1$?

16. Because the value of arctan(1) is

    ```
    > arctan(1);
    ```

    $$\frac{1}{4}\pi$$

    we can, in principle, use the series expansion for arctan(x) to accurately compute the value of π. Use the series for arctan(x) to construct a function that approximates $4\arctan(x)$ and evaluate it at $x = 1$ to get an estimate for the value of π. How many terms do you require in the series before the answer is accurate to even 2 decimal places? (To see what the problem is, try plotting the series approximation to arctan(x) near 1.)

17. The value of $16\arctan(1/5) - 4\arctan(1/239)$ is exactly π. Use this formula and a five term series approximation to arctan(x) about $x = 0$ to approximate π. Notice how much more accurate this approximation is than that of the previous exercise. Examine the graph of arctan(x) and the approximation from $x = 0$ to $x = 1$ to see what is happening. Even when a series converges, it may converge more quickly at some x values than at others.

18. Use a Taylor series expanson of order 1 to construct a function that computes the tangent line to the curve

    ```
    > f := (x) -> sin(2*x)/exp(x);
    ```

    $$x \mapsto \frac{\sin(2x)}{e^x}$$

    at $x = a$.

19. Graph the function cos(x) and the degree nine Taylor polynomial about $x = 0$ in a manner that indicates where the accuracy of this approximation starts to fail.

20. Determine the maximum error that could occur by using a degree five Taylor polynomial expanded about $x = 0$ to approximate $f(x)$.

    ```
    > f := (x) -> cos(2*x)*sin(x):
    ```

    for values of x in the range $-1 \leq x \leq 1$.

21. What degree polynomial would be required in estimating $f(x)$ of Exercise 20 to ensure that the error was never greater than 0.003 on the specified interval?

APPENDIX A The Computing Environment

A.1 The Student Package

The *Maple* commands used in this book are commands that already exist in *Maple* V Release 2, either as a main level command or as part of special groups of commands called packages. Though developed specifically for this course, the commands known collectively as the "student package" are included in all distributions of *Maple* including the "student editions." With the exception of some of the three-dimensional implicit plots, all of the problems solved in this book can be solved with these special student editions of *Maple*, and even those can be accomplished with slight variants on the plot options.

If a particular command is not defined in your current session, this will become apparent when you attempt to use it. *Maple* will simply echo your command back to you as its response, as in

```
> An_Undefined_command(f(x),x);
```
$$An\_Undefined\_command(f(x), x)$$

In most instances the reason for this will be that you mistyped the name of the command.

A second possibility is that the command exists but is part of a package that has not yet been defined. For example, the majority of this book has been written with the assumption that the student package has been loaded by the command

```
> with(student);
```
$$D, Doubleint, Int, Limit, Sum, Tripleint, changevar,$$
$$combine, completesquare, distance, equate, extrema,$$
$$integrand, intercept, intparts, isolate, leftbox,$$
$$leftsum, makeproc, maximize, middlebox, middlesum,$$
$$midpoint, minimize, powsubs, rightbox, rightsum,$$
$$showtangent, simpson, slope, trapezoid, value$$

Ideally, when working with this book, you should configure your *Maple* installation so that the student package is loaded automatically on starting *Maple*. This is accomplished by placing the command **with(student):** in the *Maple* initialization file.

The code for the student package was written in the *Maple* programming language by the author of this book and has become a standard part of the *Maple* system. Several of the concepts, including **combine( )**, have been adopted on a system-wide basis. This code, and the use of it in this book, represents a conscious decision that the underlying mathematical concepts — not the software — be the driving force behind any mathematical discussion. The purpose of the code is to adapt the software to the interactive computing needs that are dictated by the course. Its role is to provide the essential mathematical primitives at the right level of refinement. We have purposely kept any discussion of programming to a minimum.

The success of this approach is partially indicated by the extent to which the *Maple* student package has served as a foundation for other calculus based software projects. For example, Calculus T/L™ and Newton™ both use the student package from *Maple* to carry out some of their operations. An important contribution of this project to the general software environment has been the formalization of the role of inert functions, such as **Int( )** and **Sum( )**, and the establishment of mechanisms to manipulate and rearrange such objects. Such tools are now being used regularly by researchers to state and reformulate problems in *Maple* prior to "computing an answer." As software and user interfaces evolve, the manner in which such code is invoked may change, but the need for such primitives will continue.

The actual code can be accessed from within a *Maple* session. For example, the following commands show how Simpson's rule is implemented.

```
> with(student):
> interface(verboseproc=2):
> eval(simpson);
proc(F,dx)
local a,b,f,i,h,n,rg,x;
options 'Copyright 1990 by the University of Waterloo';
    if type(dx,equation) and type(op(1,dx),name) and
        type(op(2,dx),range) then
            x := op(1,dx); rg := op(2,dx)
    else
            ERROR('usage: simpson( f(x), x=a...b, iterations)')
    fi        fi
fi;
if not type(n,{name,numeric}) then
    RETURN('procname(args)')
fi;
a := op(1,rg);
b := op(2,rg);
h := (b-a)/n;
f := student[makeproc](F,x);
1/3*h*(f(a)+f(b)+4*Sum(f(a+(2*i-1)*h),i = 1 .. 1/2*n)+2*
        Sum(f(a+2*i*h),i = 1 .. 1/2*n-1))
    end
```

A.1 The Student Package

The only *Maple* code used in this book that is not part of the standard *Maple* system is listed below.

Code to Implement Newton's Algorithm
The following code was used in Chapter 3.

```
newton := proc(f,startvalue) local x0,x1;
    x0 := startvalue;
    while abs(f(x0) ) > .001 do
        x1 := x0 - f(x0)/D(f)(x0);
        x0 := x1;
        userinfo(1,trace,`current approximation: `, x0);
    od;
    RETURN(x0);
end:
```

Code to Represent Line Segments
This code was used in Chapter 4 to generate the plot data needed for line segments.

The first routine generates a small box around the individual points to make them more visible. The second uses the first routine to generate appropriate rectangles and the actual line segment, for use by the **plot( )** command.

```
PltPoint := proc(A:list) local a,b,x;
    x := .05;
    a := A[1]; b := A[2];
    fi;
    [[a-x,b],[a,b-x],[a+x,b],[a,b+x],[a-x,b]]
end:

PltLine := proc(a,b)
    [a,b], PltPoint(a), PltPoint(b);
end:
```

These are used directly by the plot command, as in

```
> plot( { PltLine([1,2],[3,5]) } );
```

Code to Draw a Circle
The following code generates the 3D plot corresponding to a circular cross section orthogonal to the *x* axis.

```
xcircle := proc(xval,yval,zval,r)
local x,y,t,n,angle,pt;
n := 31;
angle := 2*Pi/n ;
pt := unapply( [xval, yval + r*cos(t), zval + r*sin(t)],t );
PLOT3D( POLYGONS( [seq( evalf(pt(t*angle)),t=0..n)] ));
end:
```

These plots can be combined with other plots by using commands of the form

```
> plots[display3d](...);
```

A.2 Production Notes

This book was produced using a combination of Latex and *Maple*. The source files for the manuscript are standard Latex source files with embedded *Maple* commands, but with no *Maple* results included. More than 3100 *Maple* commands are used. The algebraic output generated by the commands is formatted automatically in Latex using a slightly modified version of *Maple*'s **latex** command and then incorporated directly into secondary source files. Even the plots are produced by *Maple* in PostScript and then automatically included.

Index

Maple commands and key words are indicated by **bold** type style.

Maple , 1
' , 11
"" , 28
"""" , 28
π , 4
$\sqrt{}$, 344
^ , 2
" , 3
! , 3
* , 2
** , 2
/ , 2
:= , 7
; , 2
= , 7
< , 13
>
   relation , 14
   Maple prompt , 2, 48

absolute maximum , 177
absolute minimum , 177
acceleration due to gravity , 305
aircraft landing , 172
allvalues , 56
antiderivative , 261
approximate function value , 112
approximate numbers , 4
approximations , 4
arc , 131
arccos
   simplifications , 418
arc length , 398
   parametric , 428
   polar coordinates , 436
area
   polar coordinates , 435
area between curves , 282
areas between curves , 227

assign , 55
assignment statements , 7, 8
assume , 88, 158, 452
asymptotes , 35, 209
asymptotic behavior , 212
   sequences , 449
asymptotic growth , 490
average slope , 190
averaging , 363
axis of symmetry , 288

Big OH , 481
Binomial Theorem , 6, 118
bisection , 90

cardioid , 433
Cartesian coordinates , 415
Cartesian representations , 415
Cartesian slope , 420
chain rule , 171, 315
change of variables , 314, 316
changevar , 315
chonchoid , 438
circle , 32
clearing a variable , 22
collect , 136
combine , 232, 251
combine(...,trig) , 338
combining power series , 472
combining sums , 232
commands as summaries , 213
completesquare , 26, 350
completing the square , 26, 350
composition , 88
compositions of three or more functions , 146
concave downward , 202
concave upward , 202
constant functions , 89
constrained , 33

499

continuity, 83
continuity and extrema, 187
continuous, 85
continuous functions, 188
converge, 452
convergence
    absolute, 463, 465
convergence, test
    comparison, 460
    integral, 457
    ratio, 463
    root, 465
converging at the same rate, 462
convert(...,'@'), 422
convert(...,nested), 422
convex lens, 168
coords=polar, 431
cos, 44
cosine, 173
critical number, 186
critical point, 103
critical points, 198
crossing stream, 219
cross sections, 287
    circular, 385
curves above the x-axis, 247
curves below the x-axis, 248
cycloid, 418, 423
cylindrical decomposition, 392
cylindrical shells, 298, 396

D, 100
decimal point, 4
decreasing function, 102, 197
definite integral, 246
degree of polynomial, 330
derivative, 99, 100
    at a point, 131
    of a constant, 116
    of a power, 118
    of a product, 122
    of a quotient, 128
    of a reciprocal, 127
    of a sum, 122
    of exponentials, 148
    of logarithms, 150
differentials, 112
digamma, 366
Digits, 4
directional limit, 85
discontinuities, 267
discontinuity
    jump, 84
    step, 83
discontinuous, 85
display, 85
distance, 26
distortion, 178
ditto, 3
diverge, 452
division, 2

domains, 40
D operator, 325
double quote, 3
draining tank, 160
dx, 112
dy, 112

efficiency, 334
endpoints, 180
equations, 7
error, 113
error bound, 364
error estimate, 172
    Simpson's Rule, 370
    Trapezoidal Rule, 364
evalf, 4
even functions, 44
exp, 107
expand, 6, 252
expanding sums, 232
experimentation, 38
exponential growth, 106
exponents, 2
expressions, 130
expression sequence, 4
extrema, 186
extrema, 283
extreme points, 173
extreme value theorem, 187
extremum
    local, 181

factor, 6
factorial, 3
fencing, 213
Fermat's theorem, 188
Fibonacci, 447
fibonacci, 447
floating point, 4
fluid flow, 108
focal length, 168
functional composition, 137
functional equations, 156
functions, 130

global extrema, 176
global maximum, 175
global minimum, 175

hail, 168
half integer, 349
has, 28
height, 278
help, 15
horizontal tangent line, 181
horizontal tangent lines, 172

identity function, 401
implicit derivatives, 152
implicit equations, 31
implicitplot, 282, 390

implicitplot3d, 285, 383
increasing, 102
increasing function, 197
indefinite integral, 265
inequalities, 13
inert summation, 229
infinity plots, 177
inflection, 202
inscribed rectangle, 220
int, 247
Int, 247
integral
    indefinite, 265
integration by parts, 326
intermediate expression, 7
intermediate value theorem, 190
internal maximum, 181
interp, 368
intersection, 15
intuition, 213
inverse functions, 150
irrational, 5
is, 87
isolate, 8

labels, 7
ladder, 172
layer, 381
leftbox, 248
leftsum, 248
Leibnitz notation, 111, 116, 170
lens, 168
lhs, 8, 53
limit, 61
Limit, 57, 61
limit
    definition of, 61
limit at infinity, 209
limiting ratios, 132
limits involving trig functions, 131
line
    point-intercept form, 23
    standard form, 26
linear equations, 157
list, 38, 446
list, 446
local extremum, 181
logarithmic derivative, 152, 171
lower bounds, 239

Maclaurin series, 478
map, 28, 38
mapping onto an equation, 28
mathematical induction, 239, 490
maximize, 179
maximum
    absolute, 173
    local, 181
mean value theorem, 191
midpoint, 26
minimize, 179

minimize, 102
minimum
  absolute, 173
  local, 181
minimum distance, 216
minus, 15
monotonic function, 197
moving ladder, 159
moving particle, 268
moving ships, 161
multiple lines, 2
multiply, 2

named equations, 8
names, 7
naming expressions, 7
nested, 326
newton, 163
Newton, 305
Newton quotient, 115, 169, 263
Newton's algorithm, 163, 172
Newton's second law of motion, 305
nonnegative, 201
notational limitations, 139
nq, 135
numerical approximations, 4, 361
numerical expressions, 3

O, 481
odd functions, 44
oil can, 215
one-sided limit, 85
one-sided limits, 84
optimization, 173
option remember, 448
order of composition, 145
output, 49

parameterizations, 414
parametic, plot
  polar, 431
parametric curves, 413
partial fraction, 453
partial fraction decomposition, 352
particle, 168
Pascal's Triangle, 6
patterns, 115
peaks, 102
periodic, 5, 311
petal, 435
pi, 53
Pi, 4, 53
piecewise-defined, 83
plot, 10
plot options, 11
plotting, 10
point of view, 317, 419
polar coordinates, 430
POLYGONS, 395
polynomial identity, 353
polynomial interpolation, 180, 184

population, 168
population growth, 106
position, 168
power series, 468
  arithmetic, 472
  truncated, 472
powsubs, 135
previous expression, 3
problem structure, 213
Maple prompt, 48
proportional, 108
Psi, 366
pump, 172
pyramid, 380

quadratic, 184
quit, 48

radius of convergence, 470
random polynomial, 124
randpoly, 124
randpoly, 169
rate of change, 104
reciprocal, 3
recurrence relations, 447
recursive procedure definitions, 448
reduction formula, 333
reevaluation, 38
related rates, 158
repeating decimal, 5
restart, 53
reverse operations, 260
rhs, 8, 53
Rolle's theorem, 188, 190
root finding, 165
RootOf, 56
roots
  number of, 190
rose, 435
rsolve, 448

scaling, 33
sec( ), 339
secant, 342
secant line, 65, 168
second, derivative
  parametric, 424
select, 28
semicubical parabola, 400
seq, 446
series, 478
series, 451
  alternating, 465
  approximations, 481
  error terms, 483
  harmonic, 466
  infinite, 451
  power, 468
series approximations, 470
set, 15, 446
set difference, 15

ships, 172
sign changes, 198
Simpson's Rule, 366
sin, 44
sine, 131
single quotes, 11
slope, 112
  average, 190
smooth, 188
smooth functions, 188
snowball, 158
solid of revolution, 288
solve, 9, 13, 53
solved equations, 23
solving a cubic, 153
solving for *x*, 8
solving inequalities, 13
speed, 105
sqrt, 4, 5
square roots, 4
starting a session, 48
step functions, 43
student package, 230
subs, 5, 7
subscripts, 2
substitution
  sequential, 319
  simultaneous, 319
substitutions, 7
substitution versus evaluation, 38
Sum, 2
sum, 2
summations, 228, 232
sum of angles, 344
surface area, 405
  parametric, 429
surfdata, 294

tangent line, 171
  parametric, 420
taylor, 44, 480
Taylor series, 44, 480
textplot, 399
the chain rule, 137
the quadratic formula, 29
three-dimensional, 285
trapezoidal rule, 361, 363
trig identities, 334
trunc, 43
tubeplot, 292
tubeplot, 389, 396

unassign, 22, 53
unbounded, 177
undetermined coefficients, 172
unevaluated expression, 3
unevaluated return, 229
union, 15
unnamed functions, 38
upper bound, 188
upper bounds, 188, 242

user defined functions, 27

valleys, 102
value, 247
velocity, 104
verification, 316
verifying solutions, 8
vertical, 26
viscosity, 108
visualization, 301
volume of a sphere, 287, 383
volumes of geometric figures, 380

wedge of a cylinder, 395

xradius, 386

Maple®V: Release 2 - Student Edition Incredible mathematical power for only **$99**

NOW AVAILABLE FOR DOS/WINDOWS & MACINTOSH!

Performance, reliability, and the most power for your dollar — it's all yours with Maple V!

Used worldwide by mathematicians, engineers, scientists, teachers, and students, **Maple** is a powerful, interactive computer algebra system that manipulates, solves, and graphs mathematical equations and expressions. Easy to learn and easy to use, **Maple V: Release 2, Student Edition** features 3-D graphics and can be used as a tool in all your mathematics courses — calculus, differential equations, linear algebra, and beyond.

The Student Edition gives you the power to perform a wide range of symbolic and numeric computations with speed and accuracy, and features: • more than 2500 built-in mathematical functions that can be modified or extended • standard mathematical notation • a complete online help system that allows you to navigate quickly and easily from one topic to another • 3-D and 2-D graphics including contour plots and implicit plotting • a worksheet interface that gives you control of type styles and fonts and allows you to mix mathematics, text, and graphics • animation capabilities • and much more!

This version of *Maple V: Release 2, Student Edition* comes with software plus *Maple V: Getting Started*, a brief machine-specific guide that gets you up and running; with *Maple V, Release Notes* that details changes in Release 2; and the *Maple V Flight Manual*, a tutorial.

System Requirements

Macintosh with 2 MB RAM and hard disk with minimum 10-12 MB free disk space. Coprocessor optional.

DOS/Windows: 386/486 computer with 2 MB of extended RAM, minimum 10-12 MB free disk space, and 1.2 MB or 1.44 MB floppy disk drive. Coprocessor optional. Windows version is included in the box.

No risk! Order Maple V - Student Edition on our 30-day money back guarantee! Simply fill out the coupon below and return to Brooks/Cole along with your check or credit card number.
Or call toll-free 1-800-354-9706.

Yes! I want to order

_____ Maple V Release 2—Student Edition for the Macintosh (ISBN 0-534-21228-x)   _____ @ $99.00 _____

_____ Maple V Release 2—Student Edition DOS/WINDOWS (ISBN 0-534-21222-0)   _____ @ $99.00 _____

Subtotal _____

(Residents of AL, AZ, CA, CO, CT, FL, GA, IL, IN, KY, LA, MA, MD, MI, MN, MO, NC, NJ, NY, OH, PA, RI, SC, TN, TX, UT, VA, WA, WI must add sales tax.)

TAX _____

Handling $2.00

TOTAL _____

Please ship to:

Name _____
Department_____ School_____
Street Address_____
City_____ State_____ Zip_____
Office phone number (_____)_____
Office hours (please circle)   M T W Th F /Time _____

❏ Payment enclosed (check, money order, purchase order).
❏ Please charge my ❏ Mastercard ❏ Visa ❏ American Express

Account Number _____ Expiration Date_____

Signature_____

Prices subject to change without notice. We will refund payment for unshipped out-of-stock titles after 120 days and for not-yet-published titles after 180 days unless an earlier date is requested in writing by you.

Detach and return to: **Brooks/Cole Publishing Company**
511 Forest Lodge Rd.
Pacific Grove, CA 93950
(408) 373 - 0728
FAX (408) 375 -6414

NO POSTAGE
NECESSARY
IF MAILED
IN THE
UNITED STATES

BUSINESS REPLY MAIL
FIRST CLASS PERMIT NO. 358
PACIFIC GROVE, CALIFORNIA 93950-5098

POSTAGE WILL BE PAID BY:

 Brooks/Cole Publishing Company
511 Forest Lodge Road
Pacific Grove, CA 93950-5098